PRACTICAL
SPREADSHEET STATISTICS
AND CURVE FITTING
FOR SCIENTISTS
AND ENGINEERS

PRENTICE HALL LABORATORY LOTUS® SERIES

PRACTICAL SPREADSHEET STATISTICS AND CURVE FITTING
FOR SCIENTISTS AND ENGINEERS

LOUIS M. MEZEI

PRENTICE HALL
ENGLEWOOD CLIFFS, NEW JERSEY 07632

Library of Congress Cataloging-in-Publication Data

MEZEI, LOUIS M.
    Practical spreadsheet statistics and curve fitting for scientists
and engineers / Louis M. Mezei

      p. cm.—(Prentice Hall laboratory Lotus series)

    Includes bibliographical references.
    ISBN 0-13-519877-1
    1. Science—Statistical methods—Data processing. 2. Engineering—
Statistical methods—Data processing. 3. Lotus 1-2-3 (Computer
program) 4. Curve fitting. I. Title. II. Series.
Q183.9.M49 1990
507.2—dc20

90-7287
CIP

Editorial/production supervision
 and interior design: BARBARA MARTTINE
Cover design: KAREN STEPHENS
Manufacturing buyer: KELLY BEHR

© 1990 by Prentice-Hall, Inc.
A Division of Simon & Schuster
Englewood Cliffs, New Jersey 07632

The publisher offers discounts on this book when ordered
in bulk quantities. For more information, write:
    Special Sales/College Marketing
    College Technical and Reference Division
    Prentice Hall
    Englewood Cliffs, New Jersey 07632

ISBN   0-13-519877-1

Prentice-Hall International (UK) Limited, *London*
Prentice-Hall of Australia Pty. Limited, *Sydney*
Prentice-Hall Canada Inc., *Toronto*
Prentice-Hall Hispanoamericana, S.A., *Mexico*
Prentice-Hall of India Private Limited, *New Delhi*
Prentice-Hall of Japan, Inc., *Tokyo*
Simon & Schuster Asia Pte. Ltd., *Singapore*
Editora Prentice-Hall do Brasil, Ltda., *Rio de Janeiro*

To
Heddie, Becky, Heather, and Collette

At least one person's opinion of statistics is:

There are three kinds of lies: lies, damned lies, and statistics.--Mark Twain

However,

Applied correctly, statistical analyses provide objective measures of the confidence that one can have in the conclusions being drawn.--Lou

Contents

9 INTERPOLATION   183

Preface

How I Got Hooked On the Lotus Spreadsheets and Programming Language

Have you ever noticed how scientific software manufacturers corral you into a one-inch square box and then charge you megabucks for the box?

Have you ever noticed how commercial software manufacturers try to make you do it their way, or not at all?

Do you perform experiments that gather data using instruments or sensors?

Are you having a hard time finding programs that reduce and calculate your data the way that **YOU** want it evaluated?

Are you tired of buying new software, only to be directed through yet another mountain of manuals every time you want to do something a little different?

Do you get confused trying to remember which commands go with which piece of software; leaving you with a strong desire to try to find one system that would consolidate your efforts into an single way of doing things?

Trying to find an amiable, all-encompassing answer to the above problems prompted me to look into spreadsheet programs. At first, I was awed at the number of things that I thought I would need to learn. What's worse, all of the books written on spreadsheets were written from a business standpoint. After buying about 16 books on Lotus® spreadsheets and trying to weed through such foreign topics as "rate of return," "sales forecasts," "commission schedules," and the like, I became dismayed. But, I saw a glimmer of hope. Lotus 1-2-3® and Symphony® seemed to have all of the features that I needed to reduce and graph my data. All that I needed to do was to transform the information in my books from a business focus to a scientific standpoint and try to find a way to use the features in a way that would address the kinds of programming and spreadsheet activities that a scientist could benefit from.

My incentives were numerous. I was buying expensive software that was designed by people who had never worked in a laboratory environment. The software was never a good "fit" for how things are really done in laboratories. I had doubts that software designers had ever caught on to the fact that every experiment performed in a laboratory setting is unique. This uniqueness is especially true in research laboratories. In addition, data reduction tends to be highly variable. In fact, most researchers look at their data from many different angles . . . the number of angles varying with the inverse of the data's quality.

To make a long story short, I have found Lotus spreadsheets to be a cost effective way to beat the problems associated with purchased software. I was able to find a method to talk to instruments directly from a spreadsheet, most often **WITHOUT the use of Lotus Measure**[tm]. Once the data was in the spreadsheet, Lotus solved all of my data reduction and graphics problems.

Lotus is a very rich environment that you too can use to solve your data acquisition, reduction, and graphics problems. You do not need to learn the hundreds of Lotus program commands and special-purpose formulae to use it effectively. You also do not need to spend a lot of time learning how to connect an instrument to a personal computer. My earlier book, **Laboratory Lotus®: A Complete Guide To Instrument Interfacing** teaches you quick and easy ways to get data into your spreadsheet from your instruments. This book, **Prentice Hall Laboratory Lotus® Series**[tm]**: Practical Spreadsheet Statistics and Curve Fitting for Scientists and Engineers,** teaches you efficient and flexible ways to quickly perform data reduction on data, whether the data are from instruments or not.

From robots to doughnuts, (and everything in between), if your central mission is the collection and reduction of data, **Laboratory Lotus®: A Complete Guide To Instrument Interfacing** and **Prentice Hall Laboratory Lotus® Series**[tm]**: Practical Spreadsheet Statistics and Curve Fitting for Scientists and Engineers** can help you. Armed with the first book, you can control just about everything from a robot that automates the most sophisticated genetic engineering project to a texture analyzer that measures the adhesion, bloom strength, crispness, and tackiness of your doughnuts to make sure that they are done and have not become stale. And that can be very important at coffee break time.

This book will help you make your data come alive with sophisticated statistical and curve fitting data reduction, even if you have just typed the data into the spreadsheet. When your reports contain statistical characterizations, distribution analyses, matrix mathematics, linear and curvilinear regressions, nonlinearly fitted curves, spline interpolations, ANOVA tables, nonparametric statistics, etc., you can issue your reports and draw your conclusions with increased confidence.

That is, after reading this book and learning some very clever programming techniques, you will be able to write **your own** piece of software tailored specifically to reducing **your** data in just a couple of hours.

Louis M. Mezei

1

Introduction

My first book in this series, *Laboratory Lotus®: A Complete Guide To Instrument Interfacing*, taught you how to merge scientific instruments, your personal computer and either a Lotus 1-2-3® or a Lotus Symphony® spreadsheet to produce a single, powerful unit that met your exact automation, data reduction, and graphics needs. Whether your data originate directly from instruments, are imported from files, or are simply typed into a spreadsheet, this book will teach you how to create tools that you can use to fulfill the second item in the above list—data reduction.

Lotus® is a wonderful tool for learning statistical and curve fitting techniques. With a Lotus spreadsheet you can immediately view the results of "experimenting" with various techniques. By building pauses into Lotus programs that automatically perform the techniques, you can also investigate the results of intermediate calculations . . . instantly. This book capitalizes on these Lotus capabilities to help you learn by doing and to see the consequences of your actions. To enhance this process, the general style used in the book provides the following information for each technique:

- A theoretical overview of the technique
- When to use the technique
- Step-by-step instructions on how to create the technique
- An explanation of the details of how the spreadsheet and/or program works
- An explanation of how to interpret the results

A high degree of attention to detail in this book will help you to learn the mechanics behind both the techniques and how to implement them. In the process, you will learn professional programming skills that will allow you to implement very sophisticated spreadsheets and/or Lotus programs that will automate data reduction applications for your experiments.

SOME USEFUL DEFINITIONS

This section contains definitions of a few of the basic terms that will be used throughout this book. These terms will be defined in more detail later, but cursory explanations will make fundamental descriptions easier prior to rigorous treatment.

- An *observation* is a single data point and the basic element of statistics.
- A *population* is a complete set of observations. That is, a population includes every data point in a certain grouping under consideration. In effect, a population is equivalent to the *universe* of all possible observations.
- A *sample* is a portion of a population selected for analysis. That is, a sample is an incomplete set of observations.
- A *parameter* is a summary measure that is computed to describe a characteristic of an entire population.
- A *statistic* is a summary measure that is computed to describe a characteristic from a sample of a population. In effect, a statistic is an estimate of a parameter.

A major facet of statistics is the process of using sample statistics to draw conclusions about true population parameters.

Procedures that compare means (the most common population parameters) fall into the broad category of *parametric statistics*. Parametric statistical procedures rest on several assumptions that may not always apply. *Nonparametric statistical procedures* compare populations without making assumptions regarding population parameters.

HOW THIS BOOK IS ORGANIZED

This book is progressive in format, but contains a certain degree of redundancy (when necessary). The remainder of this chapter is devoted to stating any assumptions I've made, and provides you with all the information you need to complete the objectives of the remaining chapters successfully.

This book is roughly divided into four major sections:

- The first section (Chapters 2 through 4) deals with some general concepts that will be used by later chapters. These concepts include table lookup, building a template to organize your data, and finding the distribution of a set of data.
- The second section (Chapter 5) deals with statistical methods that require stringent assumptions to be made before they can be applied to data and are valid only if the assumptions hold (i.e., parametric statistical procedures). The assumptions upon which these techniques are based include a requirement for the form of the

distribution of the data in the population and a requirement that observations be independent of each other.

- The third section (Chapters 6 through 9) deals with linear and nonlinear curve fitting techniques. Because many of these techniques require programs to be developed, Chapter 7 will teach you how to create Lotus programs.
- The final section (Chapters 10 and 11) is dedicated to teaching you advanced programming techniques for statistical tests. Chapter 10 covers statistical procedures that depend neither on the form of the distribution of the data in the population from which the sample is drawn nor with the parameters of a population (i.e., nonparametric statistical procedures). Chapter 11 contains statistical tests for three or more populations of data.

The following is a brief overview of each chapter:

- Chapter 2 will show you how to automate the tedious process of looking up values in tables. The tools that you will learn in Chapter 2 will be used throughout the remainder of the book because they will help you to automate calculations, determine whether the results of statistical tests are significant, locate the appropriate segment intervals to use in calculations, etc.
- Chapter 3 contains information on how to plan and organize your data for efficient reduction in your Lotus 1-2-3 and/or Symphony spreadsheet. To this end, Chapter 3 presents the concept of a "spreadsheet template." Chapter 3 will also teach you several methods that you can use to provide consideration for the number of significant figures in an experiment.
- Most common statistical tests and curve fitting procedures require that the form of the population data from which a sample is drawn for analysis be normally distributed. If the data distribution does not meet this criterion, accurate results may not be achieved and other techniques need to be used. For this reason, Chapter 4 is included to show you how to determine data distributions.
- Chapter 5 discusses the difference between populations and samples and how each is used with Lotus statistical tools. Then, a number of statistical procedures are provided to describe and compare central values and variabilities of sets of sample data.
- Chapter 6 will show you how to characterize mathematical relationships between variables and how to use the predictive powers of these characterizations. The relationships discussed in this chapter relate to *linear* relationships, although this term is somewhat misleading. A *linear* relationship means that the relationship is *linear in the coefficients*. For example, $y = a/x^2$ would be considered a linear relationship because "a" is linear, even though $1/x^2$ is not. Chapter 6 also contains information on how to choose a curve fitting equation. Because error analysis is such an important part of any curve fitting procedure, a majority of Chapter 6 is dedicated to the discussion of powerful tools that you can easily create within the Lotus environment and use to expose potential problems.

- Chapters 8 through 10 require programs to be developed. Chapter 7 will teach you all of the programming skills that you will need to know. A comparison to BASIC language is made for those who know BASIC. However, the focus is on Lotus programming commands and techniques. The chapter also stresses professionalism in creating spreadsheets and programs. If you do not have a copy of *Laboratory Lotus®: A Complete Guide To Instrument Interfacing,* then it is important for you to read and understand the information given in Chapter 7 before proceeding to chapters which follow.

- Not all data can be fitted and characterized with linear techniques. Chapters 8 and 9 will show you how to handle those situations, along with improving your programming skills. Chapter 8 uses the Marquardt's Compromise method. This method is an iterative method that is both very fast and reliable. For difficult curves, Chapter 9 will present three interpolation methods: linear, cubic, and cubic spline.

- Chapter 10 will provide you with spreadsheet templates and Lotus programs that will automate nonparametric statistical tests (tests that do not depend on the distribution or parameters of a population). The emphasis is on how to create your own applications, because all known statistical tests cannot be presented. The example procedures provided in this chapter include the Runs, Sign, Median, Wilcoxon Signed-Rank, Rank Sum, Kruskal-Wallis, Spearman Rank-Correlation, and Kolmogorov-Smirnov tests.

- If you are trying to analyze more than two groups of data, then you will want to investigate Chapter 11. Chapter 11 presents *AN*alysis *Of* *VA*riance (ANOVA) techniques. ANOVA is a very powerful technique that you will discover many applications for. Lotus makes ANOVA calculations easy and removes any hesitations you may have had about not using ANOVA.

- Appendices A, B, and C present concise explanations of the Lotus operators, special purpose functions, and program commands used throughout the book. The remainder of the appendices provide useful statistical tables.

All of the chapters in this book point out ways to avoid some of the common pitfalls found in Lotus. They also show you how to avoid problems arising from quirks in Lotus' operation, and what to do if you fall into a trap.

If you are an experienced Lotus user, this book will show you how to push Lotus to its limits by taking worn out, old Lotus tools that have been collecting dust and use them in innovative ways to perform some very clever tasks.

DISCLAIMER

Lotus will give you no guarantees and neither can I. However, I can assure you that I have made a strong effort to ensure that all of the programs that I present to you in this book work. I have tried all of them in both Lotus 1-2-3 and Symphony on an IBM® XT™, AT® and Model 80. However, the examples are just examples. Although all of the examples can be used directly in your laboratory, they are meant to teach you certain specific concepts about how to create data reduction tools and are not necessarily the most efficient way to program for your particular situation.

ASSUMPTIONS ABOUT YOUR ABILITIES

I wrote this book for the person who has access to a copy of Lotus 1-2-3 and/or Symphony and has at least a beginner's knowledge of how to use it. A fundamental discussion of Lotus spreadsheets and their components can be found in subsequent sections of this chapter. A more thorough discussion of these topics can be found in one of the references listed at the end of this chapter.

I also assume that you have a real desire to learn how to perform some basic programming techniques to solve your data handling challenges.

Finally, I do not intend for this book to be a comprehensive text on derivations or on a deep theoretical basis for Statistics, Linear Regression, Calculus, etc. Books on these topics abound and you can refer to them if you need assistance. This book is a practical guide on how to implement these mathematical tools to real-life data and on the mechanism of how the tools perform.

ASSUMPTIONS ABOUT YOUR SOFTWARE AND PERSONAL COMPUTER

Software

I developed and tested all of the examples in this book using Lotus 1-2-3 (Releases 2.0, 2.01, 2.2, and 3.0) and Symphony (Versions 1.1, 1.2, and 2.0). Because the Lotus macro commands and @functions are fully established, there should not be any changes in how they are invoked or how they perform (other than additions required to support the multiple layering of spreadsheets). For this reason, the information and examples given in this book should be generally applicable to all future releases of the two Lotus programs.

Because of their similarities to Lotus, the information provided in this book should also be applicable to Quattro™ (Borland International®, Inc.), VP Planner™ (Paperback Software® International, Inc.), and Excel (Microsoft® Corp).

If you haven't purchased a Lotus spreadsheet program yet, I would recommend Symphony. Lotus 1-2-3 contains a spreadsheet, graphics, and a crude database. Symphony, also contains these three environments plus word processing, telecommunications, windowing capabilities and real-time graphics. Additionally, Symphony can run DOS programs directly from a spreadsheet.

Although 1-2-3 and Symphony have many features in common, especially in the spreadsheet area, Symphony is much more powerful than 1-2-3. Lotus Development has enhanced the spreadsheet and graphics functions for Symphony and has rewritten the database functions to make this part of the program more powerful and easier to use. Symphony can easily keep track of samples and/or results for a single lab or several different labs.

Personal Computer Equipment

The computers that I used were all IBM®-based. I have a Personal System/2® Model 80, an AT®, and an XT™. All of the programs worked on all three computers. The XT proved a little slow in a couple of instances, so I used plug-in accelerator boards to boost the speed of the XT.

The first board that I used was an Orchid Technology PCturbo 286e plug-in accelerator board. (The PCturbo 286e is available from Orchid Technology, 45365 Northport Loop West, Fremont, CA, 94538.) The speed increase provided by the PCturbo 286e was dramatic. The Pcturbo 286e model that I own is an 8 MHz, zero wait-state 80286 CPU with cache memory. Because the board also uses the XT's existing 8088 microprocessor to handle input and output, the board runs rapidly through calculations and video displays. One drawback to the PCturbo 286e is that it is not compatible with EGA monitors.

The second board I used was an Intel® Inboard™ 386/PC in the XT. (The Inboard 386/PC is available from Intel Corporation, 5200 NE Elam Young Parkway, Hillsboro, OR 97124-9987.) This board has a 80386 microprocessor working at 16 MHz and provides all of the advanced features of the 80386 microprocessor. This capability means that the Intel Inboard is fully compatible with Lotus 1-2-3, Release 3, and that the board can perform calculations up to 10 times faster than the XT alone. Additionally, you can have up to 4 megabytes of high performance 32-bit extended memory for efficiency when using very large spreadsheets. More importantly, the Inboard is fully compatible with EGA monitors.

For a direct comparison of various computer and plug-in accelerator board configurations, see Chapter 9 (Table 9-1).

Drawing from my experience, you should be able to use any IBM personal computer or close compatible that supports Lotus. Use one with a speed at least as fast as an eight megahertz AT. The difference in price between an AT and an XT is inconsequential, and the AT's speed will make calculations much faster. If you already have an XT or compatible, you may want to consider a coprocessor board for some applications.

Hard Drive.   A hard drive is a necessity. Do not try to use either Lotus program without a hard drive. The current cost of hard drives will make it one of your best investments. The difference in price between a 10 and a 30 or 40 megabyte hard drive is so insignificant that I recommend you buy **at least** a 30 as you will soon surpass 10 megabytes.

RS-232 Communications Port.   If you will also be interfacing instruments to your computer, the last necessity is at least one serial RS-232 communications port. If you have two however, two serial ports can automate two instruments. And, if you will be buying a serial printer, the second serial port is an absolute must for interfacing instruments because the printer will occupy the first serial port.

Graphics Equipment.   If you plan on using graphics, make sure that your monitor and printer can handle graphics. With new releases of Lotus products, an EGA or VGA monitor is well worth the extra cost.

Math Coprocessor. Math coprocessors are specialized microprocessors that perform floating-point calculations. Without a math coprocessor, your computer performs floating-point calculations using integer mathematics and requires many calculations to produce a single floating-point result. If a math coprocessor is present, floating-point calculations can be executed with a single instruction, thus improving the speed with which your spreadsheet recalculates and programs perform. This speed increase is usually between 10% to 50%, but depends on the application. Scientific data reduction applications typically involve large macro programs and/or trigonometric functions. These two categories are the most common to benefit from the addition of a math coprocessor.

Expanded Memory. If you are going to be dealing with a **very large** amount of data, you could run out of computer memory. You may want to consider adding expanded memory to your computer. Lotus®, Intel®, and Microsoft® Corporations have jointly developed a specification that allows programs to use memory above the standard 640K maximum. Memory boards conforming to this specification are referred to as *Expanded Memory Specification* (EMS) boards. Lotus 1-2-3, Releases 2 and 3 and Symphony can take advantage of the memory on expanded memory boards. These boards are marketed by Intel, Quadram®, AST®, and many others. You can use up to four megabytes of EMS memory from a Lotus spreadsheet.

If expanded memory is present in your computer, Lotus will use the EMS memory to store formulae, floating-point (decimal) numbers, and labels. Integers are stored in standard memory. The following table outlines the type of memory (CONVentional or EMS) where various types of information reside.

	1-2-3 2.0/2.01/2.2	SYMPHONY 1.1/1.2	SYMPHONY 2.0
INTEGERS	CONV	CONV	CONV
DECIMAL NUMBERS	EMS	EMS	EMS
LABELS (TEXT IN CELLS)	EMS	EMS	EMS
MACRO PROGRAMS	EMS	EMS	EMS
FORMULAE	EMS	EMS	EMS
@FUNCTIONS	EMS	EMS	EMS
RANGE NAMES	CONV	CONV	EMS
BLANK (ERASED) CELLS	CONV	CONV	CONV
ADD-IN APPLICATION PROGRAMS	CONV	CONV	CONV
SETTINGS SHEETS	CONV	CONV	EMS
LOTUS 1-2-3/SYMPHONY	CONV	CONV	CONV
PRINT SETTINGS	CONV	CONV	EMS
LOTUS TEXT (MESSAGES)	CONV	CONV	EMS
MEMORY RESIDENT PROGRAMS	CONV	CONV	CONV

As you can see, there are subtle, but very important, differences in how the various Lotus programs utilize expanded memory.

EMS memory is not nearly as fast as conventional memory. Thus, the use of EMS memory can profoundly affect the speed of your program. This difference in speed is especially evident in Lotus 1-2-3, Release 3 and Symphony, Release 2 where your programs will execute approximately 35% slower if you are using expanded memory. (For a direct comparison, see Chapter 9, Table 9-2.)

Lotus claims that you can utilize up to four megabytes of expanded memory. However, this statement is misleading. For each "segment" ("packet") of four active cells that are stored in expanded memory, Lotus allocates about 16 bytes of conventional memory (i.e., about 4 bytes of conventional memory are consumed for each EMS cell). This conventional memory is used for control purposes and establishes a link to the expanded memory location where each cell's information is located. The amount of unused conventional memory, therefore, will place limitations on the number of cells of data that your Lotus program can support (no matter how much memory each cell takes up in expanded memory).

The following table will give some rough guidelines. It assumes that you have a computer with 640K of conventional memory and at least 640K of EMS memory, storage of floating-point (decimal) numbers, and no memory-resident, print spooler, add-in application, etc., programs present:

	1-2-3 2.0/2.01	SYMPHONY 1.1/1.2	SYMPHONY 2.0
# CELLS WITHOUT EMS	34500	24500	23000
# CELLS WITH EMS	84000	68500	74000

Thus, EMS memory can help you collect massive amounts of data, but be forewarned that EMS memory is not nearly as fast as standard memory and it will slow down your spreadsheet and programs.

Unless you anticipate collecting more than 23,000 data points in your experiments, I strongly recommend disabling the expanded memory. Your spreadsheets will run about 25% to 35% faster for nearly all operations. You can disable your expanded memory by removing the command lines in the CONFIG.SYS file that specify the EMS device driver (i.e., the ".SYS" file) and then rebooting your computer.

There is one last EMS quirk that you need to be aware of. If you are using Lotus with a co-processor accelerator board (such as the Orchid Technologies PCturbo 286e), Lotus spreadsheets WILL NOT be able to access the personal computer's EMS memory in accelerator mode. If you have added one of these boards to an old IBM PC/XT to try to improve its speed, you will need to choose between having access to the personal computer's EMS memory and speed.

However, Coprocessor boards that have their EMS memory on daughter boards (such as the Intel Inboard 386/PC) will allow Lotus to access EMS memory on the daughter board.

Extended Memory.   Extended memory (also known as "Protected Mode Memory") is memory above 1024 kilobytes (1 megabyte). The 80286 and 80386 microprocessors can access extended memory directly. The 8088 microprocessor cannot.

Only certain programs can utilize extended memory. Currently in the Lotus family of products, only Lotus 1-2-3, Release 3, can utilize extended memory. In fact, Release 3 **requires** extended memory. A stock PC or XT computer without a co-processor board cannot access extended memory, because they have an 8088 microprocessor. It is for this reason that Release 3 cannot be used on a PC or XT computer unless a co-processor board has been installed.

Extended memory has two very valuable virtues. The first virtue is that extended memory is much faster than expanded memory. This feature makes program execution and spreadsheet recalculation much faster. Furthermore, unlike expanded memory, extended memory does not require an allocation of conventional memory for control purposes. Thus, the maximum cell limits described in the preceding section are virtually eliminated with extended memory.

A SIMPLE EXPLANATION OF LOTUS, MACRO COMMANDS, AND @FUNCTIONS

Lotus 1-2-3 and Symphony

Lotus 1-2-3 and Symphony both have over 600 commands. This magnitude tends to overwhelm and discourage new users. What new users fail to understand is that they need to know only a handful of commands to get started.

Consider all of the programming commands available as if they were tools in a toolbox. To perform a certain task, look up the name of the tool and use it. The more commands that you are aware of, the more tools you have at your disposal. **Half the battle of programming is knowing what is possible.** To get a grasp of the available Lotus tools, skim the manuals accompanying the Lotus software, browse through Appendices A through C at the end of this book, or look over some of the books listed as references at the end of this chapter.

A number of the most powerful commands are not documented in the Lotus manuals and many of the reference books listed. This book uses many of these undocumented features of the Lotus programming language. It also describes many ways to use the commands in programs...in both conventional and unconventional manners.

Macro Commands and @Functions

A set of instructions in a format and language that Lotus can understand is called a Lotus "*macro*" program. The language that is used is called the "Lotus Command Language" and the instructions are called "macro commands".

There are six broad categories of Lotus macro commands. They are

- *System commands*, which allow you to control the screen display and computer's speaker.
- *Interaction commands*, which allow you to create interactive macros that pause for the user to enter data from the keyboard.
- *Program flow commands*, which let you include branching and looping in your program.

- *Cell commands*, which transfer data between specified cells and/or change the values of cells.
- *File commands*, which work with data in DOS files.
- *Menu commands*, which allow you to design and manage menus.

Special-purpose Lotus formulae are called "*@functions*". Lotus contains numerous @functions. Each @function can extend your calculating power beyond simple arithmetic and text handling operations. There are several broad categories of Lotus @functions: mathematical, text, logical, financial, statistical, database, and date/time.

In general, macro programs use both macro commands and @functions to carry out the kinds of processes that your program implements to perform data reduction chores. Your goal is to plan and write efficient macro programs. Efficient programs

- get specific work accomplished in a specific order.
- specify the exact calculations that need to be carried out and when.
- manipulate data in the cells of your spreadsheet.

Planning ahead is perhaps the most important concept of efficient programming. Knowing what you need to do, what you can do, and when you need to do it is over 90% of the data reduction game.

YOUR KEYS TO SUCCESS

If you work through this entire book, you should be well equipped to create your own data reduction applications. The real secret to becoming an expert, however, is to actually work through the examples contained within this book. You will be able to get a much better feel for the structure of a program or methodology if you type the examples into a spreadsheet. Hands-on experience is the best way for you to learn. More importantly, if you have experimental data that can be analyzed using a data reduction method being described in a chapter, you should use the data to learn more about the technique.

After you have a general idea of what each example template or Lotus program can do, you should work with any features that you don't fully understand. Do so by making variations to see the effects of your changes. Place a copy of the formulae that are within Lotus program commands into cells of the spreadsheet and see how they update as the program proceeds. Frequently, after you try using a specific Lotus command or formula and see it in action, the explanation in the book will become clearer and you will learn something important about the statistical or numerical method.

One thing that you will discover when reading through this book is that there are often many solutions to a single problem. There often isn't a single, best way to achieve a programming goal with software. I will usually present only one or two ways to address each programming challenge. However, with a programming language as versatile as this one, there are usually *many* ways to implement almost any task. Some ways are better than others, depending on the circumstances and the application. Sometimes you will be surprised to find that what you thought was a very inefficient method, is actually the most efficient. Do not be afraid to investigate several alternative programming schemes to accomplish a particular task and then choose the best.

SUMMARY OF WHAT HAS BEEN ACCOMPLISHED

For this book, there are four broad categories of statistics. They are:

- *Descriptive statistics* that use simple Gaussian formulae.
- *Parametric statistics*, based on rigid assumptions about the population from which the sample data are drawn.
- *Nonparametric statistics*, which do not depend on the distribution or parameters of a population.
- *Analysis Of VAriance comparisons* of means for three or more groups of data.

There are three broad categories of curve fitting. They are:

- *Linear regression* (polynomial, logarithmic, exponential, etc.)
- *Nonlinear iteration.*
- *Interpolation* (linear, cubic, and cubic spline).

This book will teach you the basics for all of these categories and give you numerous examples of how to apply them to scientific experimental data.

WHAT'S NEXT?

The next chapter will show you how to automate the process of looking up values from tables. This automation is very valuable because it eliminates a common, tedious, time-consuming, and error-prone task. More importantly, the tools developed in the next chapter will provide a greater degree of automation for programs developed later in this book.

BEGINNER USERS' REFERENCES

ADAMIS, EDDIE, *Command Performance: Lotus 1-2-3*. Bellevue, WA.: Microsoft Press, Div. of Microsoft Corp., 1986.

MEZEI, LOUIS, *Laboratory Lotus®: A Complete Guide To Instrument Interfacing*. Englewood Cliffs, NJ.: Prentice Hall, Inc., 1989.

MILLER, STEVEN, *Lotus Magazine: The Good Ideas Book*. Reading, MA.: Addison-Wesley Publishing Co., Inc., 1988.

The staff of Lotus Books, *The Lotus Guide To Learning Symphony Macros*. Reading, MA.: Addison-Wesley Publishing Co., Inc., 1986.

The staff of Lotus Books, *The Lotus Guide To Learning Symphony Command Language*. Reading, MA.: Addison-Wesley Publishing Co., Inc., 1985.

STARK, ROBIN, *Encyclopedia of Lotus 1-2-3*. Blue Ridge Summit, PA: Tab Books, Inc., 1987.

ADVANCED USERS' REFERENCES

FENN, DARIEN, *Symphony Macros And The Command Language*. Indianapolis, IN.: Que Corporation, 1987.

ZUCKERMAN, MARCIA, DYRUD, ANNE, and POSNER, JOHN, *The Lotus Guide to 1-2-3 Advanced Macro Commands*. Reading, MA.: Addison-Wesley Publishing Co., Inc., 1986.

CAMPBELL, MARY, *1-2-3 Power User's Guide*. Berkeley, CA.: Osborne McGraw-Hill, Inc., 1988.

2

Lookup Tables

Anyone who has used statistics for very long knows how much time is spent looking up values in tables. These values are usually used in later calculations, used to determine whether the results of a certain statistical test are significant, etc. The importance of accuracy in determining a correct value from a large matrix of values goes without saying. This requirement means that extreme caution must be used when determining a value and makes looking up values tedious and time-consuming. It is therefore desirable to have a way to be able to quickly get the appropriate value from a table, while having confidence that the value is correct.

Lotus® spreadsheets provide you with a very powerful environment and tools that will accomplish this task. This chapter will teach you how to establish a table lookup system so that you can immediately begin looking up values from tables and save valuable time.

LOTUS LOOKUP TABLES

Lotus defines a *lookup table* as a range of cells in a worksheet that contains a table of values from which you can select the value you are interested in. In form and function, a lookup table is no different from a table that you see in an appendix at the end of a statistics book.

In addition to the "values" section of a Lotus lookup table, there is also an identifying row or column (or both) at the top and/or left of the table entries. Once you have created a table of values with an *identifier,* you can find the value you are interested in. The two Lotus lookup functions that accomplish this task are the @HLOOKUP (for horizontal lookup table) and the @VLOOKUP (for vertical lookup table) function. These @functions search an identifier (row or column, respectively) and retrieve a value from a table based on the position of the identifier that meets the criterion specified in the lookup @function.

However, you can also combine these two @functions to first scan horizontally to find the column you want, then search vertically to find the row you want, or vice versa. The table value obtained will be based on the column and row offsets.

A SIMPLE LOOKUP EXAMPLE

Most statistical tables are *two-way,* meaning that they contain several rows down and several columns across. However, to make it easier to explain how to manage data in these tables, let us first look at how to manage data in a single column before moving on to a more complicated example.

Figure 2-1 shows a vertical lookup table. Although Lotus spreadsheets can perform lookups in either horizontally or vertically arranged tables; in concept, horizontal and vertical lookups are identical. Vertical lookup will be discussed in detail in this section and the subtle differences exhibited by horizontal lookup will be discussed in the next section.

Figure 2-1 contains a column of critical values for the *t*-distribution at $p = 0.10$. Along the left column of the table (Column A) there are numeric values representing intervals (Degrees of Freedom, in this example). The range of these identifier values will be used as @function arguments and will be used to uniquely identify the row for the value that is to be returned. That is, Lotus will use Column A when searching for the value that you want to find. These identifiers are the comparison values for the table.

```
--------A---------B---------C--------D---------E--------F--------G-----
 1              CRITICAL VALUES FOR t-DISTRIBUTION
 2
 3 DF \ ALPHA      0.1     0.05    0.025    0.01    0.005
 4          1     3.078
 5          2     1.886
 6          3     1.638
 7          4     1.533
 8          5     1.476
 9          6      1.44
10          7     1.415
11          8     1.397
12          9     1.383
13         10     1.372
14         30      1.31
15
16
17
18
19              DEG_FRED         10
20
21              VVAL          1.372
22
23
24
```

Figure 2-1

To create the example lookup table, type in the values as you see them in Figure 2-1. Using the arrow keys, move the cell-pointer (the reversed video, highlighted cell) to the appropriate cells and type the relevant text or number. Use numerals instead of labels whenever you create column or row headings that will be used in calculations. That is, do **not** precede numerals with a single-quote, double-quote, caret sign (", ', or ^) or spaces. By typing in numerals rather than labels, the headings will be in an appropriate form for analysis by Lotus. If labels are used, Lotus would treat them as character strings, assign their values to zero and not return proper calculations.

Next, move the cell-pointer to the cell containing the "DEG__FRED" label. Issue the / Range Name Label Right command. (In Symphony®, you can issue the {Menu} command by either pressing the slash key (/) or the F10 key.) When prompted to "Enter Label Range:", press return. This command will give cell C19 the name "DEG__FRED". Then, move the cell-pointer to cell A4. Issue the /Range Name Create command. When prompted for the range name, type T__TABLE and press Return. When prompted for the range, press period (.), move the cell-pointer to cell F14 and press Return. This command will give cells A4..F14 the name "T__TABLE".

The @VLOOKUP function requires three arguments to operate. The format of the @VLOOKUP function is

```
@VLOOKUP (TEST, RANGE, COLUMN_OFFSET)
```

The TEST argument can be a numeric value, cell name, formula, or @function. Lotus compares this value against the values in the identifier column of the lookup table (Column A, beginning at cell A4 in this example).

At a minimum, RANGE must specify all identifier cells. However, it is more common to include all of the cells in the lookup table. This convention is perfectly acceptable as long as the identifier cells occupy the left-most column of RANGE. (When preparing two-way lookup tables, specifying the entire lookup table is usually necessary because two-way tables require both @VLOOKUP and @HLOOKUP functions.)

RANGE may contain cells that have values, references to other cells, formulae, or @functions that return values or labels. The COLUMN__OFFSET is the number of columns to the right or left of the identifier column that contains the result of the @VLOOKUP function. The first column of the table (the identifier column) has an offset of 0, the column immediately to the right of this column has an offset of 1, and so on. That is, the first actual row or column of data values begins with 1. With Symphony, if you specify a negative offset, Lotus will access results to the left of the identifier column and none of the offset columns need to be within RANGE.

In a vertical lookup table, the cells in the identifier (left-most) column of the lookup table define numeric brackets. The @VLOOKUP function works by searching down the identifier column. If Lotus finds a value equal to TEST, it stops on that row. If Lotus does not find a value equal to TEST, it stops at the first value greater than TEST and then backs up one cell to the row headed by the preceding comparison value.

This procedure may seem a little confusing, so let me say it another (but slightly less accurate) way: In essence, Lotus searches for the first value that is **greater** than the TEST entry argument. When a cell is found that meets the criteria, Lotus stops at that row. The result of the function will be in the row of the table immediately **above** the row that contains the first identifier value that is greater than the TEST value.

Once Lotus has identified the row that contains the result, it uses the offset argument to determine the column that contains the result.

To illustrate lookup, type the following formula into cell C21:

`@VLOOKUP(DEG_FRED,T_TABLE,1)`

Next, move the cell-pointer to cell C19. Type in some test values. For example, type a 1 in cell C19 and press Return. A 3.078 will appear in cell C21. That is, the value in the lookup table one column to the right of cell A4 will be returned. Likewise, when a 25 is placed in cell C19, a 1.372 will appear in cell C21 and when a 30 is placed in cell C19, a 1.31 will appear in cell C21.

If you type a value into cell C19 that is less than the first value in the identifier column, the @VLOOKUP function will return the value ERR. For example, if you try to find the value for zero Degrees of Freedom, @VLOOKUP will return ERR.

If you type a value into cell C19 that is greater than the largest value in the identifier column, @VLOOKUP will assume that the result of the function is in the last row of the table. For example, if you enter a value of 31 into cell C19, @VLOOKUP will return 1.31.

HORIZONTAL LOOKUP

The @HLOOKUP function is nearly identical to the @VLOOKUP function in every way. The only difference is that @HLOOKUP looks up entries from horizontal lookup tables. The format of the @HLOOKUP function is

`@HLOOKUP(TEST,RANGE,COLUMN_OFFSET)`

In a horizontal lookup table, the identifier values are in the top-most row of RANGE and the values are returned from the cell that is at the intersection of the column that meets the criterion.

In a fashion analogous to the @VLOOKUP, the @HLOOKUP function works by searching left to right across the identifier row. If Lotus finds a value equal to TEST, it stops on that column. If Lotus does not find a value equal to TEST, it stops at the first value greater than TEST and then backs up one cell to the column headed by the preceding comparison value.

Figure 2-2 shows some additions to Figure 2-1. These additions illustrate a horizontal lookup table. More importantly, the additions will be used in the next section as part of a two-way lookup table. In that section, a horizontal search will be made to determine the column containing information for a specified confidence

```
--------A---------B--------C--------D--------E--------F--------G-----
 1                 CRITICAL VALUES FOR t-DISTRIBUTION
 2
 3  DF \ ALPHA      0.1       0.05      0.025      0.01      0.005
 4        1        3.078
 5        2        1.886
 6        3        1.638
 7        4        1.533
 8        5        1.476
 9        6        1.44
10        7        1.415
11        8        1.397
12        9        1.383
13       10        1.372
14       30        1.31
15
16  1/PROB->        10        20        40       100       200
17  COLUMN           1         2         3         4         5
18
19            DEG_FRED         6
20            PROB           0.1
21            VVAL          1.44
22            HVAL             1
23
24
```

Figure   2-2

interval. This information will be used as the column offset for an @VLOOKUP. The horizontal table in Rows 16 and 17 will provide the needed offset.

Both @VLOOKUP and @HLOOKUP are based on searches down the identifier to find the first value that is **greater** than the TEST entry argument. For this reason, the values in the identifier **MUST** be entered in ascending order. Unfortunately, most statistical tables list significance levels in the opposite order. For example, in Figure 2-1, the order is 0.1, 0.05, 0.025, etc. For this reason, either a re-arrangement of the columns or a conversion of the column headings is necessary. It is usually more convenient to convert column headings to an acceptable form in a lookup table than to re-arrange a table. The conversion can be made by taking the reciprocal of the horizontal column headings (e.g. 1/B3). In this example, the conversion formulae are in Row 16. As you can see, this conversion provides whole numbers and places the numbers in ascending order.

Create the example horizontal lookup table as follows:

1. Type +1/B3 into cell B16.
2. Issue the /Copy command to copy cell B16 to cells C16..F16. When prompted for the range to copy from, type B16 and press Return. When prompted for the range to copy to, type C16..F16 and press Return.
3. Type 1 through 5 into cells B17 through F17, respectively.
4. Type PROB and HVAL into cells B20 and C22, respectively.
5. Move the cell-pointer to the cell containing the "PROB" label. Issue the /Range Name Label Right command. When prompted to "Enter Label Range:", press Return. This command will give cell C20 the name "PROB".

6. Move the cell-pointer to cell B16. Issue the /Range Name Create command. When prompted for the range name, type KEY__TABLE and press Return. When prompted for the range, press period (.), move the cell-pointer to cell F17. Press Return. This command will give cells B16..F17 the name "KEY__TABLE".

To illustrate horizontal lookup, type the following formula into cell C22:

@HLOOKUP (1/PROB, KEY_TABLE, 1)

Note that 1/PROB is used in the formula. This correction converts the user input value to the format used in the lookup table.

Move the cell-pointer to cell C20. Type in some test values. For example, type a 0.1 in cell C20 and press Return. A 1 will appear in cell C22. That is, the value in the lookup table **one** row below cell B16 will be returned. Likewise, when a 0.025 is placed in cell C20, a 3 will appear in cell C22 and when a 0.005 is placed in cell C20, a 5 will appear in cell C22.

You are now ready to create a two-way lookup table.

TWO-WAY LOOKUP TABLES

Figure 2-3 shows a conventional lookup table. It contains more of the table for the values of the t-distribution. Creating a fully functional two-way lookup table is easy, but takes some time to type in the values. However, this time will be more than offset by the time you will save looking up values in a table each time you need them and by

```
--------A---------B--------C--------D--------E--------F--------G-----
 1              CRITICAL VALUES FOR t-DISTRIBUTION
 2
 3 DF \ ALPHA      0.1      0.05     0.025     0.01     0.005
 4         1      3.078    6.314    12.706    31.821    63.657
 5         2      1.886    2.92      4.303     6.965     9.925
 6         3      1.638    2.353     3.182     4.541     5.841
 7         4      1.533    2.132     2.776     3.747     4.604
 8         5      1.476    2.015     2.571     3.365     4.032
 9         6      1.44     1.943     2.447     3.143     3.707
10         7      1.415    1.895     2.365     2.998     3.499
11         8      1.397    1.86      2.306     2.896     3.355
12         9      1.383    1.833     2.262     2.821     3.25
13        10      1.372    1.812     2.228     2.764     3.169
14        30      1.31     1.697     2.042     2.457     2.75
15
16 1/PROB->       10       20        40       100       200
17 COLUMN          1        2         3         4         5
18
19         DEG_FRED       6
20         PROB        0.025
21         VVAL        1.44
22         HVAL           3
23         VALUE       2.447
24
```

Figure   2-3

the peace of mind you will have by knowing that the values being used in your analyses are correct.

Begin by locating the table that you want to use in a reference or text book. Then, type the values into a range of cells large enough to hold the values you want to include. Be sure to include the numeric values that represent the intervals or reference points (e.g., significance levels and Degrees of Freedom) across the top row and down the left column. You should also be sure to include a second identifier table if the identifiers are in descending order. This addition will provide an index to the appropriate column in the main table.

When you position the lookup table, prepare for the future. For example, you may want to make the lookup table available to other spreadsheets. If this case applies to you, then the most convenient position to start in is cell A1 and you should include only one lookup table per spreadsheet. Then, when you want to include a table in another spreadsheet, you can position the cell-pointer to an open area of the new spreadsheet and use the /File Combine (Lotus 1-2-3®) or {SERVICES} File Combine command to add the lookup table.

Modify the existing example lookup table as follows:

1. Type the new data into the cells.
2. Type VALUE into cell B23.
3. (This step is optional if the horizontal lookup identifier row is not part of the two-way lookup table) Change the size and starting position of the T__TABLE named range. Issue the /Range Name Create command. When prompted for the range name, type T__TABLE and press Return. When prompted for the range, press {ESC}ape, move the cell-pointer to cell A3, press period (.), move the cell-pointer to cell F14, and press Return.

To illustrate two-way lookup, type the following formula into cell C23:

```
@VLOOKUP (DEG_FRED, T_TABLE, @HLOOKUP (1/PROB, KEY_TABLE, 1) )
```

This formula is the one you will need for your two-way statistical tables. Type in some combinations of test values into cells C19 and C20 and view the value returned in cell C23. When a number is placed into one of the cells, Lotus will begin by evaluating the @HLOOKUP function. As discussed above, the @HLOOKUP will return the column offset to be used by the @VLOOKUP. Then, Lotus will evaluate the @VLOOKUP and return the result in the column specified by @HLOOKUP.

CLEANING UP THE TABLE

The equations and text in cells B21..C22 (VVAL and HVAL) were teaching/error-checking tools and are no longer needed. So, erase them by issuing the /Range Erase command (Lotus 1-2-3) or the /Erase command (Symphony). When prompted for the "Range To Erase", type B21..C22 and press Return. You can then move cells B23..C23 upward by issuing the /Move command.

Also, the more formulae in your cells, the slower a spreadsheet will recalculate. If you recall, cells B16..F16 contain active formulae. To convert the formulae to their values, you can issue the /Range Values command. When prompted for the "Range To Copy From:", type B16..F16 and press Return. When prompted for the "Range To Copy To:", type B16..F16 and press Return. This action will convert the formulae to their number equivalents and place permanent numbers into the cells, thus overwriting the formulae that were originally in the cells.

After verifying the values in the table and testing the formulae, store the spreadsheet. To store the file in Symphony, issue the {SERVICES} File Save command (F9FS). In Lotus 1-2-3, the command is / File Save (/FS). When the program prompts for a name, use one that is descriptive. For this example, "T__Test" would be appropriate. Press Return.

SUMMARY OF WHAT HAS BEEN ACCOMPLISHED

In this chapter, you learned how to create a lookup table and how to nest the Lotus @VLOOKUP and @HLOOKUP functions to specifically access data in the table. You have also learned some basic menu commands and have been introduced to the use of formulae and @functions in Lotus spreadsheets.

This method of retrieving information from tables is very powerful because you cannot only manually type in criteria and view the appropriate data, but you can also use the information in other @functions and macro programs. For example, just as the @HLOOKUP was "nested" in the @VLOOKUP function to provide information to the @VLOOKUP, you can nest both of these functions into another @function, formula, or macro command to provide information to them. This capability will provide a high level of automation to your data reduction spreadsheets.

WHAT'S NEXT?

The next chapter is a review of the concept of a template and how to construct a template to organize your data. This information was presented in detail in *Laboratory Lotus®: A Complete Guide To Instrument Interfacing*. The chapter will give you a better understanding of the applications presented throughout the remainder of this book.

Chapter 3 will also provide you with a variety of methods that you can use to provide consideration for the number of significant figures in an experiment.

3

Spreadsheet Templates

Before you start analyzing your data or writing data reduction programs, you must create a spreadsheet template to hold the data.

A spreadsheet template is simply a portion of your spreadsheet that you set aside to hold data in an organized manner. Raw data can be input manually (by typing), imported from a file, or input directly into the cells of the spreadsheet using the concepts documented in *Laboratory Lotus®: A Complete Guide To Instrument Interfacing*. The template automatically formats the data and/or performs calculations. The template usually contains only labels and formulae. You can view a template as a kind of "skeleton" into which your data are placed. If you create a template before you start programming, you will find it much easier to create the program that manipulates the data.

With the templates that you create for your data, you can change and experiment with the raw data. If you have formulae in the template, they will instantly recalculate and display the results of any changes you make. This capability is extraordinarily powerful when working with experimental data and provides you with enough flexibility to be able to thoroughly examine the data.

The design of the example template in this chapter is simple. Calculations need not be simple averages, standard deviations, maximum values, minimum values, etc. (as they are in the example). They can be formulae of immense complexity. Nearly all of the scientific formulae that you use can be placed into a spreadsheet template. For example, you can obtain logarithms, exponentials, exponents, trigonometric functions, and statistical functions. (See Appendix A for an overview of mathematical calculations and Appendix B for an overview of Lotus® @functions.) You can even set up templates for matrix mathematics, polynomial linear regression analyses, integrations, and differentiations.

DESCRIPTION OF AN ASSAY TEMPLATE

Figure 3-1 contains a simple example template created for an assay that determines the DNA concentration of a sample.

Fluorescence spectroscopy is a valuable tool in many scientific disciplines,

```
--------A--------B-------C--------D--------E--------F--------G--------H----
 1                    FLUORESCENT DNA ASSAY DATA
 2 DATE RUN:
 3                 ***STANDARD CURVE DATA****                   X    REGRESS
 4 NG DNA/ML ASSAY1  ASSAY2   ASSAY3   ASSAY4  AVERAGE  PREDICT  CURVE
 5     0                                         ERR      ERR    0.0
 6    25                                         ERR      ERR    0.0
 7    50                                         ERR      ERR    0.0
 8    75                                         ERR      ERR    0.0
 9   100                                         ERR      ERR    0.0
10   250                                         ERR      ERR    0.0
11   500                                         ERR      ERR    0.0
12   750                                         ERR      ERR    0.0
13                 =========================================
14                      Regression Output:
15                 | Constant
16                 | Std Err of Y Est
17                 | R Squared
18                 | No. of Observations
19                 | Degrees of Freedom
20                 |
21                 | X Coefficient(s)
22                 | Std Err of Coef.
23                 =========================================
24                 ***SAMPLE FLUORESCENCE DATA****
25 SAMPLE ID
26 COMMENTS
27
28 SAMPLE A1
29        2
30        3
31 SAMPLE B1
32        2
33        3
34 SAMPLE C1
35        2
36        3
37
38                 ***SAMPLE CONCENTRATION DATA****
39 SAMPLE A1  ERR       ERR       ERR       ERR       ERR       ERR       ERR
40        2   ERR       ERR       ERR       ERR       ERR       ERR       ERR
41        3   ERR       ERR       ERR       ERR       ERR       ERR       ERR
42 SAMPLE B1  ERR       ERR       ERR       ERR       ERR       ERR       ERR
43        2   ERR       ERR       ERR       ERR       ERR       ERR       ERR
44        3   ERR       ERR       ERR       ERR       ERR       ERR       ERR
45 SAMPLE C1  ERR       ERR       ERR       ERR       ERR       ERR       ERR
46        2   ERR       ERR       ERR       ERR       ERR       ERR       ERR
47        3   ERR       ERR       ERR       ERR       ERR       ERR
48           ****SUMMARY OF SAMPLE CONCENTRATION DATA****
49      AVG   ERR       ERR       ERR       ERR       ERR       ERR       ERR
50       SD   ERR       ERR       ERR       ERR       ERR       ERR       ERR
51      %CV   ERR       ERR       ERR       ERR       ERR       ERR       ERR
52      VAR   ERR       ERR       ERR       ERR       ERR       ERR       ERR
53      MIN   ERR       ERR       ERR       ERR       ERR       ERR       ERR
54      MAX   ERR       ERR       ERR       ERR       ERR       ERR       ERR
55    RANGE   ERR       ERR       ERR       ERR       ERR       ERR       ERR
56    X-2SD   ERR       ERR       ERR       ERR       ERR       ERR       ERR
57    X+2SD   ERR       ERR       ERR       ERR       ERR       ERR       ERR
```

Figure   3-1

including biochemistry, molecular biology, enzymology, immunology, pharmacology, environmental analysis, clinical pathology, and food analysis. The example template presented in this chapter is one in which a fluorescent dye binds to DNA. When the dye binds, a fluorescence appears at 458 nanometers. A standard curve is generated using a set of known standards and a linear regression is performed on the curve. Unknown samples are assayed in replicate and their concentrations are determined from the regression. This method is a convenient way to determine the amount of DNA in a sample.

Assay data usually fit nicely into a grid (matrix) wherein each row or column of data will contain replicates for each sample being analyzed. Data from most other instruments or experiments also fit into convenient matrices or formats. Take a moment to think about how you could organize the data from your instrument or experiment to fit into a matrix. **Matrices are the cornerstones of good templates**. Because a matrix is such a major part of a template, if you make an effort to organize the matrix that holds your data now, then it will be easier to work with the data in the future.

The example template in Figure 3-1 contains four matrices and is divided into five sections. These sections are typical of those found in many templates of applications that are used for assaying samples.

THE STANDARD CURVE SECTION

The first section of an assay template is the Standard Curve section. This part of the template should be set up to hold the standarization data from an instrument.

In this example, the template is set up to receive data from a fluorimeter when the standard curve is being run with known amounts of DNA. It allows for quadruplicate determinations; to obtain more replicates you can add more columns. If you want fewer replicates, you could use the template as it appears in the example. No calculation will be adversely affected by fewer data points.

The column at the right of the raw data section (entitled "AVERAGE") contains active formulae that give the averages of the raw data points. These are the Lotus @AVG formulae and are described in more detail in the next chapter.

The column entitled "X PREDICT" gives the concentration values that the known amounts would have given if they were calculated from the regression curve. The "X PREDICT" column provides a good visual check for how well the regression curve "fits" the data. That is, this column will allow you to compare the values predicted from a regression curve to actual values. When actual values closely agree with predicted values, your regression analysis is more accurate, thus making the predictions for your samples more accurate. The "REGRESS CURVE" column provides an analogous comparison for fluorescent units.

THE REGRESSION SECTION

The next section of an assay template is the Regression section. Both Lotus spreadsheets have the ability to perform a broad range of regression analyses on data. This example shows only simple linear regression. As you will see in future chapters, the

Lotus regression program is very powerful and will support almost any type of regression, including exponential, polynomial, and log regression analyses.

The Regression section sets up an area in the template to receive data from the Lotus regression program. However, when the numbers for the regression parameters are returned from the regression program, the program places text into the spreadsheet along with the numbers. The text appearing in the box of Figure 3-1 originated from the regression program. (Detailed definitions and descriptions of how to make full use of the Lotus regression programs are given in Chapter 6.)

If you were to enter your own text in this area, it would be overwritten. You can get around this problem by building a box in the format that you want, picking an area of the spreadsheet outside the viewing area, directing the regression program to place the regression numbers into that area, and placing formulae into the cells in the template that refer to the corresponding cell in the regression area.

For example, if your regression area had its "constant" cell set to Z5, you would place +Z5 in the cell of your template that corresponds to this piece of data.

If you recall, the last two columns of the Standard Curve section (the ones entitled "X PREDICT" and "REGRESS CURVE") relate to the Regression section. Later, when you create the template for this example, you will place Lotus equivalents to the formulae $x = (y - b)/m$ and $y = mx + b$, respectively, into the cells of these columns. (When you create the template for your application, use equations that appropriately describe the data for *your* standard curve.)

When a linear regression is performed on the "NG DNA" and "AVERAGE" columns in this example, the "X PREDICT" and "REGRESS CURVE" columns update using the "Constant" (intercept) and "X Coefficient" (slope) from the linear regression. Using the results, you can see how closely each of the points on your standard curve agree with those predicted by the regression analysis.

THE RAW DATA AND SAMPLE DATA SECTIONS

The final two sections of an assay template are the Raw Data section and the Sample Data section. The Raw Data section is set up to receive the Fluorescent Units data for the samples. In the example, space is included for triplicate determinations on each of three separate samplings of a DNA preparation. This section will certainly be different for other experiments and/or applications. Thus, the beauty of a spreadsheet is that you can change it at will to fulfill any purpose.

The Sample Data section contains cells for the DNA concentration calculated from the linear regression curve. These cells correspond 1:1 with the Raw Data cells and contain formulae that refer to their counterparts in the Raw Data section. As was the case in the Standards section, the formulae use the X Coefficient and Constant from the regression analysis to determine the DNA concentration for a sample. The formulae are a re-arrangement of the familiar $y = mx + b$ formula and are the Lotus equivalents to $x = (y - b)/m$. Again, this formula may change for your particular situation.

You may want to merge your Raw Data and Sample Data sections so that the raw data and final results are in alternating columns. The format you choose for presentation is a matter of preference, but separating out the sample concentrations into a single matrix makes it more convenient to perform further statistical analyses.

SUMMARY SECTION

The formulae in the bottom section of the example template calculate the average, standard deviation, coefficient of variation, variance, minimum, maximum, range (max-min), and the mean plus/minus two standard deviations. These calculations are performed on the final DNA concentrations. Statistics on final values are usually better indicators of the true accuracy and precision of data, because they take into account the sensitivity, linearity, etc., of the assay and its standard curve.

The formulae in this summary section are based on Lotus @functions. These @functions are described in detail in future chapters, but a synopsis is given below for cursory explanation:

Abbreviation	Statistic	Lotus @Function
AVG	Average	@AVG
SD	Standard Deviation	@STD
%CV	Coefficient of Variation	Formula:+(B48*100)/B47
VAR	Variance	@VAR
MIN	Minimum	@MIN
MAX	Maximum	@MAX
RANGE	Range	Formula:@MAX-@MIN
X-2SD	Mean minus 2 Std Devs.	Formula:@AVG-2*@STD
X+2SD	Mean plus 2 Std Devs.	Formula:@AVG+2*@STD

For reasons that will be discussed in Chapter 5, corrections are made in this example template to the Lotus standard deviation and variance @functions (@STD and @VAR) to give *sample* statistics (i.e., for data that are obtained as a subset from a larger population).

PLANNING YOUR TEMPLATE

To make the best use of a spreadsheet's power, take time to **plan** your template. Plan it with the intention that the template will serve as a report for your notebook records. If you are also automating data collection, take into consideration that the template's primary purpose is to serve as a recipient of data from your instrument and that the template must work in harmony with the program that obtains the data. The time that you spend planning your template will be more than offset by the time that you will save when you begin programming. If you do not properly plan your worksheet, you may have to make changes that will require inserting rows, deleting columns, moving data to different cells, etc. Re-designing templates and rewriting programs take a lot of time and are prone to errors.

When you design a template, determine the location of each piece of data. Pay particular attention to whether the data would be better presented horizontally across the shreadsheet or vertically down. From this initial design consideration, you will be able to determine the locations for any macro programs that you plan to use. Coordinating the positions of the two is important because the template and its interactive program must form an efficient unit that exchanges data with the greatest harmony and safety. Remember always that a well planned template is the key to efficient programming.

You will also want to organize the areas of a template so a program can efficiently place data into them. For example, you may want separate sections for standards responses, regression analysis, unknown sample responses, and final results. In addition, remember that experiments increase in size as projects proceed and new ideas are generated. If you plan your template for expansion from the start, you can reserve room on the spreadsheet for extra rows and columns that can be used to hold extra data.

Sometimes, a little planning will permit you to design a template that can serve dual purposes. For example, you may be able to get day-to-day performance data for quality control purposes from the same template that holds your experimental data. Or, if you work in a centralized assay laboratory that performs sample analyses from several other laboratories, your need to issue individualized reports (that contain only the data that a submitting lab is interested in) can be met by laying out a template that can be individualized. (You can use the /Worksheet Column Hide command (Lotus 1-2-3®) or the /Width Hide command (Symphony®) to hide data from other submitting laboratories and just print the appropriate data for the lab in question.)

There are four major style guidelines to consider when designing a template:

- The more formulae that you have in cells, the slower the spreadsheet will recalculate. If you have several hundred formulae in the template, every time the spreadsheet recalculates it will take a substantial amount of time. Therefore, if you need to place large data summaries in the spreadsheet, assign as much of the summarization process to the program as is practical.

- Take the user's ability into account. Use headings and labels that are easily understood. Use as little jargon as feasible. If you use non-descriptive terms, it may confuse your audience. This confusion will lead to inefficient usage and errors that will waste time and may cause serious errors in the data reduction (and, therefore, the interpretation of the experiment).

- Do not place too many different types of experiments in one spreadsheet. Crowded layouts can be cumbersome and confusing for the user. Try to keep the template as small and uncluttered as possible.

- Keep all related data together. Try to segregate raw and summary data. This separation will make it more convenient to access specific categories of data when you perform future statistical analyses.

OVERVIEW OF CREATING TEMPLATES

Creating any template is very easy after you have planned it. You type in text headers and formulae, create named ranges, and specify formats for specific cells.

Because a template is nothing more than a completed "skeleton" without data, test the template by manually entering characteristic data into it to make sure that it works correctly. Then, after all of the labels, formulae, and programs have been entered and the template is working correctly, store it. **Store the spreadsheet template without the data**. With a "clean slate" template, you will not have to worry about deleting entries for each subsequent use.

You should also create copies of a spreadsheet on several floppy disks and store them in separate locations. These backup copies will be a safeguard against having to re-create the spreadsheet if the current one is damaged or lost.

CREATING THE TEMPLATE FOR THIS EXAMPLE

The following instructions will teach you how to create the template shown in Figure 3-1. The steps for your template will be essentially the same. Creating this template will teach you some important concepts that you can apply when you create **your** template.

The column titles are centered. The worksheet will automatically perform the centering task if you issue the following commands. In Symphony, issue the /Settings Label-Prefix Center command (/SLC). In Lotus 1-2-3, the equivalent command is /Worksheet Global Label-Prefix Center (/WGLC).

Using the arrow keys, move the cell-pointer to the appropriate cells and type in the relevant text or number.

Use numerals instead of labels whenever you create column or row headings that will be used in calculations or graphics. This convention will format the headings in an appropriate form. If labels are used, Lotus treats them all as character strings, assigns their values to zero, and does not give proper calculations or graphics displays.

To create the box, use a single quote and a series of equal signs for the horizontal lines (' = = =). Use a single quote and shift-backslash (') for the vertical lines. More instructions are listed below.

1. Into cell F5, type the formula to calculate the average Fluorescent Units value in Row 5:

   @AVG (B5. . E5)

   Issue the /Copy command to copy the formula in cell F5 to Rows 6 through 12. When you are prompted for the cells to copy from, type F5 and press Return. When prompted for the cells to copy to, use the arrow keys to move to cell F6, type a period (.), and use the down arrow key to move to cell F12. Press Return.

ERR will appear in all of the cells because the ranges on which their calculations depend do not have values yet. This display will appear in the cells of your template that contain certain Lotus @functions whenever there are no data within the @function's specified range. It will also occur in cells that depend on other cells whose @functions evaluate to ERR. These displays are normal. In Chapters 9 and 11, techniques to eliminate ERR messages will be presented. The methods use the Lotus @IF function.

2. Using the / Range Name Create command, create the following ranges:

Range Name	Range Of Cells
DATERUN	B2
STDX	A5..A12
STDF	F5..F12
STATS	C14
SLOPE	E21
INTERCEPT	F15
REPORT	A1..H57

3. Into cell G5, type +(F5-$INTERCEPT)/$SLOPE (the formula for calculating the DNA concentration from the standard's average fluorescence value in cell F5 and the slope/intercept from the regression analysis). Into cell H5, type +$SLOPE*A5+$INTERCEPT. NOTE: Using the dollar sign before a cell name will allow you to *fix* the column and row coordinates. That is, if you were to copy a formula in a down direction without the dollar sign, the row coordinates for SLOPE and INTERCEPT would change by the number of rows that offsets the final cell from the original cell. However because dollar signs precede $SLOPE and $INTERCEPT, they will be copied with no change in their cell designations.

4. Issue the / Copy command to copy cells G5..H5 to cells G6..H12. (i.e., When prompted for the range to copy from, type G5..H5 and press Return. When prompted for the range to copy to, type G6..G12 and press Return.) The formulae will be copied to cells G6..H12.

5. Into cell B39, type +(B28-$INTERCEPT)/$SLOPE (the formula for calculating the DNA concentration from the fluorescence value in cell B28 and the slope/intercept from the regression analysis).

6. Issue the / Copy command to copy from cell C37 to cells B39..H47.

7. Into cell B49, type @AVG(B39..B47), the Lotus formula for the average (mean).

8. Corrections to the Lotus standard deviation formula will be explained in Chapter 5. For now, just type the following formula into cell B50:

@STD(B39..B47)*@SQRT(@COUNT(B39..B47)/(@COUNT(B39..B47)-1))

9. Into cell B51, type +(100*B50)/B49, the formula for the coefficient of variation.

10. Corrections to the Lotus variance formula will be explained in Chapter 5. For now, just type the following formula into cell B52:

    @VAR(B39..B47)*(@COUNT(B39..B47)/(@COUNT(B39..B47)-1))

11. Into cell B53, type @MIN(B39..B47), the Lotus formula to find the minimum value in the range.
12. Into cell B54, type @MAX(B39..B47), the Lotus formula to find the maximum value in the range.
13. Into cell B55, type +B54-B53, the formula for finding the range of values.
14. Into cell B56, type +B49-2*B50, the formula for the mean minus two standard deviations.
15. Into cell B57, type +B49+2*B50, the formula for the mean plus two standard deviations.
16. Use the copy command to copy the formulae in cells B49 through B57 to columns C through H. Issue the / Copy command. When prompted for the "Cells To Copy From", press ESCape, use the arrow keys to move to cell B49, type a period (.), and use the down arrow key to move to cell B57. This sequence will highlight all of the cells to copy. Press Return. When prompted for the "Cells To Copy To", use the arrow keys to move to cell C49, type a period (.), and use the right arrow key to move to cell H49. Press Return.
17. Set the source for the print function to REPORT. In Lotus 1-2-3, issue the / Print Printer Range command. In Symphony, issue the {Services} Print Settings Source Range command. When prompted for the range, type REPORT and press Return. Press {ESC}ape until you return to the spreadsheet.

SIGNIFICANT FIGURE AND ROUNDING CONSIDERATIONS

You will want to include significant figure considerations in your template. That is, if the data that come from your instrument or experiment are good only to a certain number of significant figures, then you would probably not want to display data past that number of digits.

Most laboratories have significant figure and rounding guidelines that they follow for data reduction. However, depending on the discipline, the level of accuracy/precision required, and the level of significance that the rules play in the accuracy of the data to be reviewed, there is great variability in the rules that are employed. Because of the importance of significant figures and rounding in obtaining accurate values in data reduction, this section will begin with a quick review of some of the most common rules.

One of the most common rules for determining the number of significant figures to use is as follows:

> For all mathematical operations (addition, subtraction, multiplication, division, and exponentiation), retain the equivalent of two more places of figures than present in the single observed value.

For example, if the observed values are expressed as four significant figures (e.g., 345.1), then use enough places in the computation to express the numbers to six significant figures (e.g., 345.100).

This rule can cause difficulty when squaring operations are being performed because it can lead to a doubling of the number of significant figures (e.g., $0.8^2 = 0.64$; $0.88^2 = 0.7744$; $0.888^2 = 0.788544$; and so on). Therefore, an alternative rule (which is safer) retains at least twice the number of significant figures as present in the original data. For example, you would retain at least eight significant figures for a value like 60.64.

A common significant figure rule for quantitative chemistry states that

- When expressing experimental quantities, retain no more digits beyond the first uncertain one.
- During addition and subtraction, retain only as many decimal digits in the result as there are in the component with the least number of decimal places.
- During multiplication and division, the answer should have a relative uncertainty that is of the same order as that of the least precise component (e.g., the uncertainty of the answer should lie between one-fifth and twice that of the least precise component).
- The log of a number should have as many digits to the right of the decimal as there are significant digits in the number.
- The mean or average value of a set of numbers should have the same number of significant digits as the observations upon which it is based. (To be more rigorous, the number of significant digits to retain should be such that the uncertainty in the mean corresponds to its standard deviation.)

Since the advent of the personal computer, an alternative approach has become popular. Because it is more convenient to retain the maximum number of places that the computer is capable of working with, this approach allows the computer to perform its calculations with full precision and no significant figure considerations are made until the final values are obtained. Many scientists use this approach because they feel that there are advantages to carrying along full precision during calculations and that this practice prevents the propagation of rounding errors. There are, however, some significant disadvantages to using the maximum number of places (especially in the final reported value). For example, when one number is compared to another, they may be judged as "not equal" on the basis of digits well beyond the precision of the assay (e.g., 210.2678956 is not equal to 210.2678957 as far as Lotus is concerned, even though digits beyond 210.3 are not significant). As we will see later, using the maximum number of digits can have a profound effect on the accuracy of some statistical tests.

The elimination of one or more digits to the right of a number to obtain the desired number of significant figures is called *rounding*. Unlike the variability found in significant figure rules, there is less variability in the rules for rounding. One set of rules is:

- If the residue is greater than 5, increase the last digit by one.
- If the residue is less than 5, retain the last digit unchanged.
- If the residue is exactly 5, retain the last digit unchanged if even, or increase by one if odd.

Regardless of the significant figure and/or rounding rules that your laboratory has adopted, Lotus has tools which make it easy for you to implement your rules.

The most elementary situation is one in which you always retain the same number of decimal digits. This rule is a subset of the one used in the quantitative chemistry approach given above. For example, suppose you wanted to retain one significant digit in the decimal position of data that has been received from the fluorimeter. There are two ways to implement this system. The first method affects only the **display** of significant digits and does not give consideration to significant digits during intermediate calculations.

To illustrate this method, format the "AVERAGE" and "REGRESS CURVE" regions of the Standard Curve section of the template to one digit. In Symphony, issue the / Format Fixed (/FF) command. In Lotus 1-2-3, issue the / Range Format Fixed (/RFF) command. When prompted for the number of digits, type 1 and then press Return. When prompted for the range, type F5..G12 and press Return.

Notice that we did not use this method to format the entire spreadsheet. The reason for not formatting the entire spreadsheet is: When a cell is formatted, Lotus allocates memory for the cell whether it contains data or not. Therefore, if you format several thousand cells, a great deal of memory will be used up, thus limiting the amount of data that you can place into your spreadsheet. You may even run out of memory. Formatting large numbers of cells also tends to slow the recalculation of the spreadsheet.

If you want to change the number of digits for the entire spreadsheet, you could change the global default. The command that allows you to change the default format is the / Worksheet Global Format Fixed (/WGFF) command (Lotus 1-2-3) or the /Settings Format Fixed (/SFF) command (Symphony). This command does not use up memory because it is a default. However, if you want to display cells with significant figures other than the new default, you will need to reformat those cells using the instructions given above.

Regardless of the number of digits displayed, Lotus 1-2-3 and Symphony perform calculations with a numeric precision comparable to that of most mainframe computers. That is, regardless of the width of the spreadsheet columns or the formatting of the cells, up to 15 decimal places of precision are retained during calculations.

Under these circumstances, performing calculations on scientific data can increase the number of significant figures so that the resulting number is greater than that which the experimental precision can support. However, using this method may be contrary to your set of rules. That is, you may feel that for your data, merely formatting the display to a certain number of digits gives a false sense of trustworthiness, has a material affect on accuracy during careful work, and/or can lead to apparently incorrect results. The following example shows how performing a summation on identical numbers (i.e., placing a 1.45 into all four cells) can lead to apparently incorrect results in cells that have been formatted.

General Format	Fixed, 0 Decimal Place Format
1.45	1
+1.45	+1
2.90	3

To implement your own rules and/or establish a rigorous consideration for significant digits during calculations, you can use the Lotus @ROUND function. This @function helps you manage the difference between internal storage precision, intermediate calculation precision, and displayed precision. Used by itself, @ROUND permits you to alter the precision of numbers to **a specific number of decimal places** or to round whole numbers to **a specific number of integer places**. (However, @ROUND, by itself, does not allow you to specify **a constant number of significant digits**.)

The rules that Lotus follows are: Digits less than 5 are rounded down (towards zero), while digits 5 or greater are rounded up (toward the next higher number). The format of the @ROUND function is:

```
@ROUND(number,places)
@ROUND(@function,places)
@ROUND(formula,places)
@ROUND(cell,places)
```

That is, the number to be rounded may be a number, an @function, a formula, or reference to a cell containing a number or formula. For example,

```
@ROUND(21.723,1)
@ROUND(@AVG(B39..B47),1)
@ROUND((B49+C49)/2,0)
@ROUND((B28-$INTERCEPT)/$SLOPE,1)
@ROUND(B37,1)
```

If a calculation is specified in an @ROUND function, the calculation will be performed (with full internal precision) and the value returned will be rounded to the specified number of places.

The number of places can be either a negative or positive number between -15 and $+15$. If the number of places is negative, Lotus will round the integer portion of the number. Rounding to the nearest hundred requires a -1 as the number of places and rounding to the nearest thousand requires a -2. If the number of places is positive, Lotus will round the decimal portion of the number. If the number of places is zero (0), Lotus will round the number to an integer (i.e., it is similar to the @INT function; but rounds, instead of truncating). For example,

Calculation	Value Returned
@ROUND(99.47,-1)	100
@ROUND(99.47,0)	99
@ROUND(99.47,1)	99.5
@ROUND(-99.57,0)	-100

The last example shows the rounding of a negative number. In all cases, rounding a number means rounding away from zero, no matter if the number is positive or negative. That is why -100 is returned and not -99.

To implement a different system of rounding, you can create your own formula to fit your rules. For example, recall that the @ROUND function rounds fractional numbers depending on whether the digits are less than 5 or greater than/equal to 5. The @INT function always rounds fractions down to the next whole number. If you always need to round fractions up to the next whole number, you should use a formula similar to the following:

```
@ROUND(cell+0.499999,0)
```

To round to a different number of decimal places, simply move the decimal point in the 0.499999. For example, to round up to one decimal place, use 0.0499999. To control the sensitivity of the rounding up process, vary the number of nines. The fewer nines that you use, the less sensitive the process is to very small fractions. When rounding negative numbers with this formula, the values will be reduced in absolute value. For example, -2.2 will be rounded to -2, not -3.

To round to a **specific number of places**, thereby providing for a constant number of significant digits (not just the number of decimal places), you need to construct a special formula. The fundamental structure of the formula is:

```
@ROUND(cell/10^@INT(@LOG(cell)),NUM
            +@IF(cell>=1,-1,0))*10^@INT(@LOG(cell))
```

This formula rounds any positive number, @function, or formula to the correct number of specified significant figures (NUM). In this example, the "cell" designation stands for the cell containing the number to format. The formula first divides the number by the exponent that the number would have if it were converted to scientific notation, $(10^{\wedge}@INT(@LOG(cell))$. The @IF function determines if the number is greater than or equal to 1. If it is greater than or equal to 1, NUM is decreased by 1 for the @ROUND function. If the number is less than 1, the formula leaves it unchanged. Finally, the formula multiplies the number resulting from the @ROUND function by the exponent to convert it back to the correct number of digits. Because the formula is based on a logarithmic function, it works with positive numbers only. (Because the log of a negative number or zero is undefined, Lotus would return ERR.)

As previously explained in this chapter, using digits beyond those that are significant can adversely affect tests for equality. In later chapters, equality testing is performed on the data in Figure 3-1. For this reason, you should edit the formulae that calculate the sample DNA concentrations in the template so that the data are restricted to four significant figures. Do so by typing the following formula into cell B39:

```
@ROUND(((B28-$INTERCEPT)/$SLOPE)/10^@INT(
        @LOG(((B28-$INTERCEPT)/$SLOPE))),
            4+@IF(((B28-$INTERCEPT)/$SLOPE)>=1,-1,0))*
                10^@INT(@LOG(((B28-$INTERCEPT)/$SLOPE)))
```

Issue the /Copy command to copy this formula from cell B39 to cells B39..H47.

To summarize, to obtain full precision in your calculations, but display fewer decimal places, use the /Range Format Fixed (/RFF) or the /Worksheet Global Format Fixed (/WGFF) command (Lotus 1-2-3). For Symphony, use the /Format Fixed (/FF) or the /Settings Format Fixed (/SFF) command. Lotus will **display** only the number of decimals specified, but will **preserve** all digits during calculations. If you need to strictly adhere to significant figure rules during calculations, use @ROUND functions or prepare special formulae (based on @ROUND) as outlined above. The use of these measures is especially important if you will be testing the data for equality. If you need only the integer portions of numbers and want the numbers truncated, use the @INT function.

TEST THE TEMPLATE

Now, take the time to see how the template works. Type the numbers shown in Figure 3-2 into cells B5 through E12 and B28 through H36. For now, also type in the output shown in the Regression box. (Using the Lotus Regression program will be discussed in Chapter 6.)

The formulae and @functions in the template will automatically update after each number is added. Figure 3-2 shows the template after the example numbers have been added.

Actually, by typing in representative data you will be performing an even more important task—testing the template.

Testing a template *before* you start programming is very important. If the template does not perform correctly alone, you cannot expect it to perform correctly when it is interacting with a program or with other templates.

As a general rule, it is best to test a template or a program as much as possible **before** going on to the next programming task. This action makes it easier to identify and track down errors. That is, if you wait until after you write your program, there will be many program lines and interdependencies in your template. These relationships will make it much more difficult to isolate errors. However, if you test as you build, you can verify that each step works correctly before you add another layer of complexity.

When you are finished testing the template, store it in a file called EXAMPLE. (Although you would not normally store a template with its data, the data in this template will be used as a basis for examples in future chapters of this book.)

```
--------A--------B-------C--------D--------E--------F--------G--------H----
  1                      FLUORESCENT DNA ASSAY DATA
  2  DATE RUN:31-OCT-88
  3                   ***STANDARD CURVE DATA****                  X    REGRESS
  4  NG DNA/ML ASSAY1    ASSAY2    ASSAY3    ASSAY4   AVERAGE  PREDICT  CURVE
  5          0    93.9      92.9      94.3      94.3      93.9     0.7    92.7
  6         25   130.4     134.6     133.4     135.9     133.6    26.9   130.7
  7         50   174.5     171.8     173.1     174.1     173.4    53.1   168.7
  8         75   202.9     204.5     204.2     205.2     204.2    73.3   206.7
  9        100   235.0     242.2     242.7     239.9     240.0    96.9   244.7
 10        250   473.9     462.0     469.3     473.2     469.6   248.0   472.7
 11        500   847.1     854.2     850.4     862.2     853.5   500.6   852.6
 12        750  1236.4    1236.4    1226.0    1235.2    1233.5   750.6  1232.6
 13                 =======================================
 14                      Regression Output:
 15               Constant                    92.74541
 16               Std Err of Y Est             3.441756
 17               R Squared                    0.999939
 18               No. of Observations                 8
 19               Degrees of Freedom                  6
 20
 21               X Coefficient(s)   1.519749
 22               Std Err of Coef.   0.004814
 23                 =======================================
 24               ***SAMPLE FLUORESCENCE DATA****
 25  SAMPLE IDXX102-7 XX102-7   XX102-7   XX102-7   XX102-7  XX102-7 XX102-7
 26  COMMENTS                   SAMPLE PRECISION STUDY
 27
 28  SAMPLE A1  428.8     410.1     413.6     409.3     428.8    433.5   423.7
 29         2   433.5     407.2     394.3     418.1     410.1    407.2   405.7
 30         3   423.7     405.7     416.5     410.5     413.6    394.3   416.5
 31  SAMPLE B1  410.1     414.0     410.2     408.4     409.3    418.1   410.5
 32         2   415.8     405.5     405.8     421.3     410.1    415.8   416.9
 33         3   416.9     400.5     411.6     402.3     414.0    405.5   400.5
 34  SAMPLE C1  401.9     418.7     410.1     424.5     410.2    405.8   411.6
 35         2   406.4     415.8     412.2     416.0     408.4    421.3   402.3
 36         3   408.9     420.4     408.9     418.9     408.9    412.2   418.9
 37
 38               ***SAMPLE CONCENTRATION DATA****
 39  SAMPLE A1  221.1     208.8     211.1     208.3     221.1    224.2   217.8
 40         2   224.2     206.9     198.4     214.1     208.8    206.9   205.9
 41         3   217.8     205.9     213.0     209.1     211.1    198.4   213.0
 42  SAMPLE B1  208.8     211.4     208.9     207.7     208.3    214.1   209.1
 43         2   212.6     205.8     206.0     216.2     208.8    212.6   213.3
 44         3   213.3     202.5     209.8     203.7     211.4    205.8   202.5
 45  SAMPLE C1  203.4     214.5     208.8     218.3     208.9    206.0   209.8
 46         2   206.4     212.6     210.2     212.7     207.7    216.2   203.7
 47         3   208.0     215.6     208.0     214.6     208.0    210.2   214.6
 48         ****SUMMARY OF SAMPLE CONCENTRATION DATA****
 49       AVG   212.8     209.3     208.2     211.6     210.5    210.5   210.0
 50        SD     7.0       4.4       4.2       4.7       4.2      7.4     5.2
 51       %CV     3.3       2.1       2.0       2.2       2.0      3.5     2.5
 52       VAR    49.1      19.8      17.5      22.2      17.6     54.9    26.9
 53       MIN   203.4     202.5     198.4     203.7     207.7    198.4   202.5
 54       MAX   224.2     215.6     213.0     218.3     221.1    224.2   217.8
 55     RANGE    20.8      13.1      14.6      14.6      13.4     25.8    15.3
 56     X-2SD   198.8     200.4     199.9     202.0     202.1    195.7   199.6
 57     X+2SD   226.9     218.2     216.6     221.0     218.8    225.3   220.3
```

Figure   3-2

To store the file in Symphony, issue the {Services} File Save command (F9FS). In Lotus 1-2-3, the command is / File Save (/FS). When the program prompts for a name, type "EXAMPLE" and press Return.

Finally, erase the artificial data. In Symphony, issue the / Erase (/E) command. In Lotus 1-2-3, issue the / Range Erase (/RE) command. When prompted for the range to erase, press ESCape, arrow over to cell B5, type a period (.), arrow over to cell E12, and press Return. Repeat the process to erase cells B28 through H36. These commands will return your spreadsheet to an empty template. Store the empty template in a file named "DNA".

CREATE AND ERASE CELLS FOR MAXIMAL SPEED

There is another benefit to testing a template. The slowest process during data input is the process of creating new cells in a template. Once created, the process of filling the cells with data is relatively quick. Likewise, overwriting old data in the cells with new data is a relatively slow process. The fastest way for data input and reduction macro programs to place data in cells is to have the cells already created, but empty. You can set up this condition by forcing Lotus to create the cells at the time that you make up your original template.

Thus, you would fill the cells with "dummy" data during your testing session and then erase the data. Erasing existing cells clears the previous data, but leaves the cell's memory space and attributes intact. This process is faster than creating and filling new cells and is also faster than replacing existing data in cells.

If you have large numbers of identical cells in a matrix, you can use the /Data Fill (Lotus 1-2-3) or the /Range Fill (Symphony) command to fill your template with numbers. Then you should erase the numbers and use the empty cells to receive your data.

However, do not over-anticipate and create all of the cells in a spreadsheet. Even though the cells do not contain data, they still use up memory. By creating spreadsheets full of empty cells, you are likely to run out of memory and you may actually slow the performance of your spreadsheet if you create too many cells. You will also store bigger files on your disk. Therefore, create only the cells that you actually plan to use in your template.

BACKING UP YOUR WORK

I recommend storing a spreadsheet template (without data) as soon as you have all of the components in place and again after you have tested the template. In fact, I recommend storing it often during creation. That way, if you have a power failure or a computer lockup, you will not lose the work you have already completed. If you have a large macro program and spreadsheet, you may want to store it under different names and on several floppy disks.

Before you store a template, take a moment to restore any pertinent settings to their original default values. For example, take a minute to reset the label alignment settings in the current spreadsheet. Recall that we set the label alignment to "center" for column headers. For macro programs, you will want to left-justify the text. As before, the worksheet will automatically perform the justification. In Symphony, issue the /Settings Label-Prefix Left command (/SLLQ). In Lotus 1-2-3, the equivalent command is /Worksheet Global Label-Prefix Left (/WGLL).

To store the file in Symphony, issue the {Services} File Save command (F9FS). In Lotus 1-2-3, the command is / File Save (/FS). When the program prompts for a name, type one that is descriptive. For this example, I chose the name "DNA". Press Return.

After the first time that you save a spreadsheet, the program will detect the file's presence. It will prompt you to confirm that you want to overwrite the old file. At this point, choose the Replace option for Lotus 1-2-3 or the Yes option for Symphony.

SUMMARY OF WHAT HAS BEEN ACCOMPLISHED

Before moving on to descriptions of the @functions and macro programs that interact with the data in this spreadsheet, review some of the goals that you have achieved. First, you have created an organized template (skeleton), to hold your data. This superstructure will facilitate manipulation and presentation of the data.

In anticipation of macro programs, you have learned how to create named cells and ranges to hold data from an experiment. These named cells and ranges allow you to use names in the commands of data reduction programs; the programs are therefore easier to read. These named cells and ranges also serve to *marry* the template to a macro program.

This chapter also presented a number of ways to set the number of significant figures displayed and/or used in calculations. As we shall see throughout the remainder of this book, setting the number of significant figures becomes especially important when tests for equality are to be performed on the data.

WHAT'S NEXT?

The next chapter describes how to check data to determine the frequency distribution that is being followed. This step is the first in using statistics for characterizing data.

4

Checking Distributions

When summarizing raw data (especially data from an assay that is new to your laboratory), you should always start by determining the distribution of the population from which the sample data is being taken. A frequency distribution is a representation of the relationship between categories (or classes) of the data and the frequency of events belonging to each category. From this summary, you can determine whether a certain distribution is being followed and the type of statistical analyses to use. Most commonly used statistical techniques are based on the assumption that the distribution of values follows the symmetrical, bell-shaped (so-called *normal* or *Gaussian*), curve shown in Figure 4-1. A distribution summary ensures that your data fit this criterion; if they do not, you can pick other statistical tests accordingly.

This chapter shows you how to perform the appropriate analyses in the Lotus® environment. It uses the template and data from Chapter 3 as an example.

LOTUS FREQUENCY DISTRIBUTION TOOLS

Lotus has features that work together to facilitate analyses. The most important of these features is one that makes it easy to create a *frequency table*. You can set up as many categories (*bins*, in Lotus terminology) for your data as you want; each bin being minimum and maximum values that define a range. You can then assign Lotus to scan a particular range of your database and fill in the count for each category. The resulting frequency distribution table tells you how many entries fall into each bin.

Figure 4-2 shows an addition to the bottom of the template created in Chapter 3 (Figure 3-1). This additional template has been set up for a distribution analysis. As you can see, you must perform some preliminary work before you implement the

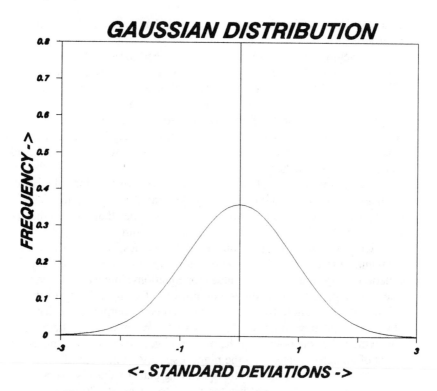

Figure 4-1

```
-------A--------B--------C--------D--------E--------F--------G--------H----
38                  ****SAMPLE CONCENTRATION DATA****
39  SAMPLE A1   221.1   208.8   211.1   208.3   221.1   224.2   217.8
40        2     224.2   206.9   198.4   214.1   208.8   206.9   205.9
41        3     217.8   205.9   213.0   209.1   211.1   198.4   213.0
42  SAMPLE B1   208.8   211.4   208.9   207.7   208.3   214.1   209.1
43        2     212.6   205.8   206.0   216.2   208.8   212.6   213.3
44        3     213.3   202.5   209.8   203.7   211.4   205.8   202.5
45  SAMPLE C1   203.4   214.5   208.8   218.3   208.9   206.0   209.8
46        2     206.4   212.6   210.2   212.7   207.7   216.2   203.7
47        3     208.0   215.6   208.0   214.6   208.0   210.2   214.6
48

-------A--------B--------C--------D--------E--------F--------G--------H---
59               ****FREQUENCY DISTRIBUTION ANALYSIS****
60         MULTIPLIER    BIN    FREQ CUM. FREQ% CUM FREQ
61           0.1       201.0    2          2      3.1
62           0.2       203.6    3          5      7.8
63           0.3       206.1    8         13     20.3
64           0.4       208.7   10         23     35.9
65           0.5       211.3   15         38     59.4
66           0.6       213.9   10         48     75.0
67           0.7       216.5    8         56     87.5
68           0.8       219.0    3         59     92.2
69           0.9       221.6    2         61     95.3
70           1         224.2    2         63     98.4
71                               0
```

Figure 4-2

analysis. You can then issue the command that activates the Lotus data distribution program.

To begin a distribution analysis, you need to assemble the data into a table of contiguous cells. The data that is used in the example originates from cells B39..H47 of Figure 3-1. As you can see, the matrix of data already conforms to this criterion and demonstrates the far-reaching benefits of taking the time to plan ahead when designing templates. For example, if the concentration data had been placed in alternating columns with the raw data, you would have been required to copy the information to a new set of cells, which were contiguous. This copying would take additional time and can lead to mistakes.

Next, you must decide on the cutoff values for the bins. The cutoff values must be in ascending order, with each value defining the upper limit of the bin to which it is assigned. That is, each bin will contain the values that are less than or equal to the value of the bin, but greater than the next lower bin.

Once you have decided on bin cutoff values, you need to create a bin range containing the cutoff values. These cutoff values may alternatively be specified by equations. Type the bin cutoff values (or equations) into a blank area of your template. The bin range must be a single column of cells immediately to the left of a blank column of cells. These blank cells will become an output range that Lotus will use to place the frequency counts of the values into. When creating the bin range, allow for one extra cell at the bottom of the output range; Lotus will use this cell to display the count of all cells greater than the highest specified bin interval.

One convenient method for specifying cutoff values is to determine the range of an experiment's values and use the range to create 10–20 bins, with each bin being 10%-5% larger than the last. Figure 4–2 demonstrates this technique by taking equal segments between the minimum and maximum value in the range (using the Lotus @MIN and @MAX functions). To make it easier to create this system, Column C is filled with "multipliers," each multiplier being 10% larger than the last.

The fastest and most convenient way to create the multipliers in Column B is to use the /Data Fill (Lotus 1-2-3®) or /Range Fill (Symphony®) command. This command allows you to fill a range of cells with an evenly spaced series of numbers. When you issue the command, Lotus will display the prompt "Enter Fill Range:". When you see this prompt, select the range to fill (B61..B70) and press Return. Next, Lotus will prompt you to define the Start value (the value that you want Lotus to place in the first cell of the range). The default value is 0, but you should change it to 0.1. Press Return. Next, Lotus will prompt you to define the Step value (the interval that you want between each number in the range). The default value is 1, but you should change it to 0.1 to get 10% increments. Press Return. Finally, Lotus will prompt you for a Stop value. The default value is 8191. Because this value is greater than the largest value you want in the Fill range, press Return. Lotus will immediately fill the range.

Next, create the following equation in cell C61

$$\text{@MIN(B\$39..H\$47)+C61*(@MAX(B\$39..H\$47)-@MIN(B\$39..H\$47))}$$

which is equivalent to the following literal formula

$$\text{MINIMUM VALUE} + 0.1 * \text{RANGE}$$

Issue the /Copy command to copy the formula from cell C61 to cells C62..C70. By using the multipliers in Column B, the formulae will take 10% increments between the minimum and maximum values in the range. That is, the following formulae will appear in the corresponding cells:

```
@MIN(B$39..H$47)+C61*(@MAX(B$39..H$47)-@MIN(B$39..H$47))
@MIN(B$39..H$47)+C62*(@MAX(B$39..H$47)-@MIN(B$39..H$47))
@MIN(B$39..H$47)+C63*(@MAX(B$39..H$47)-@MIN(B$39..H$47))
                        :
                        :
```

which are equivalent to the following formulae:

```
@MIN(B$39..H$47)+0.1*(@MAX(B$39..H$47)-@MIN(B$39..H$47))
@MIN(B$39..H$47)+0.2*(@MAX(B$39..H$47)-@MIN(B$39..H$47))
@MIN(B$39..H$47)+0.3*(@MAX(B$39..H$47)-@MIN(B$39..H$47))
                        :
                        :
```

but are much easier to create because you are not required to type in each formula individually.

(Note that the $ in the first equation will allow copying down a column without the row specifiers changing, thereby facilitating creation of the bin table.)

Next, select the /Data Distribution (/DD) command (Lotus 1-2-3) or the /Range Distribution (/RD) command (Symphony). You will be presented with a prompt to "Enter Values Range." Create the example in Figure 4-2 by typing B39..H47 and pressing Return. You will then be prompted to "Enter Bin Range." Specify the bin range only (B61..B70)—not the column to its right. Once you have pressed Return, Lotus immediately carries out the distribution analysis and displays the results in the output cells to the right of the corresponding bin number. Using this procedure on the example data in the template resulted in the data in cells C61 to C70 of Figure 4-2.

Once a frequency distribution table has been created, you should create a graphical representation of the table. Although a line or *xy* graph can be used, this example will use a histogram. The procedure for plotting the frequency distribution is slightly different for the two Lotus programs.

In Lotus 1-2-3, issue the /Graph command to bring up the graphics menu. From that menu, choose Type. Specify a Bar graph. Next, choose X (for the *X* range), specify C61..C70, and press Return. Finish the setup by choosing A (the first *Y* range), specify D61..D70, and press Return. To view the histogram, choose View. After viewing the graph, you can save it for later printing or plotting by choosing the Save menu choice.

In Symphony, issue the /Graph command to bring up the graphics menu. From that menu, choose 1st-Settings and then Type. Specify a Bar graph. Next, choose Range and then X (for the *X* range), specify C61..C70, and press Return. Finish the setup by choosing the A (the first *Y* range), specify D61..D70, press Return, and then choose Quit twice to return to the main Graph menu. To view the histogram, choose

Preview. After viewing the graph, you can save it for later printing or plotting by choosing the Image-save menu choice.

PARAMETRIC AND NONPARAMETRIC TESTS

Figure 4-3 shows the frequency distribution for the data in Figure 4-2. This distribution closely mimics the classic Gaussian "normal" curve (see Figure 4-1).

As was introduced in Chapter 1, statistical analyses that assume a Gaussian curve are called *parametric tests*. Parametric statistics test their hypotheses about their population's parameters, such as the mean or the variance. Because parametric tests assume that the data are drawn from a normal distribution, it is usually important that your data are normally distributed before these statistical tests give valid results. (However, some parametric statistical tests are robust enough in that they will tolerate deviations from normality and still provide reasonably accurate results.)

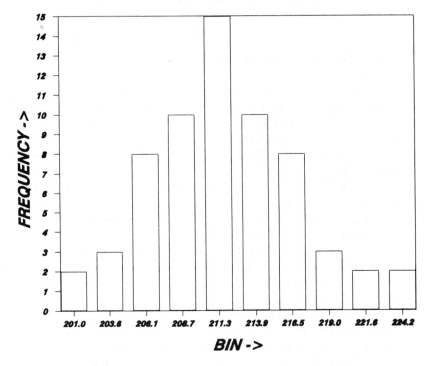

Figure   4-3

Frequently however, the data do not follow a Gaussian curve. Figures 4-4 through 4-10 show *x-y* graphs of some other possible shapes. In contrast to the symmetrical or bell-shaped Gaussian curve (characterized by equidistant observations from the central maximum having the same frequency), these shapes are not symmetrical. The following definitions describe the alternative shapes:

- Figures 4-4 and 4-5 illustrate *skewed* frequency curves. Skewed frequency curves are characterized by the tail of the curve to one side of the central maximum being longer than the other. If the longer tail occurs to the right, the curve is said to be *skewed to the right* or to have *positive Skewness*. If the reverse is true, the curve is said to be *skewed to the left* or to have *negative Skewness*.

- Figures 4-6 and 4-7 illustrate a *J shaped* and a *reversed J shaped* curve, respectively. In these cases, a maximum occurs at one end.

- Figure 4-8 illustrates a *U shaped* curve, in which there are maxima at both ends.

- Figure 4-9 illustrates a *bimodal* frequency curve, in which there are two maxima.

- Figure 4-10 illustrates a *multimodal* frequency curve, in which there are more than two maxima.

In these alternative circumstances, there is a preference to use statistical methods whose strengths do not depend on the precise shape of the distribution curve. These

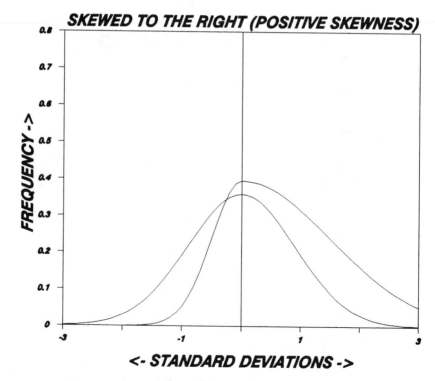

Figure 4-4

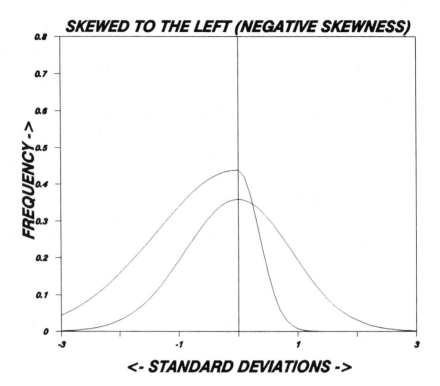

Figure 4-5

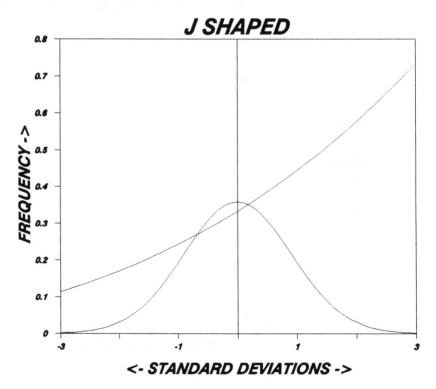

Figure 4-6

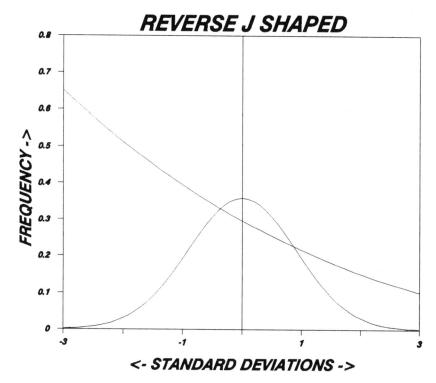

Figure   4-7

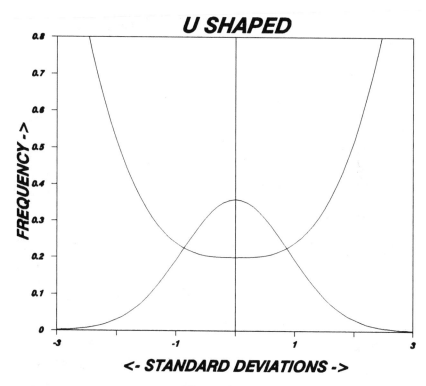

Figure   4-8

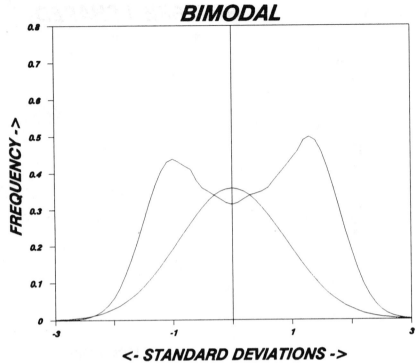

Figure   4-9

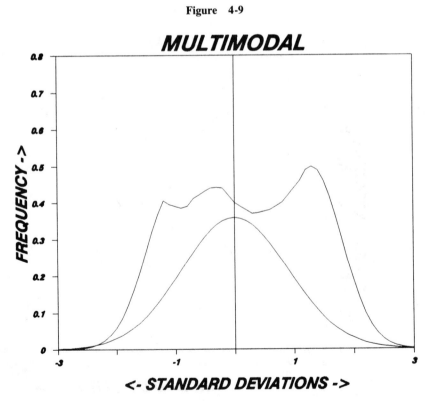

Figure   4-10

methods are based on *signs* of differences, *ranks* of measurements, and *counts* of objects or events falling into categories. Such methods do not depend heavily on explicit assumptions about population parameters or distribution. Therefore, they may be applied regardless of the kinds of distribution that the data exhibit. Because they are not based on the parameters of the normal distribution, they are called *nonparametric* statistics. Nonparametric methods are discussed in Chapter 10.

TESTING FOR NORMALITY OF DISTRIBUTION

Following from the discussion above, it is very important to determine whether your data are normally distributed. One method of empirically testing this assumption is to use the chi-square test for Goodness-of-Fit. Unfortunately, the chi-square test should be used only if there are at least five data points that fall into each bin of a frequency distribution table. This criterion is usually difficult to meet for fewer than 50 observations. Laboratory experiments more commonly have a smaller set of data. With fewer than five data points per bin, you should use the Kolmogorov-Smirnov test for Goodness-of-Fit. (The Kolmogorov-Smirnov test is presented in Chapter 10.) A quick and simple visual way to determine whether a set of data is consistent with the assumption of normality is to plot a *percent cumulative frequency curve*. Before elaborating on the chi-square test, let us first discuss the percent cumulative frequency curve and two other measures of normality (Skewness and Kurtosis).

The percent cumulative frequency curve can be easily added to your template, once you have prepared the frequency distribution table. Cells E61 through F70 in Figure 4-2 pertain to the cumulative frequency curve for the example data. Cells E61 through E70 contain formulae that calculate the cumulative frequency for each bin, i.e., the total number of measurements less than or equal to the current bin's frequency measurement. This comparison is calculated for each bin by forming a summation of the frequency for all previous bins.

Cells F61 through F70 give the percent cumulative frequency. The percent cumulative frequency is calculated using a Lotus equivalent to the following formula:

% cumulative frequency = 100 $*$ cumulative frequency / (n+1)

Use the following protocol to create the example in Figure 4-2.

1. Type +D61, into cell E61.
2. Type +E61+D62, into cell E62. (This formula will add the current row's frequency to the number in the previous row.)
3. Copy cell E62 to cells E63..E70. (Steps 1, 2, and 3 will form a summation series, thereby determining the cumulative frequency for each row/bin.)
4. Into cell F61, type +E61*100/(@SUM(D61..D70)+1). (These formulae will determine the percent cumulative frequency for each row.)
5. Copy cell F61 to cells F62..F70.

6. If you are using **Lotus 1-2-3,** issue the /Graph command to bring up the graphics menu. From that menu, choose Type. Specify an XY graph. Next, choose X (for the *X* range), specify C61..C70, and press Return. Finish the setup by choosing A (the first *y* range), specify F61..F70, and press Return. To view the graph, choose View. After viewing the graph, you can save it for later printing or plotting by choosing the Save menu choice.

7. If you are using **Symphony,** issue the /Graph command to bring up the graphics menu. From that menu, choose 1st-Settings and then Type. Specify an XY graph. Next, choose Range and then X (for the *X* range), specify C61..C70, and press Return. Finish the setup by choosing A (the first *y* range), specify F61..F70, press Return, and then choose Quit twice to return to the main Graph menu. To view the graph, choose Preview. After viewing the graph, you can save it for later printing or plotting by choosing the Image-save menu choice.

Figure 4-11 shows the graph for the example in Figure 4-2. Graphing the data from your experiment will give you a rough idea of the distribution of the data. If data come from a normal distribution, the graph will result in an S-shaped curve. Because the data in this example fit this criteria, they are probably normally distributed.

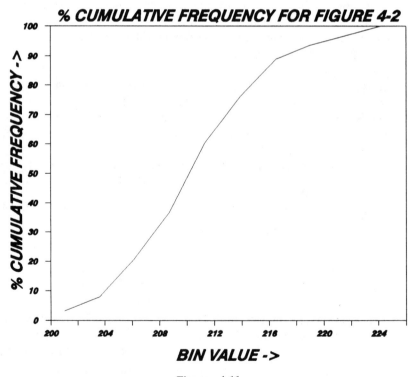

Figure   4-11

Figure 4-12 shows a graph of data that are multimodal. As you have probably surmised, the percent cumulative frequency graph will only expose gross deviations from normality. There are two other quick methods that provide better characterization of how well frequency curves approximate the normal distribution. These methods are the subjects of the next four sections.

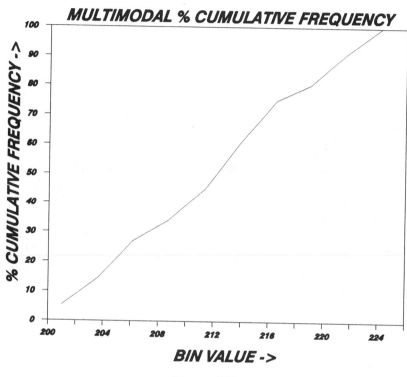

Figure 4-12

QUANTIFYING SKEWNESS AND KURTOSIS

Skewness is the degree of asymmetry of a distribution (i.e., how well the symmetry of a data distribution approximates a normal curve). If the frequency curve of a distribution has a longer "tail" to the right of the central maximum than to the left, the distribution is said to be *skewed to the right*. If the reverse is true, the curve is said to be *skewed to the left*.

The measure of Skewness is sometimes called the third moment and will take on a value of zero when the distribution is a completely symmetric, bell-shaped curve. A positive value indicates that the data are clustered more to the left of the mean with most of the extreme values to the right. A negative value indicates clustering to the right.

There are several methods that you can use to characterize Skewness. For example, you can determine the mean, and either the median or mode of the distribution. From this information, the Skewness can be calculated. (See Chapter 5 for a discussion of how to determine the median and mode.) However, the most satisfactory measure of Skewness employs the third moment about the mean.

The rth *moment about the mean, $\bar{X}$,* is defined in statistics books as follows

$$m_r = \frac{\Sigma (X - \bar{X})^r}{N}$$

You are probably already very familiar with the first two moments about the mean. The first moment about the mean is

$$m_1 = \frac{\Sigma (X - \bar{X})^1}{N}$$

which is the arithmetic mean of the deviations (and should be zero if the mean has been calculated correctly). The second moment about the mean is

$$m_2 = \frac{\Sigma (X - \bar{X})^2}{N}$$

which is the variance (the square of the standard deviation).

Similarly, the third moment about the mean is the average of the cubes of the deviations of the various values from the mean and is calculated using the following equation:

$$m_3 = \frac{\Sigma (X - \bar{X})^3}{N}$$

The Skewness can be calculated from either of the following formulae:

$$\text{Skewness} = m_3/s^3$$
$$\text{Skewness} = m_3/v^{3/2}$$

where, s is the standard deviation and v is the variance of the data.

Kurtosis is the degree of peakedness or flatness of a distribution, usually compared to a normal (Gaussian) distribution. The normal distribution is neither very peaked nor very flat-topped and has a Kurtosis of zero. If a distribution is more peaked (narrow) than a normal distribution, it will have a positive value for the Kurtosis and is called *leptokurtic*. A leptokurtic curve is shown in Figure 4-13. If a distribution is relatively flat-topped, it will have a negative value for the Kurtosis and is called *platykurtic*. A platykurtic curve is shown in Figure 4-14.

Kurtosis is sometimes called the fourth moment. The fourth moment about the mean is given by the following equation:

$$m_4 = \frac{\Sigma (X - \bar{X})^4}{N}$$

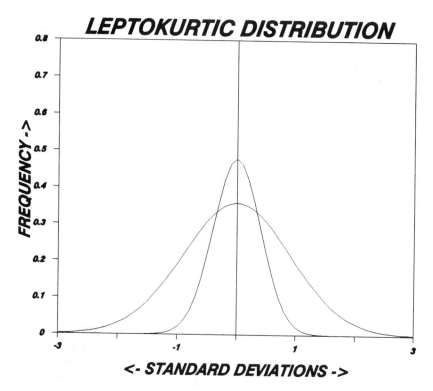

Figure   4-13

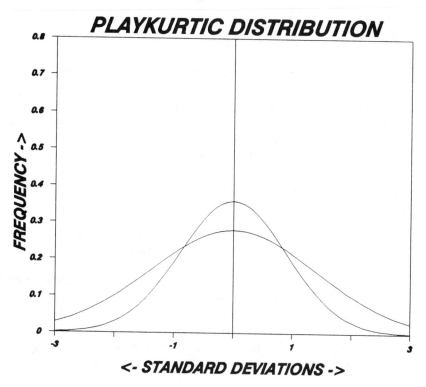

Figure   4-14

One measure of Kurtosis uses the fourth moment about the mean and can be calculated from either of the following equations

$$K = m_4/s^4$$
$$K = m_4/v^2$$

where, s is the standard deviation and v is the variance of the data.

CREATING A TEMPLATE TO CALCULATE SKEWNESS AND KURTOSIS

Figure 4-15 shows a template that calculates all four moments, Skewness, and Kurtosis. It has been arranged in such a way that some other statistical analyses can be conveniently performed (such as mode and median determination, which will be described in Chapter 5). The following will describe how to create a template of this type.

To preserve the original raw data, copy the raw data into the new template and arrange it in a single column. However, you cannot use the /Copy command to accomplish this task. If you recall from Chapter 3, the data displayed in these cells are the result of formulae that have been placed in the cells. The formulae used were of the following format:

```
@ROUND((((B28-$INTERCEPT)/$SLOPE)/10^@INT(
        @LOG(((B28-$INTERCEPT)/$SLOPE))),
            4+@IF(((B28-$INTERCEPT)/$SLOPE)>=1,-1,0))*
                10^@INT(@LOG(((B28-$INTERCEPT)/$SLOPE)))
```

Copying formulae from one range of cells to another will copy the formulae, not their values. Any cells referred to in the formula (e.g., B28), will be copied so that they refer to a cell positioned relative to the new cell. For example, copying the formula from cell B39 to cell A78 would lead to a change in the references from cell B28 to cell A67, which is incorrect.

Likewise, in Chapter 5 we will sort the data. To sort data you will need values in the cells, not formulae or @functions.

For these two reasons, you need to use the /Range Value command to copy the **contents** of the cells. That is, the /Range Value command will convert formulae and @functions in a range of cells to their current values and place these **values** into the new range of cells.

After the /Range Value command is issued, you will be prompted to "Enter Range To Copy FROM". For this example, specify B39..B47 and press Return. When prompted to "Enter Range To Copy TO", type A78 and press Return. Repeat to copy the data from ranges C39..C47, D39..D47, and E39..E47 to ranges starting at cells A87, A96, and A105, respectively. (Note that I did not use all of the data in this example so that I could fit the example on a single page.)

```
---------A--------B-----------C------D------E--------F--------G--------H----
 73                    ****SKEWNESS AND KURTOSIS ANALYSIS****
 74
 75           TEST FOR SKEWNESS              TEST FOR KURTOSIS
 76 RAW DATA SORTED DATA OCCURENCES     DIFF    DIFF^2   DIFF^3    DIFF^4
 77
 78    221.1                           10.6    112.1    1186.3   12558.7
 79    224.2                           13.7    187.3    2563.5   35084.9
 80    217.8                            7.3     53.1     386.8    2818.3
 81    208.8                           -1.7      2.9      -5.0       8.6
 82    212.6                            2.1      4.4       9.1      18.9
 83    213.3                            2.8      7.8      21.6      60.3
 84    203.4                           -7.1     50.6    -360.0    2561.1
 85    206.4                           -4.1     16.9     -69.6     286.4
 86    208.0                           -2.5      6.3     -15.9      39.9
 87    208.8                           -1.7      2.9      -5.0       8.6
 88    206.9                           -3.6     13.1     -47.2     170.6
 89    205.9                           -4.6     21.3     -98.2     453.2
 90    211.4                            0.9      0.8       0.7       0.6
 91    205.8                           -4.7     22.2    -104.7     493.8
 92    202.5                           -8.0     64.2    -514.7    4124.5
 93    214.5                            4.0     15.9      63.3     252.5
 94    212.6                            2.1      4.4       9.1      18.9
 95    215.6                            5.1     25.9     131.6     669.2
 96    211.1                            0.6      0.3       0.2       0.1
 97    198.4                          -12.1    146.7   -1777.7   21534.5
 98    213.0                            2.5      6.2      15.4      38.2
 99    208.9                           -1.6      2.6      -4.2       6.8
100    206.0                           -4.5     20.4     -92.0     415.1
101    209.8                           -0.7      0.5      -0.4       0.3
102    208.8                           -1.7      2.9      -5.0       8.6
103    210.2                           -0.3      0.1      -0.0       0.0
104    208.0                           -2.5      6.3     -15.9      39.9
105    208.3                           -2.2      4.9     -10.9      24.0
106    214.1                            3.6     12.9      46.1     165.4
107    209.1                           -1.4      2.0      -2.8       4.0
108    207.7                           -2.8      7.9     -22.3      62.7
109    216.2                            5.7     32.3     183.8    1045.3
110    203.7                           -6.8     46.4    -316.4    2155.7
111    218.3                            7.8     60.6     472.0    3675.2
112    212.7                            2.2      4.8      10.4      22.8
113    214.6                            4.1     16.7      68.2     278.8
114
115 AVERAGE        210.5            MOM1     MOM2     MOM3      MOM4
116 STD_DEV          5.309           0.0     27.4     47.2    2475.2
117 SKEWNESS         0.3156
118 MEDIAN                                  KURTOSIS   3.1147
119                                         (K-3)      0.1147
120
121
122
```

Figure   4-15

Next, issue the /Range Name Create command to create a range for the data. This name will make the creation of formulae easier and more readable. In the example, the range A78..A113 was given the name RAW__ DATA.

Type the text into the appropriate cells. Then, issue the /Range Name Label Right command to create range names for the average and standard deviation (AVERAGE and STD__ DEV, respectively). Issue the /Range Name Label Down command and highlight cells E115..H115 to create range names for the moments.

Type the following formulae into the appropriate cells:

Formula	Cell
@AVG (RAW__ DATA)	B115
+A78-$AVERAGE	E78
+E78^2	F78
+E78^3	G78
+E78^4	H78
@AVG (E78. . E113)	E116

Type the following formula for converting Lotus' population standard deviation into a sample standard deviation into cell B116:

@STD (RAW__ DATA) *@SQRT (@COUNT (RAW__ DATA) / (@COUNT (RAW__ DATA) −1))

(This formula will be explained in detail in the next chapter.) Issue the /Copy command to copy the following cells:

From	To
E78..H78	E79..H113
E116	F116..H116

Type the following formula into cell B117 for the calculation of Skewness:

+MOM3 / STD__ DEV^3

Into cell G118, type the following equation to calculate Kurtosis:

+MOM4 / STD__ DEV^4

For the normal distribution, $K = 3$. For this reason, the Kurtosis is sometimes defined by $(K - 3)$, which is positive for a leptokurtic distribution, negative for a platykurtic distribution, and zero for the normal distribution. Into cell G119, type the following equation to compare the Kurtosis to that of a normal distribution:

+G118−3

INTERPRETING THE EXAMPLE TEMPLATE

The example data in Figure 4-15 give a Skewness of 0.3156 and a Kurtosis of 0.1147.

Because the Skewness Coefficient is positive, the distribution is skewed positively, i.e., to the right. Because a normal distribution has a Kurtosis of 3 and the Kurtosis of this set of data is slightly greater than 3 (by 0.1147), the distribution is slightly leptokurtic with respect to the normal distribution (i.e., slightly more peaked than the normal distribution).

Combinations of indices are sometimes useful. One such combination is

$$3 * (\text{SKEWNESS})^2 - 2 * (\text{KURTOSIS}) + 6$$

The result of this equation is zero for a normal distribution, positive for a binomial or Poisson distribution, and negative for distributions that are extremely peaked. If you were to type the equation into the template of Figure 4-15, the result would be 0.06927.

SKEWNESS AND KURTOSIS CAUTIONS

You must use caution when using third and fourth moment tests. The fourth moment (and to some extent, the third) is restricted in its usefulness unless a large number of data points is used in the determination. If N is not large, the value of the peakedness can be unreliable because of its high sensitivity to fluctuations in the tail regions of the distribution.

If this criterion is met, these higher moments can often be very useful for studies of the properties of distributions and in arriving at theoretical distributions for fitting observed data.

GOODNESS-OF-FIT

The remainder of this chapter is dedicated to an example that ties together most of the important concepts of this, and the previous two, chapters. More importantly, this section shows you how to integrate Lotus tools (such as lookup tables, spreadsheet templates, data distributions, @functions, formulae, etc.) to form powerful systems that solve common statistical problems. This particular example creates a template to determine how well the normal curve is representative of a distribution of data.

Because this book is not intended to cover every known statistical test or how to apply all statistical tests to every possible situation, this example is important. The example begins to teach you how to choose Lotus tools and assemble them into a reusable system that solves **your** statistical problems. So, you will know what to do when you begin to develop your own system based on other statistical tests.

The statistical test described in this section is the *chi-square Test for Goodness-of-Fit* and is illustrated by Figures 4-16 and 4-17. One of the most common uses of the chi-square distribution is to test the hypothesis that a random variable has a specified theoretical statistical distribution. The term *Goodness-of-Fit* refers to the comparison of an observed sample frequency distribution with that of a theoretical frequency distribution.

The three most important functions describing expected distributions in science and engineering are:

- The Gaussian (normal) distribution.
- The Binomial distribution.
- The Poisson distribution.

```
--------A--------B--------C--------D--------E--------F--------G--------H----
 38                  ****SAMPLE CONCENTRATION DATA****
 39  SAMPLE A1    221.1    208.8    211.1    208.3    221.1    224.2    217.8
 40         2     224.2    206.9    198.4    214.1    208.8    206.9    205.9
 41         3     217.8    205.9    213.0    209.1    211.1    198.4    213.0
 42  SAMPLE B1    208.8    211.4    208.9    207.7    208.3    214.1    209.1
 43         2     212.6    205.8    206.0    216.2    208.8    212.6    213.3
 44         3     213.3    202.5    209.8    203.7    211.4    205.8    202.5
 45  SAMPLE C1    203.4    214.5    208.8    218.3    208.9    206.0    209.8
 46         2     206.4    212.6    210.2    212.7    207.7    216.2    203.7
 47         3     208.0    215.6    208.0    214.6    208.0    210.2    214.6
 48
--------A--------B--------C--------D--------E--------F--------G--------H----
125             ****CHI_SQUARE GOODNESS-OF-FIT TEST****
126
127                   MEAN    210.4238    STD    5.364871
128
129     FREQUENCY DISTRIBUTION      Z      GAUSS      AREA    EXPECTED
130  MULTIPLIER    BIN    FREQ    FOR CLASS    AREA    FOR CLASS    FREQ        CHI
131     0.2      203.6      5      -1.28    0.3997
132     0.4      208.7     18      -0.32    0.1255    0.2742    17.2746    0.030461
133     0.6      213.9     25       0.64    0.2389    0.3644    22.9572    0.181774
134     0.8      219.0     11       1.61    0.4463    0.2074    13.0662    0.326734
135     1        224.2      4       2.57    0.4949    0.0486     3.0618    0.287484
136                          0
137
138               N=        63                         CHI_SQR      0.83
139
```

Figure   4-16

If you have at least five data points per bin, the chi-square test for Goodness-of-Fit will help you decide whether the results of an experiment are governed by one of these distributions. Just as importantly, the chi-square test will **quantitate** how well the theoretical distributions (normal, binomial, Poisson, etc.) fit the experimental data. If you have fewer than five data points per bin, you should use the Kolmogorov-Smirnov test presented in Chapter 10. (Note that the data in this example do not meet this criterion. Nonetheless, the data will be used to illustrate the mechanics of setting up a chi-square test, thus allowing for brevity in the template and, therefore, a focus on the more important feature of the template—implementation of the chi-square test.)

Perhaps the most important application of the chi-square distribution in this context is testing whether or not a random variable is normally distributed. This test is important because virtually all of the Lotus statistical procedures presented in this book make the assumption that data are normally distributed. In this section, we shall make a comparison of the example DNA assay distribution to the normal (Gaussian) distribution.

The chi-square Goodness-of-Fit test in this example is based on the difference between the frequencies within ranges of the DNA samples and the expected range frequencies if the sample variation conforms to the normal distribution. Thus, the template must compute the expected frequencies of the DNA data as if the distribution of the data were normal. Then, the expected distribution is compared with the actual distribution.

You may be wondering why frequencies within ranges are used. The reason for this strategy is simple: The data are continuously variable; so the expected number of measurements equal to any one value would be extremely low (especially when the 15-digit precision of the computer is taken into account). Therefore, it is better to work

--------R--------S--------T--------U--------V--------W--------X--------Y--------Z--------AA------AB----

AREAS UNDER THE STANDARD NORMAL CURVE FROM 0 TO z

z	0	1	2	3	4	5	6	7	8	9
0.0	0.0000	0.0040	0.0080	0.0120	0.0160	0.0199	0.0239	0.0279	0.0319	0.0359
0.1	0.0398	0.0438	0.0478	0.0517	0.0557	0.0596	0.0636	0.0675	0.0714	0.0754
0.2	0.1179	0.1217	0.1255	0.1293	0.1331	0.1368	0.1406	0.1443	0.1480	0.1517
0.3	0.1179	0.1217	0.1255	0.1293	0.1331	0.1368	0.1406	0.1443	0.1480	0.1517
0.4	0.1554	0.1591	0.1628	0.1664	0.1700	0.1736	0.1772	0.1808	0.1844	0.1879
0.5	0.1915	0.1950	0.1985	0.2019	0.2054	0.2088	0.2123	0.2157	0.2190	0.2224
0.6	0.2258	0.2291	0.2324	0.2357	0.2389	0.2422	0.2454	0.2486	0.2518	0.2549
0.7	0.2580	0.2612	0.2642	0.2673	0.2704	0.2734	0.2764	0.2794	0.2823	0.2852
0.8	0.2881	0.2910	0.2939	0.2967	0.2996	0.3023	0.3051	0.3078	0.3106	0.3133
0.9	0.3159	0.3186	0.3212	0.3238	0.3264	0.3289	0.3315	0.3340	0.3365	0.3389
1.0	0.3413	0.3438	0.3461	0.3485	0.3508	0.3531	0.3554	0.3577	0.3599	0.3621
1.1	0.3643	0.3665	0.3686	0.3708	0.3729	0.3749	0.3770	0.3790	0.3810	0.3830
1.2	0.3849	0.3869	0.3888	0.3907	0.3925	0.3944	0.3962	0.3980	0.3997	0.4015
1.3	0.4032	0.4049	0.4066	0.4082	0.4099	0.4115	0.4131	0.4147	0.4162	0.4177
1.4	0.4192	0.4207	0.4222	0.4236	0.4251	0.4265	0.4279	0.4292	0.4306	0.4319
1.5	0.4332	0.4345	0.4357	0.4370	0.4382	0.4394	0.4406	0.4418	0.4429	0.4441
1.6	0.4452	0.4463	0.4474	0.4484	0.4495	0.4505	0.4515	0.4525	0.4535	0.4545
1.7	0.4554	0.4564	0.4573	0.4582	0.4591	0.4599	0.4608	0.4616	0.4625	0.4633
1.8	0.4641	0.4649	0.4656	0.4664	0.4671	0.4678	0.4686	0.4693	0.4699	0.4706
1.9	0.4713	0.4719	0.4726	0.4732	0.4738	0.4744	0.4750	0.4756	0.4761	0.4767
2.0	0.4772	0.4778	0.4783	0.4788	0.4793	0.4798	0.4803	0.4808	0.4812	0.4817
2.1	0.4821	0.4826	0.4830	0.4834	0.4838	0.4842	0.4846	0.4850	0.4854	0.4857
2.2	0.4861	0.4864	0.4868	0.4871	0.4875	0.4878	0.4881	0.4884	0.4887	0.4890
2.3	0.4893	0.4896	0.4898	0.4901	0.4904	0.4906	0.4909	0.4911	0.4913	0.4916
2.4	0.4918	0.4920	0.4922	0.4925	0.4927	0.4929	0.4931	0.4932	0.4934	0.4936
2.5	0.4938	0.4940	0.4941	0.4943	0.4945	0.4946	0.4948	0.4949	0.4951	0.4952
2.6	0.4953	0.4955	0.4956	0.4957	0.4959	0.4960	0.4961	0.4962	0.4963	0.4964
2.7	0.4965	0.4966	0.4967	0.4968	0.4969	0.4970	0.4971	0.4972	0.4973	0.4974
2.8	0.4974	0.4975	0.4976	0.4977	0.4977	0.4978	0.4979	0.4979	0.4980	0.4981
2.9	0.4981	0.4982	0.4982	0.4983	0.4984	0.4984	0.4985	0.4985	0.4986	0.4986
3.0	0.4987	0.4987	0.4987	0.4988	0.4988	0.4989	0.4989	0.4989	0.4990	0.4990
3.1	0.4990	0.4991	0.4991	0.4991	0.4992	0.4992	0.4992	0.4992	0.4993	0.4993
3.2	0.4993	0.4993	0.4994	0.4994	0.4994	0.4994	0.4994	0.4994	0.4995	0.4995
3.3	0.4995	0.4995	0.4995	0.4996	0.4996	0.4996	0.4996	0.4996	0.4996	0.4997
3.4	0.4997	0.4997	0.4997	0.4997	0.4997	0.4997	0.4997	0.4997	0.4997	0.4998
3.5	0.4998	0.4998	0.4998	0.4998	0.4998	0.4998	0.4998	0.4998	0.4998	0.4998
3.6	0.4988	0.4988	0.4999	0.4999	0.4999	0.4999	0.4999	0.4999	0.4999	0.4999
3.7	0.4999	0.4999	0.4999	0.4999	0.4999	0.4999	0.4999	0.4999	0.4999	0.4999
3.8	0.4999	0.4999	0.4999	0.4999	0.4999	0.4999	0.4999	0.4999	0.4999	0.4999
3.9	0.5000	0.5000	0.5000	0.5000	0.5000	0.5000	0.5000	0.5000	0.5000	0.5000

Figure   4-17

with expected numbers in intervals (e.g., $a < x < b$). That is, we must divide the range of possible values into bins.

The procedure for choosing bins depends somewhat on the nature of the experiment being dealt with. Rules for determining bins are given in most statistics reference books. One method is to use the mean and standard deviation. In this case, from four to eight bins are set up. If four bins are used, the following scenario is used:

Bin	Bin Range
1	$x < (\text{mean} - \text{std})$
2	$(\text{mean} - \text{std}) < x < \text{mean}$
3	$\text{mean} < x < (\text{mean} + \text{std})$
4	$(\text{mean} + \text{std}) < x$

If eight bins are used, the following scenario is used:

Bin	Bin Range
1	$x < ($mean $- 1.5 *$ std$)$
2	$($mean $- 1.5 *$ std$) < x < ($mean $-$ std$)$
3	$($mean $-$ std$) < x < ($mean $- 0.5 *$ std$)$
4	$($mean $- .5 *$ std$) < x < $ mean
5	mean $< x < ($mean $+ 0.5 *$ std$)$
6	$($mean $+ 0.5 *$ std$) < x < ($mean $+$ std$)$
7	$($mean $+$ std$) < x < ($mean $+ 1.5 *$ std$)$
8	$($mean $+ 1.5 *$ std$) < x$

However, I have found the scheme described in Figure 4-2 (i.e., taking equal divisions between the minimum and maximum values) to be very convenient and quite reliable. For chi-square Goodness-of-Fit testing however, I usually divide the range into five bins, each bin being 20% higher than the last (see Figure 4-16). The equation that implements this system is the same as the one presented earlier in this chapter

@MIN($CHI__DATA)+A131*(@MAX($CHI__DATA)-@MIN($CHI__DATA))

where, CHI__DATA is defined as cells B39..H47.

Having divided the range of values into bins, the number of measurements that fall within each bin can be determined using the /Data Distribution (Lotus 1-2-3) or /Range Distribution (Symphony) command. We denote this number by O_k (where O stands for "observed number" and k is the bin number).

Next, the expected number of measurements (E_k) in each bin (k) is calculated using a table of areas under the standard normal curve. The calculation of the expected numbers (E_k) is straightforward. The probability that any one measurement will fall in a bin interval $a < x < b$ is just the area under the Gauss curve between $x = a$ and $x = b$. However, the bins are not numerically in standard Gaussian form. That is, the table of areas is based on a normally distributed standard deviation and is centered on a mean of zero. The bins have been set up as 20% increments between the minimum and maximum values (i.e., linear). For this reason, a calculation must be made on each bin to convert the bin to the correct scale. The "z" values thereby obtained are the transformation values of the upper limits for each bin so that the table can be used. The following formula shows the format that is used in cells D131..D135 to find the z values:

@ROUND((B131-$MEAN)/$STD,2)

To clarify the use of this equation, suppose you had chosen the alternate system of using the mean and multiples of the standard deviation to set the ranges. The equation for the bin formed by the mean minus two standard deviations would become

$$Z = \frac{(\text{mean} - 2*\text{std}) - \text{mean}}{\text{std}} = -2$$

illustrating that z is just the number of standard deviations away from a zero mean (because it is already in standard Gaussian form).

Using the z value and a lookup table, the area under the curve is determined. Figure 4-17 is the lookup table and can be found in any statistics reference book. It is prepared in the same way as the one in Chapter 2, except for the absence of a KEY__TABLE. If you recall, a key table was necessary for the t-distribution because of the reversed order of the identifiers (descending). The key table was also necessary because a row was needed to specify the column offset for an @VLOOKUP. Fortunately, the Gaussian table has column headings that can be used directly. That is, the numbers are in ascending order and are integers. Therefore, an @HLOOKUP with the following format can be constructed to give the correct column offset:

```
@HLOOKUP (TEST, $GAUSS, 0) +1
```

where, GAUSS is defined as R3..AB43 for this example. (The entire table is used in the definition because it contains identifiers for both horizontal and vertical lookups.)

The two key components of this @HLOOKUP are the 0 row offset and the "+1". An offset of zero is a special case for @HLOOKUP (and @VLOOKUP). It returns the value in the identifier row (column). In Figure 4-17, these values are zero through nine. Because the data columns are one through ten, a "+1" is added to make the correction, eliminating the need for a second lookup table.

The values in the lookup table are divided as follows:

- Each horizontal row in the table represents the values within a single decimal digit (e.g., #.#).
- The vertical columns within each row in the table represent the second decimal digit (e.g., ?.?#).

For example, to find the area under the curve for $z = 1.62$, you would look down the table to the row that contains 1.6 at the left. You would then look across the table to the column that contains 2 at the top. This lookup sequence means that you must divide every number that you want to look up into two parts. The first part contains the #.# section, and the second part contains the hundredths decimal digit. The following set of nested @functions is the appropriate one to find the first part of the number (and thereby, the test for the row identifier):

```
@ABS (@VALUE (@STRING (D131, 2) ) )
```

This formula begins by changing the number in cell D131 into a string with two decimal digits. This number of digits is necessary to prevent rounding. Next, the string is converted back to its numerical value and the absolute value is found.

To find the second part of a number (and thereby, test for a column offset), the following formula is used:

```
@ABS (@VALUE (@RIGHT (@STRING (D131, 2) , 1) ) )
```

This formula begins by changing the number in the cell to a string of characters with two digits past the decimal. The @RIGHT function is then used to return only one character, the last one. This character is then converted back to its numerical value and the absolute value is found. Thus, this formula returns a single integer from zero to nine.

The lookup formulae in cells E131..E135 are straightforward and have the following format:

```
@VLOOKUP (@ABS (@VALUE (@STRING (D131, 2) ) ) , $GAUSS,
    @HLOOKUP (@ABS (@VALUE (@RIGHT (@STRING (D131, 2) , 1) ) ) , $GAUSS, 0) +1)
```

From cells E131..E135, the areas for each bin interval are determined. As stated earlier, the expected number of measurements (E_k) in each bin (k) is calculated as the area under the Gauss curve between $x = a$ and $x = b$. Subtracting the area at point b from the area at point a will give the area under the Gauss Curve. This area subtraction is the function of the formulae in cells F132..F135. These formulae find the areas under the normal curve between successive values of z. The following rules are observed:

- When corresponding zs have the same sign, the value is obtained by subtracting the previous area from the current one in Column E.
- When zs have opposite signs, the value is obtained by adding the previous area to the current one in Column E. This addition calculates the area about the origin.

The formulae in cells F132..F135 have the following format:

```
@ABS (@IF (D131<0#AND#D132>=0, E132+E131, E132-E131) )
```

Thus, if the previous cell was negative and the current cell either positive or zero, then the areas are added. If both cells are either negative or positive, the areas are subtracted.

To find the expected frequency (E_k) in each bin (k), these probabilities are multiplied by the total number of measurements, (63 in this example). The expected frequency is calculated in cells G132..G135 according to equations of the following format:

```
+F132*@COUNT ($CHI__DATA)
```

The @COUNT in this equation returns the actual number of cells containing data in the CHI__DATA range. By using this @function, a degree of flexibility is added to the template. That is, if a different number of data points is imported into the template, the formulae will automatically update to reflect the change.

A chi-square value is calculated using the following formula

$$\text{chi-square} = \sum \frac{(O_k - E_k)^2}{E_k}$$

The Lotus equivalent of this formula for each bin is found in cells H132..H135. For example, in cell H132 the equation is

```
+(C132-G132)^2/G132
```

To find the total chi-square, a summation is made on these values. The Lotus @SUM function in cell G138 has the following format:

```
@SUM(G132..G135)
```

This example gives a chi-square value of 0.83. We must then decide on how well the observed numbers O_k compare with the expected numbers E_k. To state the scenario simply: If chi-square is zero, the observed and theoretical frequencies agree exactly. If chi-square is greater than zero, they do not agree exactly. The larger the value of chi-square, the greater the discrepancy between observed and expected frequencies.

To determine the significance of the chi-square value, cutoff values are found by looking up their corresponding values in a standard chi-square table. The values used are based on the Degrees of Freedom and levels of significance.

In general, the number of Degrees of Freedom (DF) in a statistical calculation is defined as the number of observed data points minus the number of parameters computed from the data and used in the calculation. For chi-square problems, the number of observed data is the number of bins. Therefore,

$$DF = n - c$$

where n is the number of bins and c is the number of parameters that had to be calculated from the data in order to compute the expected numbers E_k. The number c is often called the number of constraints. In this example, the number of constraints is 3 (total number of data points, the minimum value, and the maximum value). Therefore, the Degrees of Freedom for this template is:

$$DF = 5 \text{ bins} - 3 \text{ constraints} = 2$$

On the other hand, if the standard deviation method were to be used to set up bins, the number of constraints would also be 3. One Degree of Freedom is lost by specifying the sample size, and two Degrees of Freedom are lost due to the estimation of the population mean by the sample mean and the population standard deviation by the sample standard deviation.

A rigorous definition for the number of Degrees of Freedom (DF) is given by the following:

- If you are working with very large numbers of samples (i.e., populations), $DF = k - 1$. Note that 1 is subtracted because you will be using the number of data points.

- If expected frequencies can be computed only by estimating m population parameters from statistics using a small number of samples, $DF = k - m$; where m is the number of known constants used in determining the expected frequencies.

Based on 2 Degrees of Freedom and a confidence limit of 0.50, the percentile value for chi−square of 1.39 is found in a standard reference book. Because the chi-square value determined in this template (0.83) is less than the percentile value, the data in the template do indeed follow the normal (Gaussian) distribution.

Before we create the template, let me give you a word of caution when using the chi-square Goodness-of-Fit test. You should look with suspicion upon circumstances where the chi-square test is **too close to zero**, because it is rare (but possible) that observed frequencies agree too well with expected frequencies. To examine such situations, determine whether the computed value of chi-square is less than the chi-square cutoff value at $p = 0.05$ or $p = 0.01$. The agreement at these two levels of significance should clarify whether the fit is "too good".

The chi-square cutoff value for $p = 0.05$ and 2 Degrees of Freedom is 0.103. Thus the chi-square value determined in this example (0.803) indicates that the fit is not so good as to be unbelievable.

CREATING THE EXAMPLE CHI-SQUARE TEMPLATE

To create the chi-square template, type the multipliers in cells A131 through A135. Next, type in the template text headings and the GAUSS lookup table (Figure 4-17). Activate the following ranges by issuing the /Range Name Create command and specifying the following range names and cells:

Range Name	Cell Range
CHI__DATA	B39..H47
MEAN	D127
STD	F127
GAUSS	R3..AB43

Complete the template by typing the following formulae into the specified cells of the template:

1. Type the formula for finding the average of the data into cell D127

   @AVG (CHI__ DATA)

2. Type the formula for finding the standard deviation of the data into cell F127

   @STD (CHI__ DATA) *@SQRT (@COUNT (CHI__ DATA) / (@COUNT (CHI__ DATA) −1))

3. Type the formula for setting up the bins into cell B131 and copy it to cells B132..B135

   @MIN ($CHI__ DATA) +A131* (@MAX ($CHI__ DATA) −@MIN ($CHI__ DATA))

4. Type the formula for $z-$transformation into cell D131 and copy it to cells D132..D135

```
@ROUND((B131-$MEAN)/$STD,2)
```

5. Type the formula for using the z values to find the area under the Gaussian curve from the GAUSS table into cell E131 and copy it to cells E132..E135

```
@VLOOKUP(@ABS(@VALUE(@STRING(D131,2))),$GAUSS,
    @HLOOKUP(@ABS(@VALUE(@RIGHT(@STRING(D131,2),1))),$GAUSS,0)+1)
```

6. Type the formula to determine the area under the Gauss curve for the bin interval into cell F132 and copy it to cells F133..F135

```
@ABS(@IF(D131<0#AND#D132>=0,E132+E131,E132-E131))
```

7. Type the formula to calculate the expected frequency for the bin interval into cell G132 and copy it to cells G133..G135

```
+F132*@COUNT($CHI__DATA)
```

8. Type the formula to find the chi-square difference in the observed and expected frequency for the bin interval into cell H132 and copy it to cells H133..H135

```
+(C132-G132)^2/G132
```

9. Type the formula to find the sum of the individual chi-squares into cell G138

```
@SUM(G132..G135)
```

10. Issue the /Data Distribution (Lotus 1-2-3) or /Range Distribution (Symphony) command to recalculate the new bin values. When prompted to "Enter Values Range", specify CHI__DATA. When prompted to "Enter Bin Range", specify B131..B135.

This template has its data in a 7x9 matrix. If your data do not totally fill a matrix, then you will need to create another template in an open section of your spreadsheet. This new template should be arranged so that the data are in a single column. This format is required because, if you recall, when you highlight cells for data distribution analysis, all of the cells MUST contain data. Otherwise, the bin cells may contain erroneous data.

INCREASING THE NUMBER OF BINS

As previously mentioned, other bin configurations can be used to determine Goodness-of-Fit. Figure 4-18 shows a chi-square determination using the bin framework and equations shown in Figure 4-2. It has ten bins.

```
--------A--------B--------C--------D--------E--------F--------G--------H-----
38                        ****SAMPLE CONCENTRATION DATA****
39 SAMPLE A1   221.1    208.8    211.1    208.3    221.1    224.2    217.8
40        2    224.2    206.9    198.4    214.1    208.8    206.9    205.9
41        3    217.8    205.9    213.0    209.1    211.1    198.4    213.0
42 SAMPLE B1   208.8    211.4    208.9    207.7    208.3    214.1    209.1
43        2    212.6    205.8    206.0    216.2    208.8    212.6    213.3
44        3    213.3    202.5    209.8    203.7    211.4    205.8    202.5
45 SAMPLE C1   203.4    214.5    208.8    218.3    208.9    206.0    209.8
46        2    206.4    212.6    210.2    212.7    207.7    216.2    203.7
47        3    208.0    215.6    208.0    214.6    208.0    210.2    214.6
48
--------A--------B--------C--------D--------E--------F--------G--------H-----
125               ****CHI_SQUARE GOODNESS-OF-FIT TEST****
126
127                    MEAN    210.4238    STD   5.364871
128
129     FREQUENCY DISTRIBUTION      Z       GAUSS     AREA    EXPECTED
130 MULTIPLIER    BIN    FREQ   FOR CLASS   AREA   FOR CLASS   FREQ      CHI
131     0.1     201.0      2     -1.76    0.4608
132     0.2     203.6      3     -1.28    0.3997    0.0611   3.8493  0.187387
133     0.3     206.1      8     -0.80    0.2881    0.1116   7.0308  0.133604
134     0.4     208.7     10     -0.32    0.1255    0.1626  10.2438  0.005802
135     0.5     211.3     15      0.16    0.0636    0.1891  11.9133  0.799754
136     0.6     213.9     10      0.64    0.2389    0.1753  11.0439  0.098672
137     0.7     216.5      8      1.13    0.3708    0.1319   8.3097  0.011542
138     0.8     219.0      3      1.61    0.4463    0.0755   4.7565  0.648647
139     0.9     221.6      2      2.09    0.4817    0.0354   2.2302  0.023761
140       1     224.2      2      2.57    0.4949    0.0132   0.8316  1.641604
141                        0
142            N=         63                        CHI_SQR    3.55
```

Figure   4-18

As the number of data points increases in an experiment, you can increase the number of bins to slice the analysis into smaller pieces. (As long as there are at least five data points per bin.) Although this example has the same data as Figure 4-16, it has a different number of bins and a different chi-square value (3.55). However, the chi-square test gives the same final assessment. The reason for this similarity is that the number of Degrees of Freedom for determining the cutoff changes. That is, for ten bins

$$DF = 10 \text{ bins} - 3 \text{ constraints} = 7$$

Based on 7 Degrees of Freedom and a confidence limit of 0.50, the percentile value of 6.35 is found for chi-square in a standard reference book. Because the chi-square value determined in this template (3.55) is less than the percentile cutoff value, the same conclusions as Figure 4-16 are reached.

Converting the template from Figure 4-16 to Figure 4-18 is easy and demonstrates some of the advantages of the copy command and lookup tables. To perform the conversion, extend the MULTIPLIER column and re-assign the multiplier values so that the multipliers range from 0.1 to 1 in increments of 0.1. Use the /Data Fill command (Lotus 1-2-3) or the /Range Fill command (Symphony). Then, use the /Copy command to extend the formulae in Row 135, Figure 4-16 to correspond to the number of rows in MULTIPLIER. Perform this action by using the /Copy command to copy cells D135..H135 to cells D136..H140. Note that all formulae instantly update with correct values, demonstrating the usefulness of deploying lookup tables.

Next, issue the /Data Distribution (Lotus 1-2-3) or /Range Distribution (Symphony) command and specify the new bin range to recalculate the new bin values. The remaining values in the template will automatically recalculate.

Finish the editing process by changing the N and CHL_SQR formulae to span the new ranges. The new chi-square value will be displayed.

So, changing a template is easy if you design the template correctly from the start. That is, if you plan a template from the start, you can usually use the /Copy command and some minimal editing to expand the template to different formats. This demonstration shows the flexibility of spreadsheet analysis. However, there are ways that you can expand the size of a template that are even more powerful and convenient. Chapter 11 contains a section entitled "Expanding the Size of a Template", which explores a carefully pre-planned use of Move and Insert menu commands to perform this task. These commands will allow you to expand your template without having to edit formulae.

GOODNESS-OF-FIT FOR POISSON DISTRIBUTIONS

As stated above, the chi-square test can be used for determining whether data conform to other distributions. To help you understand how to modify a template to test for other distributions, consider how a template would be modified to perform a chi-square test to analyze data for Goodness-of-Fit to a Poisson distribution.

The *Poisson distribution* is useful in the study of discrete or rare events that occur in an interval of time, length, area, volume, etc. In practice, an event is considered rare if the number of trials is at least 50 ($N >= 50$) and the number of trials times the probability (p) that an event will occur in any single trial (N_p) is less than about 5. That is, the Poisson distribution is often used when there is a large number of trials (or samples) but only a small probability of an event occurring in each of the trials.

Many experiments involve observing the number of discrete or rare events in a fixed time interval or in a fixed area, length, or volume. The outcome of these experiments often has the characteristics of a Poisson process. The characteristics of the Poisson distribution are:

- The probability of a single occurrence of an event is directly proportional to the size of the interval.
- If an interval is sufficiently small, the probability of two or more occurrences of an event is negligible.
- The occurrences of an event in nonoverlapping intervals are independent.

These characteristics have made the Poisson distribution the basis for a class of valuable statistical tools in the study of radioactivity, material flaw/defect analyses, cell distributions, back mutations, mixing efficiencies, viral particle analyses, dilution studies on limited numbers of molecules, etc. However, before valid results can be obtained from this category of statistical tools, the data must conform to the Poisson distribution. Therefore, as in the case of the normal (Gaussian) distribution, you must first verify that your data conform to the Poisson distribution.

The example shown in Figures 4-19 and 4-20 show how the chi-square Goodness-of-Fit test can be applied to aliquots of a viral sample. In this example, a very large dilution was made on an original sample so that only a very few viral particles would be present in each of 99 aliquots. The number of viral particles in each aliquot were then determined using a current state-of-the-art CytoPathic Effect (CPE) method. The CPE assay provided a tally of the number of plaques (areas of cell death) on monolayer lawns of susceptible cells and the results were placed in the matrix shown in Figure 4-19.

	A	B	C	D	E	F	G	H	I	J	K
1			****NUMBER	OF	VIRAL	PLAQUES	DETECTED****				
2	0	6	2	1	2	4	3	4	4	5	4
3	2	1	4	5	2	2	0	3	2	0	2
4	1	4	2	2	4	1	5	2	1	4	1
5	5	2	0	1	3	7	2	1	2	2	3
6	2	1	3	3	1	2	5	3	5	3	2
7	1	3	5	2	0	4	3	0	1	2	4
8	7	2	1	3	2	3	1	2	3	4	1
9	4	0	3	2	2	3	4	3	4	3	3
10	2	1	2	3	4	3	2	4	2	1	2
11											

Figure   4-19

	A	B	C	D	E	F	G	H
15		****CHI_SQUARE	GOODNESS-OF-FIT	TEST****				
16		(POISSON	DISTRIBUTION)					
17		MEAN		2.57				
18								
19	FREQUENCY	DISTRIBUTION		BIN		EXPECTED		
20	BIN	FREQUENCY		FACTORIAL		FREQ	CHI	
21	0	7		1		7.610000	0.048896	
22	1	17		1		19.52464	0.326451	
23	2	29		2		25.04677	0.623953	
24	3	20		6		21.42047	0.094196	
25	4	16		24		13.73939	0.371948	
26	5	7		120		7.050112	0.000356	
27	6	1		720		3.014694	1.346403	
28	7	2		5040		1.104953	0.725016	
29	8	0		40320		0.354366	0.354366	
30		0						
31								
32	N=	99				CHI_SQR	0.55	

Figure   4-20

Figure 4-20 shows the template used to determine the chi-square Goodness-of-Fit. It is a modification of the template used in Figure 4-18. As you can see, some of the columns in Figure 4-18 are no longer needed for the calculation of the EXPECTED FREQUENCY. In this case, the EXPECTED FREQUENCY is determined from an equation having the following format:

$$@EXP(-\$MEAN)*\$MEAN^{A21}/D21*@COUNT(\$CHI\_DATA)$$

To use this equation, the bin number (Column A) and a factorial based on the bin number (Column D) are needed. The factorial is determined by placing a 1 in cell D21 and the following formula in cell D22:

```
+A22*D21
```

The formula in cell D22 is then copied to cells D23..D29 to form a series that provides factorial calculation. (Note how easy it is to form a factorial, summation, etc., in spreadsheets.)

The bin number is provided from Column A. Contrary to the normal distribution example, bins for Poisson distributions need to be equally-spaced, whole numbers. They must also reflect the rare occurrence criterion anticipated by Poisson (e.g., you could not use values of 201 to 224 as in the example of Figure 4-18). In this case, the numbers are zero through 8. These numbers provide both the number of the bin and the cutoff values for the distribution program.

The remaining formulae are the same as in Figure 4-18, updated for new ranges. To create the Poisson example, type the data and text of Figures 4-19 and 4-20 into a spreadsheet.

Activate the following ranges by issuing the /Range Name Create command and specifying the following range names and cells:

Range Name	Cell Range
CHI__DATA	A2..K10
MEAN	D17

Complete the template by typing the following formulae into the specified cells of the template:

1. Type the following formula to find the average of the chi data into cell D17:

   ```
   @AVG(CHI__DATA)
   ```

2. Issue the /Data Fill command (Lotus 1-2-3) or the /Range Fill command (Symphony) to fill cells A21..A29 with the numbers 0 through 8.

3. Type a 1 into cell D21.

4. Create a factorial series by typing the following formula into cell D22 and copying it to cells D23..D29:

   ```
   +A22*D21
   ```

5. Type the formula for the Poisson distribution function into cell F21 and copy it to cells F22..F29:

   ```
   @EXP(-$MEAN)*$MEAN^A21/D21*@COUNT($CHI__DATA)
   ```

6. Type the chi-square formula into cell G21 and copy it to cells G22..G29:

```
+(B21-F21)^2/F21
```

9. Type the formula to find the chi-square value into cell G32:

```
@SUM(G21..G29)/(@COUNT(G21..G29)-2)
```

10. Issue the /Data Distribution (Lotus 1-2-3) or /Range Distribution (Symphony) command to recalculate the new bin values. When prompted to "Enter Values Range", specify CHI_DATA. When prompted to "Enter Bin Range", specify B21..B29.

In this section, the number of constraints is 2 (total number of data points and the mean value). Therefore, the Degrees of Freedom for this template is calculated as follows:

$$DF = 9 \text{ bins} - 2 \text{ constraints} = 7$$

Based on 7 Degrees of Freedom and a confidence limit of 0.50, the percentile value of 6.35 for chi-square is found in a standard reference book. Because the chi-square value determined in this template (0.55) is less than the percentile value, the data in the template either follow the Poisson distribution or there are not enough data to reject the hypothesis. If the former is true, the Poisson equations can be used with confidence.

SUMMARY OF WHAT HAS BEEN ACCOMPLISHED

This chapter described how to implement a number of statistical tests to describe and evaluate the distribution for sets of assay data. More importantly, it demonstrated how to choose Lotus tools and assemble them into systems that can handle your statistical data reduction applications. Chi-square examples were given to illustrate the integrating of templates, Lotus data distribution, lookup tables, @functions, and formulae to form a powerful data reduction tool.

The chi-square examples also illustrated the utility of a template. That is, a well-designed template can be used over and over again to study all sorts of data. More importantly, very little modification is required to change a well-designed template for new applications.

For example, to re-use the chi-square template, you would import new data into the Raw Data section of the template. The Sample Concentration section and all other formulae dependent on the Raw Data section would instantly update. If necessary, you could issue the /Range Name Create command to change the definition of the CHI_DATA range. In this example, when prompted for the range name, type CHI_DATA. When prompted for the range, use the arrow keys to highlight the new data range and press return. Then re-issue the /Data Distribution (Lotus 1-2-3) or /Range Distribution (Symphony) command to recalculate the new bin values. The remaining values in the template will automatically recalculate and the new chi-square value will be displayed. This example illustrates the power of using templates.

WHAT'S NEXT?

Now that you know how to confirm the distribution of a set of data, you are ready for a discussion on Lotus statistical formulae. All Lotus statistical formulae are based on the normal (Gaussian) distribution. The next chapter will provide you with a rigorous explanation of these formulae, most of which have been used in the preparation of the example templates up to this point.

5

Descriptive Statistics

Collecting, organizing, summarizing, and analyzing data from experiments are important elements in the presentation of all scientific data. Laboratories produce data that are used to make important decisions. All data generated are associated with some degree of variability that affects evaluation and interpretation. It is therefore essential to have some knowledge of this degree of uncertainty. Statistics form the basis of this knowledge and are key to drawing valid conclusions and making reasonable decisions.

Statistics are also the key to condensing data into a form that at minimum makes the main features of the data clearer. If two or more sets of data are to be compared, the condensing process is even more important. This process may be accomplished qualitatively by grouping the data into frequency tables as shown in Chapter 3, but more often it is necessary to use quantitative measures to represent data. Quantitative measures are vital to deciding whether two sets of data show real differences or if the differences are due to sampling fluctuations. Quantitative measures are also central to establishing the precision of any parameter derived from the data. That is, quantitative measures will fix the limits within which the true value will lie.

The statistics most commonly used to depict these measures of distribution are:

- Measures of central value, which give the location of some central or typical value. The central value is representative of a set of data. An example of a central value is the average. Because average values lie centrally within a set of data arranged according to magnitude, average values are also called *measures of central tendency*. Examples of average values are arithmetic mean, median, mode, geometric mean, and harmonic mean.

- Measures of dispersion, which give the degree of spread of the data about the central value. Examples of dispersion are the variance and standard deviation.
- Measures of Skewness, which give the lack of symmetry of a data set's distribution.
- Measures of Kurtosis, which give the peakedness of a data set's distribution.

Skewness and Kurtosis were described in Chapter 3. The purpose of this chapter is to introduce the rest of the general statistical methods that describe and analyze a given group of data. These methods analyze data without drawing conclusions or inferences about larger groups. This phase of statistics is often referred to as *descriptive statistics*. In addition to basic descriptive statistics, tools to compare groups of data are provided (i.e., the F test and the t-test).

In Chapter 3, a template was created to hold experimental data from a fluorimetric DNA assay (see Figure 3-1). The template contained basic Lotus statistical @functions. The example template used in this chapter is the same one created in Chapter 3. In the interest of convenience, Figure 3-1 has been reproduced in this chapter and is shown as Figure 5-1.

POPULATIONS AND SAMPLES

Statistics were first applied to the affairs of state. They were used for preparing tables of figures giving births, deaths, marriages, divorces, accidents, and so on. These tables were used by governments for effective planning, ruling, and (of course) tax-collecting.

In this capacity, data that were collected were concerned with the characteristics of all of the individuals or objects involved. The totality of items under consideration is called a *population*. A summary measure that is computed to describe a characteristic of an entire population (i.e., on the complete collection of possible observations) is called a *parameter*.

However, it is often impossible or impractical to collect large numbers of data points to observe an entire population during scientific experimentation. For this reason, experiments are usually set up to collect data on a small portion of a population (called a *sample*). A summary measure that is computed to describe a characteristic from only a sample of a population (i.e., on a limited number of observations) is called a *statistic*. If a sample is representative of a population, important conclusions about the population can often be inferred from the analysis of the sample.

Lotus 1-2-3® and Symphony® statistical @functions are configured to yield **population** parameters. For example, the Lotus @STD function calculates a population standard deviation using division by the number of observations (N). When working with portions of a population, a corrected sample standard deviation is appropriate. To make the correction, you should multiply the value obtained from the Lotus® @function by the appropriate correction factor for the decreased number of

```
--------A--------B--------C--------D--------E--------F--------G--------H-----
 1                        FLUORESCENT DNA ASSAY DATA
 2 DATE RUN:31-OCT-88
 3                   ****STANDARD CURVE DATA****                    X    REGRESS
 4 NG DNA/ML ASSAY1    ASSAY2     ASSAY3    ASSAY4    AVERAGE  PREDICT   CURVE
 5        0    93.9      92.9       94.3      94.3      93.9      0.7     92.7
 6       25   130.4     134.6      133.4     135.9     133.6     26.9    130.7
 7       50   174.5     171.8      173.1     174.1     173.4     53.1    168.7
 8       75   202.9     204.5      204.2     205.2     204.2     73.3    206.7
 9      100   235.0     242.2      242.7     239.9     240.0     96.9    244.7
10      250   473.9     462.0      469.3     473.2     469.6    248.0    472.7
11      500   847.1     854.2      850.4     862.2     853.5    500.6    852.6
12      750  1236.4    1236.4     1226.0    1235.2    1233.5    750.6   1232.6
13                   ========================================
14                          Regression Output:
15                   Constant                        92.74541
16                   Std Err of Y Est                 3.441756
17                   R Squared                        0.999939
18                   No. of Observations                     8
19                   Degrees of Freedom                      6
20
21                   X Coefficient(s)   1.519749
22                   Std Err of Coef.   0.004814
23                   ========================================
24                   ****SAMPLE FLUORESCENCE DATA****
25 SAMPLE IDXX102-7  XX102-7   XX102-7   XX102-7   XX102-7  XX102-7  XX102-7
26 COMMENTS                   SAMPLE PRECISION STUDY
27
28 SAMPLE A1   428.8    410.1      413.6     409.3     428.8    433.5    423.7
29        2   433.5    407.2      394.3     418.1     410.1    407.2    405.7
30        3   423.7    405.7      416.5     410.5     413.6    394.3    416.5
31 SAMPLE B1   410.1    414.0      410.2     408.4     409.3    418.1    410.5
32        2   415.8    405.5      405.8     421.3     410.1    415.8    416.9
33        3   416.9    400.5      411.6     402.3     414.0    405.5    400.5
34 SAMPLE C1   401.9    418.7      410.1     424.5     410.2    405.8    411.6
35        2   406.4    415.8      412.2     416.0     408.4    421.3    402.3
36        3   408.9    420.4      408.9     418.9     408.9    412.2    418.9
37
38                   ****SAMPLE CONCENTRATION DATA****
39 SAMPLE A1   221.1    208.8      211.1     208.3     221.1    224.2    217.8
40        2   224.2    206.9      198.4     214.1     208.8    206.9    205.9
41        3   217.8    205.9      213.0     209.1     211.1    198.4    213.0
42 SAMPLE B1   208.8    211.4      208.9     207.7     208.3    214.1    209.1
43        2   212.6    205.8      206.0     216.2     208.8    212.6    213.3
44        3   213.3    202.5      209.8     203.7     211.4    205.8    202.5
45 SAMPLE C1   203.4    214.5      208.8     218.3     208.9    206.0    209.8
46        2   206.4    212.6      210.2     212.7     207.7    216.2    203.7
47        3   208.0    215.6      208.0     214.6     208.0            214.6
48              ****SUMMARY OF SAMPLE CONCENTRATION DATA****
49     AVG    212.8    209.3      208.2     211.6     210.5    210.5    210.0
50      SD      7.0      4.4        4.2       4.7       4.2      7.4      5.2
51     %CV      3.3      2.1        2.0       2.2       2.0      3.5      2.5
52     VAR     49.1     19.8       17.5      22.2      17.6     54.9     26.9
53     MIN    203.4    202.5      198.4     203.7     207.7    198.4    202.5
54     MAX    224.2    215.6      213.0     218.3     221.1    224.2    217.8
55   RANGE     20.8     13.1       14.6      14.6      13.4     25.8     15.3
56   X-2SD    198.8    200.4      199.9     202.2     202.1    195.7    199.6
57   X+2SD    226.9    218.2      216.6     221.0     218.8    225.3    220.3
```

Figure 5-1

Degrees of Freedom. For example, in Chapter 3, the template for Figure 3-1 was set up to correct the @STD. The equation used was

@STD (B39..B47) *@SQRT (@COUNT (B39..B47) / (@COUNT (B39..B47) -1))

which is equivalent to

$$\text{STANDARD DEVIATION} \, * \, \sqrt{N/(N-1)}$$

A similar situation exists with the @VAR variance function (described in more detail below). In Chapter 3, the following correction was applied

@VAR(B39..B47)*(@COUNT(B39..B47)/(@COUNT(B39..B47)-1))

which is equivalent to

$$\text{VARIANCE} \, * \, (N/(N-1))$$

LOTUS PARAMETRIC STATISTICAL @FUNCTIONS

The template in Figure 5-1 contains the basic Lotus® statistical @functions. These @functions were used in Chapters 3 and 4. All of the Lotus @functions assume a normal (Gaussian) distribution. Much of this chapter is dedicated to giving a more rigorous and formal description of the Lotus statistical @functions that were previously used. These @functions are @AVG, @STD, @VAR, @MIN, @MAX, and @COUNT.

Each of these @functions uses an argument within a set of parentheses. An argument can be a numeric value, single cell address, range of cells, range name, formula, or a mixture of these elements. The argument can also be a list. The list can include individual cell values separated by commas as well as ranges of cells, formulae, etc. For example,

@AVG(B39,C40,D41,E39..E47,F40)
@AVG(1.1*B39,1.7*C40,1.3*D40)
@AVG(RANGE)
@AVG(RANGE_ONE,RANGE_TWO)

are all valid arguments. Note that by using commas, the statistical @functions can be used on non-contiguous cells.

These @functions ignore all blank cells within a range. They do not, however, ignore label cells. If a label cell is within the range, the @functions treat them as being equal to zero (0). These phenomena can affect your calculations in two ways: Blank cells will not be included in the count of cells, even though you may want them to be (although this problem is not very common). Conversely, label cells you inadvertently include in an argument will be used to calculate the results and can thereby produce inaccurate results.

Therefore, you must be careful when using Lotus statistical @functions and make certain that none of the cells in your range contain labels that you do not want to include in the calculation. For example, many users use the spacebar on the keyboard to place a space in a cell to erase its contents instead of using the /Range Erase (Lotus 1-2-3) or /Erase (Symphony) command. This invisible entry is considered a label by

Lotus and is counted as a zero. **Using spaces to erase cells is an extremely dangerous practice and is not recommended**. If you have used a space bar in one of the cells, calculations based on the cell will probably be incorrect.

With these guidelines in mind, let us examine each of the Lotus @functions that can be used for scientific experimental data, along with some other useful tools. Begin by typing the data shown in Figure 5-1 into the template created in Chapter 3 (Figure 3-1).

AVERAGES AND OTHER MEASURES OF CENTRAL TENDENCY

There are three main ways to measure the central location of a set of data. They are:

- The mean (arithmetic, geometric, harmonic).
- The median.
- The mode.

All have their uses, but the arithmetic mean is by far the most important. The other measures are used in special cases.

The arithmetic mean (average) is defined as

$$\bar{X} = \sum X_i / N$$

That is, the arithmetic mean is merely the sum of the values divided by the number of values. The mean represents the center of a symmetrically or normally distributed group of values. Thus, the mean is a value that is typical or representative of a set of data. In this book, I will use the terms *average* and *mean* interchangeably.

The average row in the template (Row 49) contains the Lotus @AVG function. The Lotus @AVG function is equivalent to the arithmetic mean and is calculated by performing a summation on all of the specified values and dividing by the number of values in the group. For example, the @AVG(B39..B47) in cell B49 calculates the average of the values in the range B39..B47.

As outlined above, you can use Lotus @functions to find the average of both contiguous and non-contiguous cells. However, you can also use the spreadsheet to determine other types of averages, the most common being the weighted mean, median, mode, geometric mean, and harmonic mean.

To find a weighted mean, weighting factors can be incorporated into the arguments for an @function. Weighting factors depend on the significance or importance attached to the numbers. For example

@AVG(1.1*B39,1.7*C40,1.3*D41)

will apply the weighting factors of 1.1, 1.7, and 1.3 to cells B39, C40, and D41, respectively, and then find the average value. The weighting factors can also be mathematical calculations or other @functions.

By using weighting factors, greater emphasis can be given to observations at the upper end of the scale. This emphasis leads to a class of higher-order averages that are particularly useful to characterize distributions of properties, such as the length of polymers.

The geometric mean G of a set of N numbers is the Nth root of the product of the numbers:

$$G = (X_1 * X_2 * X_3 \cdots * X_N)^{1/N}$$

A geometric mean should only be computed when all observations are positive. The principal application of the geometric mean is to average a sequence of ratios or percentages. However, it is convenient to use the data from the DNA experiment to demonstrate the calculation. The data will also illustrate some important points about the geometric mean.

Figure 5-2 shows how to calculate the geometric means of the numbers in Figure 5-1. To create this example, type +B39 into cell B146. Issue the /Copy command to copy this formula to cells B146..G146. The values in cells B39 through G39 should now appear in cells B146 through G146. This row of cells initializes a multiplication series. To create the remainder of the multiplication series, type the following formula into cell B147:

```
+B146*B40
```

Issue the /Copy command to copy this formula to cells B147..H147. Cells B147

```
--------A--------B--------C--------D--------E--------F--------G--------H-----
38
39  SAMPLE A1    221.1    ****SAMPLE CONCENTRATION DATA****
                          208.8    211.1    208.3    221.1    224.2    217.8
40         2     224.2    206.9    198.4    214.1    208.8    206.9    205.9
41         3     217.8    205.9    213.0    209.1    211.1    198.4    213.0
42  SAMPLE B1    208.8    211.4    208.9    207.7    208.3    214.1    209.1
43         2     212.6    205.8    206.0    216.2    208.8    212.6    213.3
44         3     213.3    202.5    209.8    203.7    211.4    205.8    202.5
45  SAMPLE C1    203.4    214.5    208.8    218.3    208.9    206.0    209.8
46         2     206.4    212.6    210.2    212.7    207.7    216.2    203.7
47         3     208.0    215.6    208.0    214.6    208.0    210.2    214.6
48
```

```
--------A--------B--------C--------D--------E--------F--------G--------H-----
144
145                      ****GEOMETRIC MEAN CALCULATIONS****
146               221.1    208.8    211.1    208.3    221.1    224.2    217.8
147             49570.6  43200.7  41882.2  44597.0  46165.7  46387.0  44845.0
148             1.1E+07  8.9E+06  8.9E+06  9.3E+06  9.7E+06  9.2E+06  9.6E+06
149             2.3E+09  1.9E+09  1.9E+09  1.9E+09  2.0E+09  2.0E+09  2.0E+09
150             4.8E+11  3.9E+11  3.8E+11  4.2E+11  4.2E+11  4.2E+11  4.3E+11
151             1.0E+14  7.8E+13  8.1E+13  8.5E+13  9.0E+13  8.6E+13  8.6E+13
152             2.1E+16  1.7E+16  1.7E+16  1.9E+16  1.9E+16  1.8E+16  1.8E+16
153             4.3E+18  3.6E+18  3.5E+18  4.0E+18  3.9E+18  3.8E+18  3.7E+18
154             8.9E+20  7.7E+20  7.4E+20  8.5E+20  8.1E+20  8.1E+20  7.9E+20
155  GEOMETRIC
156     MEAN     212.7    209.3    208.2    211.6    210.4    210.4    209.9
157
158
```

Figure 5-2

through H147 now contain the products of the numbers for their appropriate columns. To find the *N*th root of these products, type the following formula into cell B156:

```
+B154^(1/@COUNT(B146..B154))
```

Issue the /Copy command to copy this formula to cells C156..H156.

As you can see, the intermediate numbers (e.g., cells B146..B154) in the template are quite large. Working with very large numbers can yield unpredictable results, especially if the numbers approach the limits of Lotus' numeric precision. For this reason, it is desirable to first convert the data to natural logarithms, perform a summation on the logarithms, and then convert back to decimal for the final result. That is, the following algorithm is used:

$$Ln\ G\ =\ \frac{1}{N} \sum Ln\ X_j$$

Figure 5-3 illustrates how to work with large numbers. To create this example,

1. Initiate a series for each column of data by typing the following formula into cell B146 and copying it to cells B147..H147:

```
@LN(B39)
```

2. Form a summation series that adds the natural log of each subsequent number in

	A	B	C	D	E	F	G	H
38				****SAMPLE CONCENTRATION DATA****				
39	SAMPLE A1	221.1	208.8	211.1	208.3	221.1	224.2	217.8
40	2	224.2	206.9	198.4	214.1	208.8	206.9	205.9
41	3	217.8	205.9	213.0	209.1	211.1	198.4	213.0
42	SAMPLE B1	208.8	211.4	208.9	207.7	208.3	214.1	209.1
43	2	212.6	205.8	206.0	216.2	208.8	212.6	213.3
44	3	213.3	202.5	209.8	203.7	211.4	205.8	202.5
45	SAMPLE C1	203.4	214.5	208.8	218.3	208.9	206.0	209.8
46	2	206.4	212.6	210.2	212.7	207.7	216.2	203.7
47	3	208.0	215.6	208.0	214.6	208.0	210.2	214.6
48								

	A	B	C	D	E	F	G	H
144				****GEOMETRIC MEAN CALCULATIONS****				
145				(LOGARITHMIC)				
146		5.3986	5.3414	5.3523	5.3390	5.3986	5.4125	5.3836
147		10.8112	10.6736	10.6426	10.7054	10.7400	10.7448	10.7110
148		16.1947	16.0010	16.0039	16.0482	16.0923	16.0351	16.0723
149		21.5361	21.3548	21.3458	21.3843	21.4313	21.4015	21.4151
150		26.8955	26.6817	26.6736	26.7605	26.7727	26.7609	26.7778
151		32.2582	31.9924	32.0198	32.0772	32.1264	32.0878	32.0885
152		37.5734	37.3607	37.3612	37.4631	37.4683	37.4157	37.4347
153		42.9032	42.7201	42.7092	42.8229	42.8044	42.7919	42.7513
154		48.2407	48.0935	48.0468	48.1917	48.1419	48.1400	48.1201
155	GEOMETRIC							
156	MEAN	212.7	209.3	208.2	211.6	210.4	210.4	209.9
157								
158								

Figure   5-3

the range by typing the following formula into cell B147 and copying it to cells B147..B154:

```
@LN(B40)+B146
```

3. Calculate the geometric mean by dividing by *N* and converting the natural logarithms back to decimal. Type the following formula into cell B156 and copy it to cells C156..H156:

```
@EXP(B154/@COUNT(B146..B154))
```

Figure 5-4 illustrates yet another way to calculate geometric mean. To show you some of the variation possible with this genre of templates, the template in Figure 5-4 calculates the geometric mean in a somewhat different manner. Individual logarithms are calculated (without forming a summation series), the average of the logarithms is taken, and the average is re-converted back to decimal. This method gives equivalent results to the summation series method and is slightly easier to create. To create this example,

1. Type the following formula into cell B146 to calculate the logarithm of cell B39 and copy the formula to cells B146..H154:

```
@LN(B39)
```

	A	B	C	D	E	F	G	H
38			****SAMPLE	CONCENTRATION	DATA****			
39	SAMPLE A1	221.1	208.8	211.1	208.3	221.1	224.2	217.8
40	2	224.2	206.9	198.4	214.1	208.8	206.9	205.9
41	3	217.8	205.9	213.0	209.1	211.1	198.4	213.0
42	SAMPLE B1	208.8	211.4	208.9	207.7	208.3	214.1	209.1
43	2	212.6	205.8	206.0	216.2	208.8	212.6	213.3
44	3	213.3	202.5	209.8	203.7	211.4	205.8	202.5
45	SAMPLE C1	203.4	214.5	208.8	218.3	208.9	206.0	209.8
46	2	206.4	212.6	210.2	212.7	207.7	216.2	203.7
47	3	208.0	215.6	208.0	214.6	208.0	210.2	214.6
48								

	A	B	C	D	E	F	G	H
144			****GEOMETRIC	MEAN	CALCULATIONS****			
145				(LOGARITHMIC)				
146		5.3986	5.3414	5.3523	5.3390	5.3986	5.4125	5.3836
147		5.4125	5.3322	5.2903	5.3664	5.3414	5.3322	5.3274
148		5.3836	5.3274	5.3613	5.3428	5.3523	5.2903	5.3613
149		5.3414	5.3538	5.3419	5.3361	5.3390	5.3664	5.3428
150		5.3594	5.3269	5.3279	5.3762	5.3414	5.3594	5.3627
151		5.3627	5.3107	5.3462	5.3166	5.3538	5.3269	5.3107
152		5.3152	5.3683	5.3414	5.3859	5.3419	5.3279	5.3462
153		5.3298	5.3594	5.3481	5.3599	5.3361	5.3762	5.3166
154		5.3375	5.3734	5.3375	5.3688	5.3375	5.3481	5.3688
155	GEOMETRIC							
156	MEAN	212.7	209.3	208.2	211.6	210.4	210.4	209.9
157								
158								

Figure 5-4

2. Determine the geometric mean by calculating the average of the logarithms. Then, convert the natural logarithms back to decimal. Do so by typing the following formula into cell B156 and copying it to cells C156..H156:

```
@EXP(@AVG(B146..B154))
```

The harmonic mean H of a set of N numbers is the reciprocal of the arithmetic mean of the reciprocals of the numbers. That is,

$$\frac{1}{H} = \frac{1}{N} \sum \frac{1}{X}$$

The harmonic mean is a useful measure when observations are expressed inversely to what is required in the average (e.g., miles per gallon). Again, it is convenient to use the data from the DNA experiment to illustrate this calculation. Figure 5-5 shows how to calculate the geometric means of the numbers in Figure 5-1. To create this example,

1. Type a formula into cell B162 to calculate the reciprocal of cell B39 and copy the formula to cells B162..H170.

```
+1/B39
```

2. Determine the harmonic mean by calculating the average of the reciprocals and inverting. Type the following formula into cell B172 and copy it to cells C172..H172:

```
+1/@AVG(B162..B170)
```

	A	B	C	D	E	F	G	H
38			****SAMPLE	CONCENTRATION	DATA****			
39	SAMPLE A1	221.1	208.8	211.1	208.3	221.1	224.2	217.8
40	2	224.2	206.9	198.4	214.1	208.8	206.9	205.9
41	3	217.8	205.9	213.0	209.1	211.1	198.4	213.0
42	SAMPLE B1	208.8	211.4	208.9	207.7	208.3	214.1	209.1
43	2	212.6	205.8	206.0	216.2	208.8	212.6	213.3
44	3	213.3	202.5	209.8	203.7	211.4	205.8	202.5
45	SAMPLE C1	203.4	214.5	208.8	218.3	208.9	206.0	209.8
46	2	206.4	212.6	210.2	212.7	207.7	216.2	203.7
47	3	208.0	215.6	208.0	214.6	208.0	210.2	214.6
48								

	A	B	C	D	E	F	G	H
160			****HARMONIC	MEAN	CALCULATIONS****			
161								
162		0.004522	0.004789	0.004737	0.004800	0.004522	0.004460	0.004591
163		0.004460	0.004833	0.005040	0.004670	0.004789	0.004833	0.004856
164		0.004591	0.004856	0.004694	0.004782	0.004737	0.005040	0.004694
165		0.004789	0.004730	0.004786	0.004814	0.004800	0.004670	0.004782
166		0.004703	0.004859	0.004854	0.004625	0.004789	0.004703	0.004688
167		0.004688	0.004938	0.004766	0.004909	0.004730	0.004859	0.004938
168		0.004916	0.004662	0.004789	0.004580	0.004786	0.004854	0.004766
169		0.004844	0.004703	0.004757	0.004701	0.004814	0.004625	0.004909
170		0.004807	0.004638	0.004807	0.004659	0.004807	0.004757	0.004659
171	HARMONIC							
172	MEAN	212.6	209.2	208.2	211.5	210.4	210.3	209.9
173								
174								

Figure 5-5

```
---------A--------B--------C--------D--------E--------F--------G--------H-----
| 177       ****COMPARISON OF AVERAGE, GEOMETRIC MEAN, AND HARMONIC MEAN****   |
| 178                                                                          |
| 179                                                                          |
| 180  AVERAGE    212.84   209.33   208.24   211.63   210.46   210.49   209.97 |
| 181  GEO MEAN   212.74   209.29   208.21   211.59   210.42   210.37   209.91 |
| 182  HARMONIC   212.64   209.25   208.17   211.54   210.38   210.26   209.85 |
| 183                                                                          |
| 184                                                                          |
| 185                                                                          |
```

Figure 5-6

Figure 5-6 shows the relationship between the arithmetic mean (average), geometric mean, and harmonic mean. The format has been increased to two digits to better illustrate the relationship. As a general rule, the geometric mean of a set of positive numbers is less than or equal to their arithmetic mean. Likewise, the harmonic mean is less than or equal to their geometric mean. The three means are equal only if all of the numbers are identical.

CALCULATION OF MEDIAN AND MODE

The median and mode are two other averages. The *median* of a set of numbers arranged in order of magnitude is the middle value or the arithmetic mean of the two middle values. Geometrically, the median is the value corresponding to the vertical line that divides a histogram into two equal parts. That is, there are equal numbers of observations greater than and less than the median. The following rules apply to the median value:

- For samples with an odd number of items, the median is the middle item.
- For samples with an even number of items, the median is the arithmetic average of the two middle items.

The median is important because many nonparametric statistical tests use it instead of the arithmetic average. The median is also a valuable test for "open ended" testing. For example, in calculating the half-life of a radioactive element, the time for complete disintegration is infinite. In this case, the median has the practical advantage because it is expressed in terms of only half the atoms-those that disintegrate first.

One other common use of the median is as a measure of the toxicity of insecticides. If a quantity of insects are exposed to a test material, they do not all die at the same time. Some may survive completely. The median life is the time that it takes for half of the insects to die and is a measure of the toxicity of the material.

The median has two features that make it a valuable measure of location for small samples. They are:

- It is very simply determined, especially if N is less than 5.
- It is not influenced by freak extreme observations.

For these reasons the median is sometimes used to summarize the outcome of routine tests that have few replicates (e.g., 3). For normally distributed data, the median gives a less accurate estimate of the center of a distribution than the arithmetic mean. For symmetric, non-normal distributions that are sharply peaked, the median may be the more reliable measure. For skewed distributions, the median is often considered the measure of choice for central tendency.

The *mode* of a set of numbers is the value that occurs with the greatest frequency. That is, the mode is the most common value. The mode may not exist, and even if it does exist, it may not be unique. The presence of two or more modes usually means that the data are not homogeneous and that two or more distributions have been combined. Multimodal distributions usually occur when data from discrete experiments (e.g., different days or instruments) are pooled together for analysis.

To find the median and/or the mode of a set of numbers, you must first sort the numbers in increasing order. To preserve the order of raw data, copy it to an open area of the spreadsheet and arrange it in a single column. Do so by issuing the /Range Value command, not the /Copy command. Because the data will be sorted, you want values in the cells, not formulae or @functions.

When prompted for the range to copy from, specify the appropriate portion of the template. When prompted for the range to copy to, move the cell-pointer to an open area of the spreadsheet and press Return.

If you recall, this procedure was used to prepare the template in Figure 4-15. We can therefore conveniently just modify the existing template for these analyses. Figure 5-7 shows the modifications.

Next, issue the /Copy command to copy the data to the next column (i.e., the column to the right of the raw data column). This column will be the one sorted. By doing so, you achieve an important goal. You preserve the original order of the data for future use in other test procedures.

Next, issue the /Range Name Create command to create a range for the data. This action will make the creation of formulae easier and more "readable". In the example, the range B78..B113 was given the name SORT.

Now, sort the data. In Lotus 1-2-3, this procedure is initiated with the /Data Sort menu command. In Lotus Symphony, it is initiated with the /Querry Settings menu command.

After issuing the appropriate command, you will be presented with a menu from which you need to specify the Data-Range and Primary-Key. In Lotus 1-2-3, choose Data-Range. In Symphony, choose Basic Database.

The Data-Range (Lotus 1-2-3) and Basic Database (Symphony) options set the range that is to be sorted. Do not include any blank rows below the range to be sorted. However, you **MUST** include a blank cell above the range. This requirement is due to a peculiarity in the Lotus sorting routine. If you do not include a blank cell above the range, the data in the first cell of the range containing data will not be included in the sort. To set the data range, move the cell-pointer to the top corner of the range to sort (the blank cell). Once the command is issued, the command will prompt "Enter sort range". Enter the range by anchoring the cell-pointer by pressing period (.), using the arrow keys to highlight the entire range of data, and pressing Return. For the example, the Data-Range is B77..B113.

```
---------A--------B-----------C------D------E--------F--------G--------H-----
73            ****SKEWNESS, KURTOSIS, MEDIAN, AND MODE ANALYSIS****         |
74
75         TEST FOR SKEWNESS                     TEST FOR KURTOSIS
76 RAW DATA SORTED DATA OCCURENCES      DIFF    DIFF^2     DIFF^3    DIFF^4
77
78    221.1     198.4     1.0 -          10.6     112.1    1186.3   12558.7
79    224.2     202.5     1.0 -          13.7     187.3    2563.5   35084.9
80    217.8     203.4     1.0 -           7.3      53.1     386.8    2818.3
81    208.8     203.7     1.0 -          -1.7       2.9      -5.0       8.6
82    212.6     205.8     1.0 -           2.1       4.4       9.1      18.9
83    213.3     205.9     1.0 -           2.8       7.8      21.6      60.3
84    203.4     206.0     1.0 -          -7.1      50.6    -360.0    2561.1
85    206.4     206.4     1.0 -          -4.1      16.9     -69.6     286.4
86    208.0     206.9     1.0 -          -2.5       6.3     -15.9      39.9
87    208.8     207.7     1.0 -          -1.7       2.9      -5.0       8.6
88    206.9     208.0     1.0 -          -3.6      13.1     -47.2     170.6
89    205.9     208.0     2.0 -          -4.6      21.3     -98.2     453.2
90    211.4     208.3     1.0 -           0.9       0.8       0.7       0.6
91    205.8     208.8     1.0 -          -4.7      22.2    -104.7     493.8
92    202.5     208.8     2.0 -          -8.0      64.2    -514.7    4124.5
93    214.5     208.8     3.0 MODE        4.0      15.9      63.3     252.5
94    212.6     208.9     1.0 -           2.1       4.4       9.1      18.9
95    215.6     209.1     1.0 -           5.1      25.9     131.6     669.2
96    211.1     209.8     1.0 -           0.6       0.3       0.2       0.1
97    198.4     210.2     1.0 -         -12.1     146.7   -1777.7   21534.5
98    213.0     211.1     1.0 -           2.5       6.2      15.4      38.2
99    208.9     211.4     1.0 -          -1.6       2.6      -4.2       6.8
100    206.0     212.6     1.0 -          -4.5      20.4     -92.0     415.1
101    209.8     212.6     2.0 -          -0.7       0.5      -0.4       0.3
102    208.8     212.7     1.0 -          -1.7       2.9      -5.0       8.6
103    210.2     213.0     1.0 -          -0.3       0.1      -0.0       0.0
104    208.0     213.3     1.0 -          -2.5       6.3     -15.9      39.9
105    208.3     214.1     1.0 -          -2.2       4.9     -10.9      24.0
106    214.1     214.5     1.0 -           3.6      12.9      46.1     165.4
107    209.1     214.6     1.0 -          -1.4       2.0      -2.8       4.0
108    207.7     215.6     1.0 -          -2.8       7.9     -22.3      62.7
109    216.2     216.2     1.0 -           5.7      32.3     183.8    1045.3
110    203.7     217.8     1.0 -          -6.8      46.4    -316.4    2155.7
111    218.3     218.3     1.0 -           7.8      60.6     472.0    3675.2
112    212.7     221.1     1.0 -           2.2       4.8      10.4      22.8
113    214.6     224.2     1.0 -           4.1      16.7      68.2     278.8
114
115 AVERAGE     210.5                 MOM1      MOM2      MOM3      MOM4
116 STD_DEV     5.309                  0.0      27.4      47.2    2475.2
117 SKEWNESS    0.3156
118 MEDIAN      209.45                       KURTOSIS    3.1147
119                                          (K-3)       0.1147
120
121
122
```

Figure   5-7

The Primary-Key (Lotus 1-2-3) and Sort-Keys Primary (Symphony) menu choices are used to specify the fields that will determine the order of the records. The primary key may be only one column at a time. After you issue the command, it prompts "Primary Sort Key". Enter the first data cell in the range to be sorted by typing or highlighting it. (For this example, the Primary-Key is cell B78). Lotus then prompts, "Sort order (A or D)". "A" stands for ascending (lowest to highest) and "D" stands for descending. Choose A.

To execute the sort, issue the Go command from the Data Sort menu in Lotus 1-2-3 or the Record-Sort command from the Query menu in Symphony. When prompted for the uniqueness of the sort, choose All. (Cells B78..B113 shown in Figure 5-7 have already been sorted.)

    To find the median of the SORT range, you must develop an equation based on @IF and @MOD functions. The following formula in cell B118 of Figure 5-7 calculates the median of the example range:

```
@IF(@MOD(@COUNT(SORT),2)=0,(@INDEX(SORT,0,@COUNT(SORT)/2)
    +@INDEX(SORT,0,@COUNT(SORT)/2-1))/2,
        @INDEX(SORT,0,@INT(@COUNT(SORT)/2))))
```

This equation tests to see if the number of values in the list is even or odd using the @MOD function. The @MOD in this equation divides the number of occupied cells in SORT by 2 and returns the remainder (modulus). If the remainder is zero, then there is an even number of values in the range and the first equation following the test is executed. The first equation uses two @INDEX functions to return the two middle values in the range and averages them. For example, if $n = 4$ the following values would be used:

$$210.1$$
$$\rightarrow 223.7$$
$$\rightarrow 228.8$$
$$233.5$$

and 226.25 would be returned. If the @MOD's remainder is non-zero, then there is an odd number of values and the second equation following the test is executed. The second equation uses one @INDEX function to return the middle value in the range. For example, if $n = 5$ the following value would be used:

$$210.1$$
$$215.8$$
$$\rightarrow 223.7$$
$$228.8$$
$$233.5$$

and 223.7 would be returned.

    It takes two columns to find the mode. As outlined above, the mode is the most commonly occurring value. Therefore, you need one column of formulae to count the number of times each value appears and a second column to call out the mode(s). In example Figure 5-7, Columns C and D are prepared as follows:

1. Type a 1 into cell C78.

2. Into cell C79, type the following:

```
@IF(B79=B78,C78+1,1)
```

3. Issue the /Copy command and copy cell C79 to cells C80..C113.

4. Into cell D78, type the following:

```
@IF(C78=@MAX($C$78..$C$113),"MODE","-")
```

5. Issue the /Copy command and copy cell D78 to cells D79..D113.

Complete the analysis by finding the mean. In the example, type the following formula into cell B115:

```
@AVG(SORT)
```

DISPERSION/VARIATION

The degree to which numerical data tend to spread about an average value is called the *variation* or *dispersion* of the data. Various measures of dispersion are available using Lotus @functions, either directly or after simple calculation. The most common measures of variation are standard deviation, variance, coefficient of variation, and range. The standard deviation and variance, which are in substance equivalent, are of great importance both theoretically and practically. The coefficient of variation is closely affiliated with standard deviation. The range, though easiest to calculate, is limited in usefulness because it provides no real insight to a group of data. Let us look at each of these measures in turn.

STANDARD DEVIATION

The standard deviation is a good indicator of the "spread" of the distribution of a population. The standard deviation of a set of N numbers is usually denoted by s and is defined by

$$s = \sqrt{\frac{\Sigma(X_i - \bar{X})^2}{N}}$$

The standard deviation is the square root of the variance and is an index of variability in the original measurement units. Although the variance has great importance in many applications of statistics, it represents quantity in square units of measurements and is therefore not a linear function. For this reason, the standard deviation is more commonly used.

Lotus defines its @STD function as a **population** standard deviation and is the root mean square of the deviations from the mean. As outlined earlier in this chapter, scientific experiments often require a redefinition of N by replacing it with $(N-1)$ in the denominator because it represents a better estimate of the standard deviation of a population from which a sample is taken. For large values of N ($N>30$), there is practically no difference between the two definitions. Because N is small in the example experiment, the template for Figure 3-1/5-1 was set up to correct the @STD. The equation used in cell B50 was

@STD(B39..B47)*@SQRT(@COUNT(B39..B47)/(@COUNT(B39..B47)-1))

which is equivalent to

$$\text{STANDARD DEVIATION} \; * \; \sqrt{N/(N-1)}$$

VARIANCE

The variance of a set of data is defined as the square of the standard deviation and is denoted by s^2. The variance is the mean squared deviation of the individual values from the average value.

The variance measures the extent to which data differ among themselves. If the values are all identical, the variance is zero. If they differ only slightly from each other, the variance is small. If they differ widely, the variance is large.

The variance is important because it is a convenient statistic whenever mathematical manipulations are required. For example, suppose you wanted to find the overall standard deviation of several samples that differ only because of sampling fluctuations. That is, samples are drawn from populations with the same standard deviation, but different means. To find the overall standard deviation, a summation is required. For reasons beyond the scope of this book, **summations cannot be performed on the standard deviation**, but can be performed on the variance. Therefore, the procedure to calculate the overall variance is:

$$V = \frac{(n_1 - 1)V_1 + (n_2 - 1)V_2 + \cdots + (n_k - 1)V_k}{N - k}$$

where n_1, n_2, etc. are the numbers of observations in the first, second, etc., samples. As you can see, this equation is a weighted mean, and each variance is weighted by its Degrees of Freedom. If all samples are of equal size, then

$$V = (V_1 + V_2 + \cdots + V_k)/k$$

From the overall variance, the overall standard deviation can be determined by taking the square root.

Lotus defines its @VAR function as a **population** variance. As outlined earlier in this chapter, scientific experiments often require a redefinition of N by replacing it with $(N-1)$ in the denominator because it represents a better estimate of the variance of a population from which a sample is taken. For large values of N ($N > 30$), there is practically no difference between the two definitions. Because N is small in the example experiment, the template for Figure 3-1 (5-1) was set up to correct @VAR. The equation used in cell B52 was

@VAR(B39..B47)*(@COUNT(B39..B47)/(@COUNT(B39..B47)-1))

which is equivalent to

$$\text{VARIANCE} * (N/(N - 1))$$

COEFFICIENT OF VARIATION

The coefficient of variation is the standard deviation expressed as a percentage of the arithmetic mean

$$\text{CV} = (\text{SD/MEAN}) * 100\%$$

Because the standard deviation and mean are both expressed in the same units, the coefficient of variation is independent of the units of measurement. Therefore, the coefficient of variation is a convenient statistic to relate the spread of observations for two or more methods that measure the same analyte, provided they have the same zero.

Perhaps more often, the coefficient of variation is used to compare the variability of observations made at widely differing concentrations or levels of response. In this context, the coefficient of variation can be an invaluable tool when dealing with assay values whose standard deviations rise in proportion to the mean.

In cell B51 of the example template (Figure 5-1), the coefficient of variation is calculated with the Lotus® version of the above formula. That is

```
+(100*B50)/B49
```

RANGE

The range is perhaps the simplest (and sometimes least satisfactory) of all measures of dispersion. The range of a set of numbers is the difference between the largest and smallest numbers in the set. Although the range is easy to calculate, it has limited usefulness for large sample sizes because it fails to give any consideration to the arrangement of the values between the two extremes. Therefore, it can be influenced by an unusually large or small value (or both), if such a value is present.

The equation used in Figure 3-1/5-1 utilized the Lotus @MIN and @MAX functions. These @functions find the minimum and maximum values in a specified range. From these values, the range is calculated by simple subtraction. The following equation was used:

```
+B54-B53
```

Alternatively, the following formula could have been used to save two rows of cells in the template:

```
+@MAX(B39..B47)-@MIN(B39..B47)
```

MEAN +/– 2 STANDARD DEVIATIONS

The within-run precision of a test is usually given as "95% limits." These limits are

$$\bar{X} +/- t * SD$$

where t is dependent on the Degrees of Freedom and is obtained from a t-distribution table found in most statistics books. For 6 to 17 Degrees of Freedom, t declines from 2.45 to 2.11. When the Degrees of Freedom exceed 30, t is close to 2.00. Therefore, the precision of a test at 95% (actually 95.45%) confidence limits is usually given as

$$\bar{X} +/- 2 * SD$$

For 99.73% confidence limits, the interval would be

$$\bar{X} +/- 3 * SD$$

For 68.27% confidence limits, the interval would be

$$\bar{X} +/- 1 * SD$$

These intervals are called the *confidence intervals of the mean* and provide assessments of the interval in which the mean of the population lies (i.e., the unknown "true" mean).

In cells B56 and B57 of the example template (Figure 5-1), the confidence limits are calculated with the Lotus version of the above 95% formula. That is

```
+B49-2*B50
+B49+2*B50
```

Alternatively, the following formulae could have been used:

```
+@AVG(B39..B47)-2*@STD(B39..B47)
+@AVG(B39..B47)+2*@STD(B39..B47)
```

COMPARISON OF VARIANCES: THE F TEST

The F test was named in honor of R. A. Fisher by George Snedecor. The F test is used to determine

- whether there is a significant difference in precision between two different methods, two analysts using the same method, etc.
- the choice of the form of a t-test equation that needs to be applied for a given **sample** variance situation.
- whether certain statistical tests can be used reliably.
- whether there is a difference in the means of three or more samples (see Analysis of Variance, Chapter 11). With the F test, the assumption is made that the variances of two normal populations (A and B) are equal. If n_A and n_B samples

are taken from those two populations and the sample variances are calculated, F is defined as the ratio of the variances:

$$F = \frac{s_A^2}{s_B^2}$$

The larger of the two sample variances is placed in the numerator. The critical value for F for any given level of significance is obtained from a standard table of F. The first step in using F tables is to establish the magnitude of the risk you are willing to take when deciding that the numerator is less precise than the denominator when, in actuality, there is no difference in precision. It is common to set this level of error (called significance level) at 5, 2.5, 1.0, or 0.5 percent. If 5 percent is chosen, it means that you are willing to be in error 5 percent of the time when deciding that the data corresponding to the numerator have poorer precision.

Next, you need to determine the number of Degrees of Freedom to use. Because both the numerator and denominator in the F test can be associated with differing Degrees of Freedom, a ratio of two variances would also be associated with two such quantities. It follows that the criterion for looking up the value is based on the Degrees of Freedom for the numerator and the denominator and is one less than the number of specimens (or observations) in the respective samples.

The critical value of F is obtained from an F-distribution table for the selected significance level by locating the number of Degrees of Freedom associated with the numerator in a column, and the number of Degrees of Freedom associated with the denominator in a row. As always, you can either look the value up in a standard statistics reference book or build a lookup table and use the Lotus @VLOOKUP and @HLOOKUP functions (applying the concepts of Chapter 2).

If the calculated value of the F ratio does not equal or exceed the critical value found in the F table, there is not enough evidence to conclude that the two sample variances differ significantly. If, however, the calculated value for F equals or exceeds the tabular value, the precision of the data corresponding to the numerator can be said to be poorer than the precision of the sample corresponding to the denominator.

The actual testing of two variances by the F test is quite straightforward in Lotus spreadsheets. You must create a cell containing the ratio of two @VAR functions for the ranges being compared. For example, suppose you wanted to compare the variance of cells B39..B47 to the variance of cells C39..C47 in Figure 5-1. Into cell B57 you would type

```
@IF (@VAR (B39. .B47) >=@VAR (C39. .C47) ,@VAR (B39. .B47) /@VAR (C39. .C47) ,
        @VAR (C39. .C47) >=@VAR (B39. .B47) )
```

or

```
@IF (B52>=C52,B52/C52,C52/B52)
```

The @IF test checks to see if the variance of Column B is greater than or equal to the variance of Column C. If it is, the variance for Column B is the numerator; if it is not, the variance for Column C is the numerator.

Although not shown in the example template, the result of the F test using either of these formulae calculates to 2.48. To explain, assume that a 5 percent level was chosen for the example. At the 5 percent significance level, 8 Degrees of Freedom for the numerator, and 8 Degrees of Freedom for the denominator, the critical value listed in an F-distribution table is 3.44. The calculated F value is 2.48. Because the calculated value is less than the critical value found from the table, you would conclude that there is insufficient evidence to support a difference in precision.

As stated at the beginning of this section, the F test for equality of variances is based on the assumption that populations are normal (Gaussian) in form. Unless a sample size is large, failure of the distribution to meet the normality assumption can lead to serious errors. However, using large sample sizes does not make it acceptable to violate the normality assumption and vice versa. Whether or not a given F test is robust depends on the specific experimental design and the nature of the deviation from normality.

COMPARISON OF MEANS: THE T-TEST

The t-test was first described by W. S. Gosset, who wrote under the pen name "Student". The t-test is a convenient method of comparing two averages for a significant difference. Rigorously stated, the t-test evaluates whether two samples were drawn from populations with the same means. Less rigorously stated, the t-test can answer the question: "Are the means the same?"

The t-test evaluates the tails of the distributions of two averages to assess whether the overlap is enough to consider the averages to be the same. This evaluation is accomplished by a modified consolidation of two variances. The kind of variance consolidation that is used depends on whether populations or samples are being studied and whether the variances are equal or unequal. The kind of consolidation also depends on whether the averages are dependent or independent.

If the same set of samples is analyzed by two methods, or by two analysts, the averages obtained are classified as *dependent*. Dependent averages occur when paired data are analyzed. The need for this type of analysis often arises during method comparisons. In this situation, several different samples are assayed by two different methods and a comparison is made on the paired values for each sample. The comparison represents an average and distribution that can be evaluated by the t-test. The performance of the t-test for paired (dependent) samples can be very conveniently accomplished using Lotus, as described in Chapter 6.

If separate (non-paired) samples are used in determining averages, the averages are classified as *independent*. Comparing independent averages occurs frequently in laboratory work. The following are just a few of the many situations where the comparison of independent means is applicable:

- Means from two different instruments, methods, analysts, etc.
- Normal ranges for clinical diagnostic tests.
- Experimental means with "true" means to verify validity of test procedures (e.g., an average with a reference value).

- Means from two different samples to study their equivalency.
- Means for data with unknown standard deviations.

For independent averages with known **population** variances, the equation for the t-test is:

$$t = \frac{\overline{X}_1 - \overline{X}_2}{\sqrt{\dfrac{s_1^2}{n_1} + \dfrac{s_2^2}{n_2}}}$$

where, n_1 is the number of observations comprising $\overline{X}_1$ with variance s_1^2 and n_2 is the number of observations comprising $\overline{X}_2$ with variance s_2^2.

The above method depends on **population** variances and is used for data that come from well-established methods, ones in which the **population** variances of the averages are well known. However, it is more common in experimental work that the **population** variances associated with two averages are not known. In these situations, **sample** variances must be calculated from the same data used to determine the averages.

The form of the t-test equation to be applied in any given sample testing situation depends on the two **sample** variances. To determine the equation to use, the two sample variances are subjected to the F test. If the two variances are not significantly different by the F test, the sample variances are termed "equal". If there is a significant difference in the variations, the sample variances are termed "unequal".

For independent averages with equal sample variances (i.e., passes the F test), the equation for the t-test is:

$$t = \frac{\overline{X}_1 - \overline{X}_2}{s_p} \sqrt{\frac{n_1 n_2}{n_1 + n_2}}$$

where, s_p is the consolidated sample standard deviation of the first and second sets of data and is defined as

$$s_p = \sqrt{\frac{(n_1 - 1)s_1^2 + (n_2 - 1)s_2^2}{(n_1 - 1) + (n_2 - 1)}}$$

For independent averages with unequal sample variances (i.e., fails the F test), the equation for the t-test is:

$$t = \frac{\overline{X}_1 - \overline{X}_2}{\sqrt{\dfrac{s_1^2}{n_1} + \dfrac{s_2^2}{n_2}}}$$

To compare an average to a theoretical or accepted value, the equation for the t-test is:

$$t = \frac{(\overline{X} - \overline{X}_a)\sqrt{n}}{s}$$

where, X_a is the accepted, true, or theoretical average.

The boundary values for t for any given level of significance can be obtained from a standard table of t. As always, you can either look the value up in a standard statistics reference book or build a lookup table using the concepts of Chapter 2.

The Degrees of Freedom for each category can be calculated as follows:

- For two independent averages with known **population** variances,

$$DF = \infty$$

- For two independent averages with equal **sample** variances,

$$DF = n_1 + n_2 - 2$$

- For two independent averages with unequal **sample** variances,

$$DF = \frac{\left(\dfrac{s_1^2}{n_1} + \dfrac{s_2^2}{n_2}\right)^2}{\dfrac{\left(\dfrac{s_1^2}{n_1}\right)^2}{n_1 + 1} + \dfrac{\left(\dfrac{s_2^2}{n_2}\right)^2}{n_2 + 1}} - 2$$

- For a comparison of an average with a theoretical or accepted value,

$$DF = n - 1$$

The t-test can give unsatisfactory results under certain circumstances. Strictly speaking, the t-test is applicable only when data are normally (Gaussian) distributed. However, the t-test can often lead to valid results if there is moderate departure from the normal distribution; but it is not valid for extremely non-normal data.

The actual testing of means by the t-test is quite straightforward in Lotus spreadsheets. You simply need to decide on the applicable t-test category and create a cell containing the Lotus equivalent to the appropriate t-test formula.

For example, suppose you had a well-established assay procedure with which you collected the data for Figure 5-1. Further, suppose you had established the variances for each sample and wanted to compare the means for cells B39..B47 to cells C39..C47 in Figure 5-1. You would then type the following formula into cell B58 to determine the t-value:

```
@ABS (@AVG (B39. .B47) -@AVG (C39. .C47) ) /@SQRT (@VAR (B39. .B47) /
        @COUNT (B39. .B47) +@VAR (C39. .C47) /@COUNT (C39. .C47) )
```

which is the Lotus equivalent of the first t-test formula in this section. The @ABS function is required because the average of cells C39..C47 may be larger than B39..B47, leading to a negative value. In this example, t = 1.347. Because the cutoff value from a standard two tailed t-test table for infinite Degrees of Freedom at a 0.05 level of significance is 1.96, the conclusion is made that there is no evidence of a significant difference between the two samples.

Alternatively, you could have prepared this equation by taking the square of the standard deviation, but because that squaring is the definition of the @VAR function, the @VAR can be used directly, thus saving you some typing.

You can obtain the values that are needed for the t-test from the template of Figure 5-1. For example,

@ABS(B49-C49)/@SQRT(B52/9+C52/9)

In this example, t = 1.270. This value is different than the one calculated above because the @VARs in the template have been corrected to the sample variance (i.e., they correct to $N - 1$ Degrees of Freedom). Using cells that do not apply this correction, the value of 1.347 is obtained.

More commonly, you will **not** know the **population** variance.

If the variances for two samples are equal (as determined by the F test), you must convert the t-test equation for equal **sample** variances to a Lotus formula. This conversion will require two cells. The first cell will contain the consolidated standard deviation of the first and second sets of data and will have the following format:

@SQRT(((@COUNT(B39..B47)-1)*@VAR(B39..B47)
    +(@COUNT(C39..C47)-1)*@VAR(C39..C47))
      /(@COUNT(B39..B47)+@COUNT(C39..C47)-2))

The second cell will contain the Lotus equivalent of the t-test formula for equal sample variances and will refer to the cell containing the above formula. For example, if the consolidated standard deviation is in cell J39, the equation will have the following format:

(@ABS(@AVG(B39..B47)-@AVG(C39..C47))
    @SQRT(@COUNT(B39..B47)@COUNT(C39..C47)
      /(@COUNT(B39..B47)+@COUNT(C39..C47))))/J39

In this example, cell J39 calculates to 5.530 and t = 1.347. Because the cutoff value from a standard two tailed t-test table for $n_1 + n_2 - 2 = 16$ Degrees of Freedom at a 0.05 level of significance is 2.12, the conclusion is made that no significant difference is evident between the two samples.

If the variances for two samples are unequal (as determined by the F test), you must convert the t-test equation for unequal sample variances to a Lotus formula. Doing so would lead to the following:

@ABS(@AVG(B39..B47)-@AVG(C39..C47))/@SQRT(@VAR(B39..B47)
    /@COUNT(B39..B47)+@VAR(C39..C47)/@COUNT(C39..C47))

In this example, t = 1.347. Because the cutoff value from a standard two tailed t-test table for

$$\frac{((49.1/9) + (19.8/9))^2}{\dfrac{(49.1/9)^2}{10} + \dfrac{(19.8/9)^2}{10}} - 2 = 14.9 = 15$$

Degrees of Freedom (rounded to the nearest whole number) at a 0.05 level of significance is 2.13, the conclusion is made that no significant difference is evident between the two samples.

Another common calculation is one in which sample averages are compared to theoretical or accepted values. For example, the "X PREDICT" cells in the template of Figure 5-1 (G5..G12) contain predicted values for standards using the slope and intercept from regression analysis. The question arises whether these points are sufficiently close to the known DNA concentrations given in Column A (A5..A12). For instance, is the experimental value for the 250 ng/mL standard close enough to the "weighed in" value?

To answer this question, you must first convert the raw data in cells B10..E10 to concentrations. In an open section of the spreadsheet, type the following conversion formula and copy it to the three cells at the right of this cell:

`+(B10-$INTERCEPT)/$SLOPE`

The Lotus equation for this category of t-test is (assuming that the conversion formulae are in cells J10..M10):

`(@ABS(@AVG(J10..M10)-A10)*@COUNT(J10..M10))/@STD(J10..M10)`

In this example, t = 2.610. Because the cutoff value from a standard two tailed t-test table for $n - 1 = 3$ Degrees of Freedom at a 0.05 level of significance is 3.18, the conclusion is made that no significant difference is evident between the experimentally determined standard and "weighed in" value.

SUMMARY OF WHAT HAS BEEN ACCOMPLISHED

This chapter provided a rigorous review of Lotus @functions as they apply to scientific applications. That is, it was stressed that if a **sample** is being studied, corrections need to be made to the Lotus @functions (which are based on **population** parameters).

This chapter also showed

- the flexibility in the arguments supplied to Lotus @functions.
- the power of nesting @functions to achieve both complex calculations and template flexibility.
- how easy it is to use the /Copy command and simple formulae to create series, summation, factorial, etc., calculations.
- the importance of going through logarithms of numbers whenever they are either very large or very small.
- the power and convenience of templates (i.e., a well-designed template allows new tests to be added to an existing superstructure, thus saving both time and effort).
- how to sort data in ascending and descending order.

- how to perform some parametric statistical analyses that determine the equiva-lency of variances (F-test) and means (t-test).
- an introduction to the scheme of converting statistical formulae readily found common text books into Lotus formulae that can solve statistical problems.

WHAT'S NEXT?

Chapter 6 begins a new section of this book—curve fitting. Chapter 6 opens this topic (the characterization of the relationship between two or more variables) by discussing linear, multiple, and curvilinear regressions. Also, the discussion of the t-test will be continued and will be applied to paired samples.

6

Linear, Multiple, and Curvilinear Regression

Previous chapters covered the organization of data, the checking of distributions, the comparison of mean values, simple description techniques for samples, etc. These discussions were based exclusively on repeated measurements of single, independent quantities.

However, one of the most common goals of experimental work is to characterize the mathematical relationships between variables and to subsequently use the predictive powers of these characterizations. The design of these experiments is optimized for finding empirical relationships between one or more variables so that the relationships can be used to estimate the value of any one variable, given values of all the others. Once determined, empirical relationships can be used to

- characterize and quantify the relationship between two variables in a standard curve (e.g., absorbance and concentration), so that a prediction of unknown values of one of the variables (concentration) can be made for given values of the other variable (absorbance), thereby converting measurement units to concentration units.
- study the sensitivity, linearity, and bias of an assay via its standard curve.
- compare the sensitivity, linearity, etc., of one method with another.
- investigate the *fit* (accuracy and precision) of a standard curve to its individual calibrators.
- determine the limits of precision and accuracy of results derived from a standard curve.
- compare "unknown" sample results of one method to another so that the accuracy of the method can be evaluated.
- study the functional dependencies between samples or assays.

- develop conversion formulae to transform units from one assay into units of other assays.
- determine yield vs. conditions so process optimization can be achieved.
- correlate one, easily measured attribute to a second (harder to measure) attribute so that predictions can be more conveniently made.
- predict the value of one variable from that of other(s).
- make predictions beyond the range of the current data.
- allow measurement of a limited number of attributes so that they can be used to predict longevity, etc.
- estimate the dependence of yield and quality of products on reaction conditions such as temperature, pressure, concentration, time of reaction, etc. (i.e., to determine the best operating conditions of a process and to assess what latitude can be tolerated in the reaction conditions).
- increase one's understanding of the chemistry and kinetics of a reaction.
- determine rate of reaction (e.g., change in absorbance per unit time) to assign units of activity to a catalyst or enzyme.
- study the joint distribution of two or more variables and the extent of the association between them.

This chapter deals mainly with the problem of estimating linear, multiple linear, and curvilinear (polynomial) relationships between variables. This process is known as *linear regression curve fitting*.

This chapter makes extensive use of the Lotus® regression program. The Lotus regression program is very powerful and will support almost any type of regression. For example, Lotus can do geometric, exponential, trigonometric, and logarithmic regression analyses.

In some cases, however, these curve fitting procedures are inadequate. For the Lotus regression program to work correctly, an equation must be linear in its parameters or transformable to be linear in its parameters. The variance of the transformed data must also be constant throughout the range being studied. If it is not, you must use nonlinear curve fitting techniques.

Nonlinear curve fitting techniques are very different from linear/polynomial curve fitting regression and require a basic knowledge of macro programming because the techniques are based on approximations, iterations, interpolations, etc. For this reason, Chapter 7 is included to teach you some basic macro programming skills and nonlinear curve fitting is discussed separately in Chapters 8 and 9.

OVERVIEW OF HOW TO FIND THE APPROPRIATE LINEARIZATION FORMULA

Although linear relationships are the most common relationships found in experimental data, many other relationships can also occur. For example, one class of relationships can be described by second, third, fourth and *n*th degree polynomials. The following is a summary of these relationships.

- *Straight line*: $Y = a_0 + a_1X$
- *Parabolic or Quadratic*: $Y = a_0 + a_1X + a_2X^2$
- *Cubic*: $Y = a_0 + a_1X + a_2X^2 + a_3X^3$
- *nth degree*: $Y = a_0 + a_1X + a_2X^2 + \cdots + a_nX^n$
- *Multiple linear*: $Y = a_0 + a_1X_1 + a_2X_2 + \cdots + a_nX_n$

Other, more complex relationships can also occur. The following are some of these other relationships:

- *Hyperbola*: $Y = \dfrac{1}{a_0 + a_1X}$ or $\dfrac{1}{Y} = a_0 + a_1X$
- *Exponential*: $Y = ab^X$ or $log\ Y = log\ a + (log\ b)X$
- *Modified exponential*: $Y = ab^X + g$
- *Geometric*: $Y = aX^b$ or $log\ Y = log\ a + b\ log\ X$
- *Modified geometric*: $Y = aX^b + g$
- *Gompertz*: $log\ Y = log\ p + b^X log\ g = ab^X + g$

  Logistic: $Y = \dfrac{1}{ab^x + g}$ or $\dfrac{1}{Y} = ab^x + g$

Figures 6-1 through 6-9 show qualitative plots of data that can fitted using each of the above categories of equations. If you have historical foundation or literature references that point to the appropriate formula to use, then you should use the formula to linearize the data prior to regression (if possible). However, if these aids are unavailable, you will need to determine the appropriate equation to use to linearize the data. The fastest and easiest way to decide on which curve fitting equation to use is to create a scatter diagram of the data. A scatter diagram will give you a **rough** idea of how the variables X and Y are related and whether one of the above formulae can be used to linearize the relationship. To qualify the formula(e) to evaluate, compare the scatter diagram qualitatively to Figures 6-1 through 6-9.

If, for example, the data in a scatter diagram can be linearized using an exponential equation (*log y* vs. *x*), then the scatter diagram will look like Figure 6-6. On the other hand, if the data of a scatter diagram can be linearized using a *log y* vs. *log x* relationship, the diagram will look like Figure 6-7 and the equation will have the form of a geometric curve. If your scatter diagram qualitatively matches more than one of these figures, try all of the appropriate formulae and choose the one that provides the best fit (as defined below).

Lotus has tools that make it easy to create scatter diagrams. For example, consider the data in Figure 6-10. To create a scatter diagram for these data, you would use the Lotus graphics environment.

If you are using Lotus 1-2-3®, issue the /Graph command to bring up the graphics menu. From that menu, choose Type. Specify an XY graph. Next, choose X (for the X range), specify A5..A12, and press Return. Then choose A (the first Y range), specify F5..F12, and press Return. Finish the setup by choosing Options

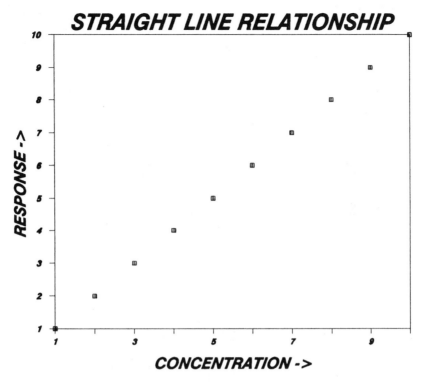

Figure   6-1

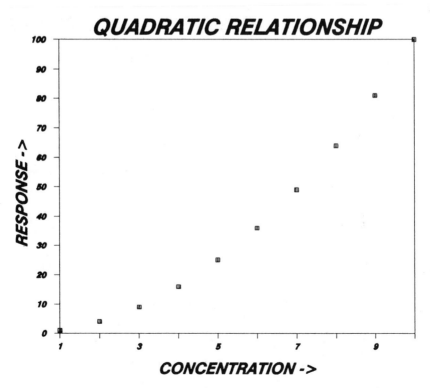

Figure   6-2

CUBIC RELATIONSHIP

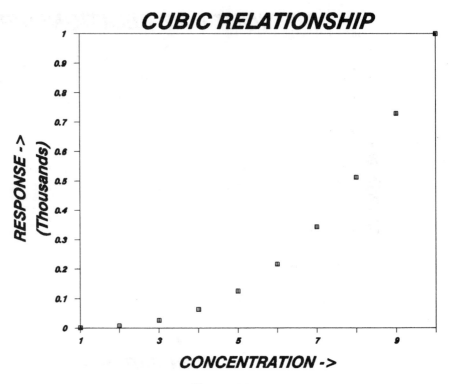

Figure   6-3

MULTIPLE LINEAR RELATIONSHIP

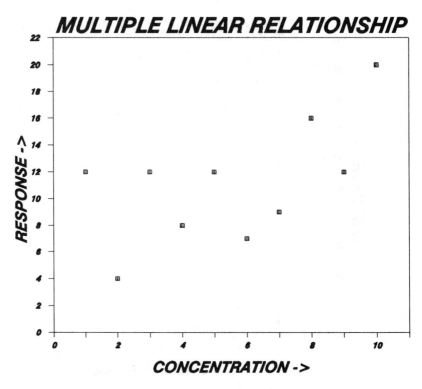

Figure   6-4

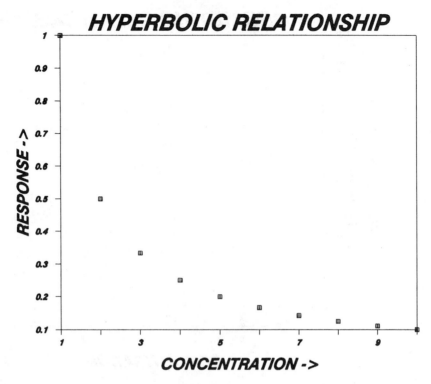

Figure   6-5

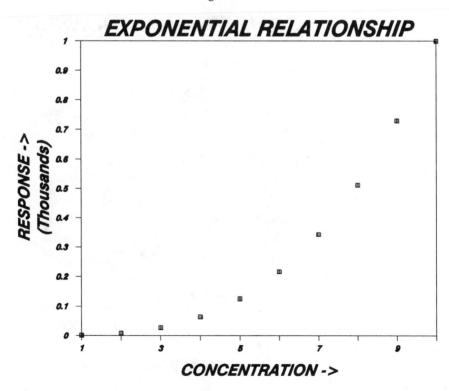

Figure   6-6

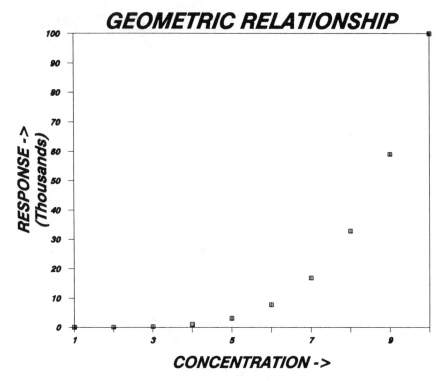

Figure   6-7

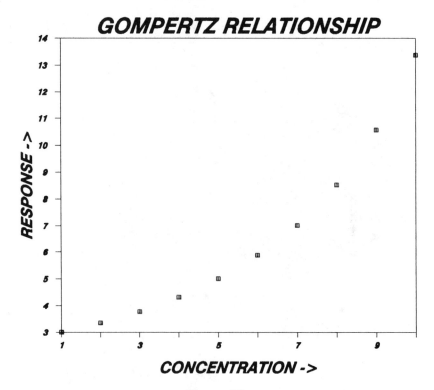

Figure   6-8

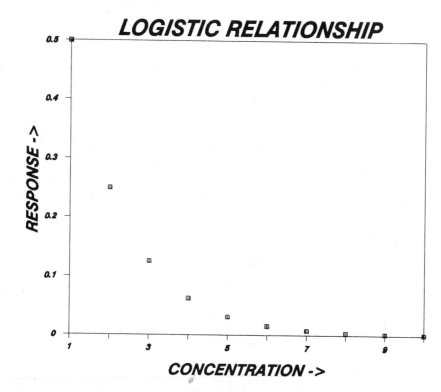

LOGISTIC RELATIONSHIP

RESPONSE ->

CONCENTRATION ->

Figure   6-9

```
--------A--------B--------C--------D--------E--------F--------G--------H-----
 1                         FLUORESCENT DNA ASSAY DATA
 2 DATE RUN:31-OCT-88
 3                  ****STANDARD CURVE DATA****                  X      REGRESS
 4 NG DNA/ML ASSAY1    ASSAY2    ASSAY3    ASSAY4    AVERAGE   PREDICT   CURVE
 5      0     93.9      92.9      94.3      94.3      93.9      0.7      92.7
 6     25    130.4     134.6     133.4     135.9     133.6     26.9     130.7
 7     50    174.5     171.8     173.1     174.1     173.4     53.1     168.7
 8     75    202.9     204.5     204.2     205.2     204.2     73.3     206.7
 9    100    235.0     242.2     242.7     239.9     240.0     96.9     244.7
10    250    473.9     462.0     469.3     473.2     469.6    248.0     472.7
11    500    847.1     854.2     850.4     862.2     853.5    500.6     852.6
12    750   1236.4    1236.4    1226.0    1235.2    1233.5    750.6    1232.6
13           =================================================================
14                         Regression Output:
15           Constant                    92.74541
16           Std Err of Y Est             3.441756
17           R Squared                    0.999939
18           No. of Observations                 8
19           Degrees of Freedom                  6
20
21           X Coefficient(s)  1.519749
22           Std Err of Coef.  0.004814
23           =================================================================
```

Figure   6-10

Format A Symbols. To view the graph, choose View. After viewing the graph, you can save it for later printing or plotting by choosing the Save menu choice.

If you are using Symphony®, issue the /Graph command to bring up the graphics menu. From that menu, choose 1st-Settings and then Type. Specify an XY graph. Next, choose Range and then X (for the X range), specify A5..A12, and press Return. Then choose A (the first Y range), specify F5..F12, press Return and then {ESC}ape. Finish the setup by choosing Format A Symbols and then choose Quit twice to return to the main Graph menu. To view the graph, choose Preview. After viewing the graph, you can save it for later printing or plotting by choosing the Image-save menu choice.

Figure 6-11 shows the scatter diagram for the example in Figure 6-10. If data come from a linear relationship, the graph will result in a straight line curve. In this case the data are linear. However, what happens if the data take the form exemplified by Figures 6-3 or 6-4 (or one of the other figures)? If the graph is not a straight line, you will need to create adjacent column(s) containing Lotus versions of the equations outlined at the beginning of this section and perform the regression analysis on the new column(s) containing the transformed X-variable(s). (More detailed instructions on this technique are given in a later section of this chapter.) After the Lotus regression analysis has been performed, create another graph to evaluate the suitability of the fitted curve.

The second evaluation graph must contain the experimental y-values overlaid by the fitted curve. To create the evaluation graph, first prepare a column that uses the

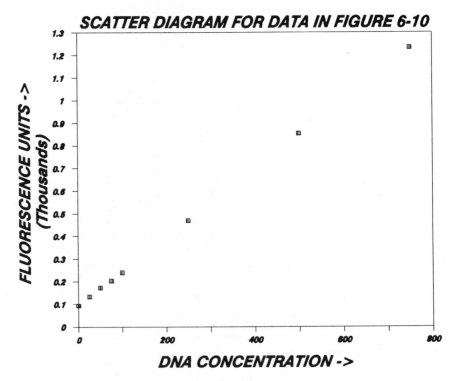

Figure   6-11

appropriate linearizing formula and the regression coefficients to convert x-values into predicted y-values. Then, plot the resulting predicted y-values (formatted to "lines") and the experimental y-values (formatted to "symbols") on the Y axis vs. the untransformed x-values on the X axis. Look qualitatively at the fitted curve to ensure that the predicted y-values are sufficiently close to the experimental y-values. If the qualitative fit looks promising, evaluate the statistics returned by the Lotus regression program to ascertain whether the fit is within the accuracy guidelines established for your laboratory. (Alternatively, you can use one of the more formal goodness-of-fit tests provided elsewhere in this book.) If the statistics do not meet your standards, then try another curve or one of the nonlinear curve fitting programs provided in Chapter 8 or 9.

For example, in Chapter 3, cells H5..H12 (in Figure 6-10) were created using the Lotus version of $a_0 + a_1 x$ ($y = mx + b$). These formulae used the regression slope and intercept to calculate predicted y-values. Setting the X range in the Lotus graphics program to cells A5..A12, the A range to F5..F12 (formatted to "symbols"), and the B range to H5..H12 (formatted to "lines") gives the plot shown in Figure 6-12. An example of a less-than-perfect fit that needs an alternative equation to be tried is shown in Figure 6-13.

One other very important point needs to be made before proceeding. Note that in all of the examples in this chapter, the X-variables are defined as DNA concentrations (i.e., column A in Figure 6-10). However, you could also treat the average fluores-

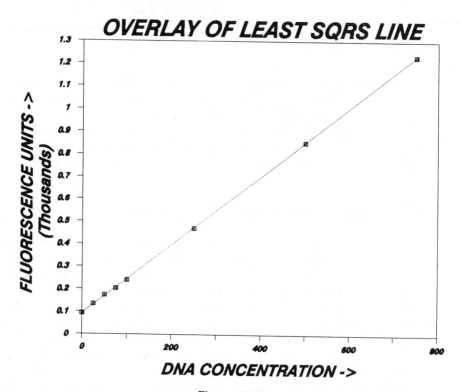

Figure   6-12

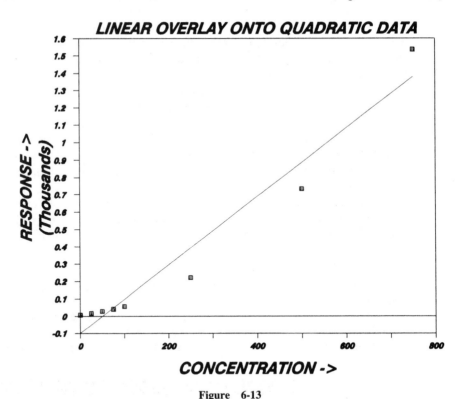

Figure   6-13

cence units data as X-variables (i.e., column F in Figure 6-10), perform regression analysis, etc., and use the regression output in a more straightforward way to calculate concentrations from standard curves. (Using response as the X-variable would be a more convenient method to use if, for example, the equation was polynomial.)

The next few sections will consider the simplest linearization equation (a straight line) so that we can focus our attention on the Lotus regression program.

OVERVIEW OF SIMPLE LINEAR REGRESSION ANALYSIS

As stated in the introduction to this chapter, the objective of curve fitting is to find the relationship between two (or more) variables. By definition, the variable whose value you are trying to determine is called the *dependent* variable. Variables that have assigned (or known) values are called *independent* variables. Thus, the goal of curve fitting is to measure the extent to which a dependent variable (or characteristic) depends upon, or is influenced by, an independent variable. This objective leads to the determination of the dependent or independent variable, given the counterpart.

To avoid confusion, in this chapter the dependent variable will be referred to as "*Y*" and the independent variable will be referred to as "*X*" (although these roles can be transposed, depending on objectives of the analysis). If more than one independent variable is used, they will be assigned the names X_1, X_2, X_3, and so on. When referring to individual experimental data pairs, lower-case letters will be used. For example, the points y_i are the Y-values of the experimental data points corresponding to the X data points x_i.

Linear regression analysis with one independent variable is called simple linear regression. Simple linear regression analysis involves the use of experimental data pairs (x_i, y_i) to find the coefficients a_0 and a_1 corresponding to the best-fit straight line $Y = a_0 + a_1X$.

There are several possible methods that can be used to assess a_0 and a_1. One of the most useful methods is called the Method of Least Squares. This method consists of finding the values of a_0 and a_1 so that the sum of the squares of the deviations between the predicted y-values and their corresponding observed y-values is minimized. In actuality, the deviations of the observed values from the regression line are minimized. The following formulae illustrate forms of the Least Squares equations that can be used for this evaluation:

$$a_0 = \text{intercept} = \frac{(\Sigma\, y_i)(\Sigma\, x_i^2) - (\Sigma\, x_i)(\Sigma\, x_iy_i)}{N\,\Sigma\, x_i^2 - (\Sigma\, x_i)^2}$$

$$a_1 = \text{slope} = \frac{N\,\Sigma\, x_iy_i - (\Sigma\, x_i)(\Sigma\, y_i)}{N\,\Sigma\, x_i^2 - (\Sigma\, x_i)^2}$$

The following formula gives the Least Squares equation that evaluates the Goodness-of-Fit of a line to the data pairs:

$$D_1^2 + D_2^2 + \cdots + D_N^2 = \sum_{i=1}^{N} (y_i - y_i(x_i))^2$$

where, y_i is the observed value and $y_i(x_i)$ is the fitted value calculated from the curve.

When this sum of squares function is a minimum, the curve is called the "best-fitting curve".

Regression is such a common mathematical procedure that both Lotus spreadsheets contain highly flexible programs that efficiently perform the complicated calculations. This feature makes linear regression straightforward because you do not have to create your own program to do it. Using the data regression commands in Lotus 1-2-3 and Symphony, you can perform linear-regression analysis simply by highlighting worksheet data ranges and selecting items from a data regression menu. The results returned by Lotus provide instantaneous details on the regression analysis. These results allow you to immediately use the information in spreadsheet projections and analyses.

ATTACHING THE STAT.APP PROGRAM TO SYMPHONY

The Lotus 1-2-3 regression function is permanently integrated into the Lotus 1-2-3 work environment and is available immediately. However, in Symphony the regression program is an add-in application and must be attached before it can be used.

To attach the STAT.app program, press {SERVICES} Application Attach, highlight STAT.app, and press Return. Press {ESC}ape until you return to the spreadsheet. The STAT.app program needs to be attached only once during a Symphony session.

USING THE LOTUS REGRESSION PROGRAM

This example describes how to perform a simple linear regression analysis on the data of Figure 6-10. (X = DNA concentration; Y = fluorescence units)

To bring up the Lotus regression menu, issue the /Data Regression command (Lotus 1-2-3) or the /Range Regress command (Symphony). To compute the regression statistics, you MUST first define the X range, the Y range, and the Output range or Lotus will issue an error message and abort the regression program. Choose the X-range, Y-range, and Output-range options in turn to specify the following

Menu Choice	Setting	Range
X-range	Independent Variable	A5..A12
Y-range	Dependent Variable	F5..F12
Output-range	Top-left Corner of Results Range	C14

Alternatively, you can specify A5..A12 and F5..F12 by their range names. If you recall, in Chapter 3 these cells were given the range names STDX and STDF, respectively.

After the appropriate specifications have been set, execute the regression analysis by choosing the Go menu option. Figure 6-10 shows the outcome of this action and is explained below.

The Intercept option allows you to compute a linear regression with a pre-set y-intercept of 0. That is, the regression program will force the regression line through the origin by adjusting the other regression parameters (e.g., the slope) to compensate. This option is rarely used for scientific data and can lead to misinterpretation of data if you are not careful.

THE REGRESSION SECTION OF THE TEMPLATE

The Regression section of a template is the area that receives data from the Lotus regression program. When the numbers for the regression parameters are returned from the regression program, the program places text into the spreadsheet along with the numbers. The text appearing in the box of Figure 6-10 originated from the regression program. Definitions of the information returned are listed below, grouped according to category.

DESCRIPTION OF THE LINE

The form of the line that is returned in the example is: Y = "Constant" + "X Coefficient" * X (i.e., $Y = a_0 + a_1 X$ or $Y = mx + b$).

The *Constant* is better known in the sciences by one of its more common names.

In various disciplines it is called the "intercept," "b," "a_0," or the "Y-intercept". In this book the Constant will be referred to as "intercept" in text and a_0 in equations. The intercept of a straight line is the Y-coordinate of the point where the line crosses the Y-axis.

An *intercept* of a set of experimental data is a constant or systematic error as opposed to a random or proportional error. It expresses itself as a persistent plus or minus deviation in the method.

In simple regression, the *X Coefficient(s)* is the slope (i.e., "a_1" or "m") of the line. If multiple, quadratic, or polynomial regressions are performed, coefficients are given for each independent variable and/or power of X (e.g., $a_1, a_2, \ldots, a_n$). The slope of a line is the amount of change in Y per unit change in X; a slope of 2 means that Y increases 2 units while X increases 1 unit.

In addition to these parameters, the regression program also returns two other pieces of information that describe the fundamental elements of regression analysis. The "No. of Observations" is the number of observations. It is simply the total number of values in the Y range.

Degrees of Freedom is defined in the same way that it has been throughout this book. It is the number of observations minus the number of constants calculated from the observations. For simple linear regression the Degrees of Freedom is computed by subtracting TWO from the number of observations.

The presence of (N-2) Degrees of Freedom is analogous to the (N-1) Degrees of Freedom that appeared in our estimate of the standard deviation of N measurements of one quantity X. If you recall, when we calculated the standard deviation, we used N measurements $x_1, x_2, \ldots, x_N$. However, before we could calculate the standard deviation, we had to use the same set of data to find the mean. This calculation left only (N-1) independent measured values and therefore, only (N-1) Degrees of Freedom remained.

Exactly analogous to this guideline, simple linear regression also makes use of N measurements, but before calculating error analyses, the two quantities a_0 and a_1 must be calculated. After this calculation, only (N-2) Degrees of Freedom remain.

For multiple or polynomial regression, the Degrees of Freedom is also defined as the number of independent measurements minus the number of parameters calculated from these measurements. For example, if linear regression is performed using a quadratic equation ($a_0 + a_1X + a_2X^2$), three quantities (a_0, a_1, and a_2) must be calculated. In this case, (N-3) Degrees of Freedom are reported by Lotus.

ERROR ANALYSIS FOR LINEAR REGRESSION

Before using regression equations, you should examine them to see how appropriate they are. In addition to graphing, many numeric tools are available that you can use to determine how well a regression equation fits the data. These tools will also allow you to quantitatively check the validity of the assumptions that you made in calculating the regression equation. Additionally, quantitative error analysis statistics can be used as tools to determine how well the results from one method agree with another.

Error analysis is of the utmost importance and can help to reveal problems in the following areas:

- If a small sample of a larger population appears to show a linear relationship strictly by chance, while in reality the relationship is nonlinear or nonexistent.
- If there is an unacceptably high scatter of the data points about the regression line.
- If a sub-optimal or incorrect linearizing formula is chosen.

These areas all relate to whether the regression equation adequately describes the relationship between X and Y. Faced with the possibility of considerable error that can be caused by these environs, you ought to find out how well the regression equation will describe your data before using the equation.

Before investigating the tools available for evaluating the Goodness-of-Fit of a regression curve, it is important to point out factors that influence the evaluation. Even in a fairly large sample from a population in which the regression is linear, a plot of the actual observations of one variable for given values of the other will not usually fall exactly on the straight line drawn using regression coefficients. These deviations are due, in varying degrees, to usual random sampling errors and can complicate the evaluation of the least squares fit.

Other factors that affect the precision and accuracy of a regression analysis are:

- the precision of the observations for each individual sample.
- the number of observations.
- the range of values of the X variable.
- the offset of the range of x-values (i.e., whether x-values are clustered at some offset from the origin).
- the extent of scatter about the regression.

Precision and sampling evaluation have been discussed in previous chapters. The subject of numerical evaluation of the extent of scatter about the regression using Lotus regression output is discussed next.

The three error analysis tools returned by the Lotus regression program are shown in Figure 6-10. They are the Std Err of Y Est, R squared, and Std Err of Coef.

The *Std Err of Y Est* is the "Standard Error Of Y Estimate" (S_{yx}) and is a measure of the certainty with which the independent values in the sample can be used to predict the dependent values. That is, S_{yx} is an estimate of the error in a single value of y calculated using the regression equation. S_{yx} is the standard deviation of the differences between observed and calculated y values and is calculated with the equation

$$S_{yx} = \sqrt{\frac{\Sigma\,(y_i - y_i(x_i))^2}{(N - 2)}}$$

where, $y_i(x_i)$ is the value predicted from the curve, y_i is the observed value, and (N-2) is the Degrees of Freedom. The (N-2) Degrees of Freedom arise from the calculated $y_i(x_i)$ values based on estimates of two constants (a_0 and a_1), and thus the requirement for subtracting 2.

S_{yx} has properties comparable to the standard deviation. For example, if lines are constructed parallel to the regression line and at vertical distances S_{yx}, $2S_{yx}$, and $3S_{yx}$

from the line, 68%, 95%, and 99.7% of the sample points, respectively, would be included between the lines. (See Figure 6-14 for a graph containing $+/- 3S_{yx}$ limits.) Also akin to the standard deviation, the larger the value of S_{yx} compared to the predicted y values, the less certain that you can be about the prediction.

One important use of S_{yx} is in the calculation of the confidence limits for the a_i coefficients of the calculated curve. This chapter contains a subsequent discussion of how to use S_{yx} in conjunction with the t-test to calculate the confidence limits that establish a band about the calculated curve that contains the true curve.

R Squared is the "coefficient of determination" or the "square of the correlation coefficient" and is a measure of the Goodness-of-Fit for a straight line regression. This statistic tells you what percentage of the variation in the dependent (Y) variable is "explained" by the variation in the independent variable (X).

The value of R squared can range from 0 to 1. If the relationship between two variables is strong, the R Squared value will be close to 1. If the relationship between variables is weak or non-existent, the value will tend to be toward zero. For example, an R Squared value of 0.721833 means that about 72% of the variation in Y is explained by the change in X.

Although the value of R squared can theoretically result in only positive numbers, there is a quirk in the Lotus regression program than can, in fact, return a

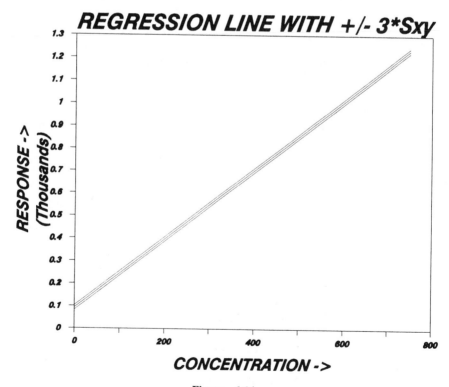

Figure 6-14

negative R squared value. This problem is a result of forcing the intercept through the origin with the Intercept Zero menu choices and having a set of fairly random numbers.

The square root of R squared (r) is called the *correlation coefficient* and is another commonly used measure of the joint distribution of variables and the degree of association between them. The correlation coefficient between the (x, y) data points is a measure of the scatter of the data about a linear line of the means.

Data that are perfectly represented by a linear least squares approximation have a correlation coefficient of $+/- 1$, depending on the sign of their slope (a_1). Uncorrelated data (random scatter) have a correlation coefficient of zero. Although it depends on the application and the size of the sample, most disciplines consider a correlation coefficient of less than 0.9 to be poor. A correlation coefficient of less than 0.9 often suggests that another curve fitting formula should be used.

The use of correlation coefficients by scientists has become more common than the use of R Squared. This popularity is somewhat unfortunate because there are some deceptive traps in using the correlation coefficient to make important decisions. Implicit in the definition of the correlation coefficient is the fact that $100r^2$ is the percentage of the total variation of the dependent variables, which is explained by their relationship to the dependent variables. That is, it is R Squared that defines the relationship, not R. Therefore, although the correlation coefficient, by itself, is an important measure of the relationship between the variables, it is R Squared that permits comparisons of the strengths of relationships. This point is where you must exercise care in interpretation. For example, if $r = 0.80$ in one study and $r = 0.40$ in another study, it would be misleading to report that the 0.80 correlation was "twice as good" as the 0.40 correlation because when $r = 0.80$, $100 (0.80)^2 = 64$ percent; while, when $r = 0.40$, $100 (0.40)^2 = 16$ percent. Thus, the percentage of the variation in the dependent variables accounted for by their relationship to the independent variables is different by **four** times. Likewise, $r = 0.60$ is nine times as strong as a relationship for which $r = 0.20$.

The following equation can be used for manually determining the sample correlation coefficient (r):

$$r = \frac{\sum x_i y_i - \dfrac{(\sum x_i)(\sum y_i)}{N}}{\sqrt{\left(\sum x_i^2 - \dfrac{(\sum x_i)^2}{N}\right)\left(\sum y_i^2 - \dfrac{(\sum y_i)^2}{N}\right)}}$$

However, it is much more convenient to use the Lotus @SQRT function to determine the square root of the R Squared cell. For example, in Figure 6-10, the correlation coefficient could be found using the following formula

@SQRT (F16)

The *Std Err of Coef* is an estimate of the standard deviation of the coefficient of an independent variable in an equation. Generally, the larger the Standard Error of Coefficient in relation to the corresponding X coefficient, the less certain that you can be about the prediction.

If you use the Lotus regression program to determine more than one coefficient (e.g., for multiple or polynomial regression), standard errors will be returned for each coefficient. The equation used for the first coefficient (i.e., the slope) is

$$S_{a_1}^2 = \frac{S_{xy}^2}{(\Sigma x_i^2) - \dfrac{(\Sigma x_i)^2}{N}}$$

The main use of the S_a values is to test each of the a_i coefficients to ensure that they are significant and that there is a good correlation between X at the ith level and the y data. That is, if the absolute value of a coefficient is significantly larger than its standard error of coefficient, then the coefficient contributes in a productive manner to the regression equation, and should therefore be used in the regression equation. Otherwise, the coefficient is not part of the equation and should be set to zero, removing the term from the equation.

To illustrate the testing concept, consider an a_1 returned by the regression program. To determine whether this term is significant (and should therefore be used in the final equation), use the following formula:

$$|a_1| > t * S_{a_1}$$

where t is a value from a standard t-table for the required confidence interval and $(N-2)$ Degrees of Freedom. If you have at least six data points, the 95% confidence interval for t is about 2.5 and decreases for more Degrees of Freedom. Therefore you can be conservative and use 2.5 times the standard error for the coefficient in question to determine whether the coefficient is significant.

You can automate the significance calculation for the example data in Figure 6-10 by placing the following Lotus formula into an open cell of the spreadsheet:

```
@IF (@ABS (E21) >2.5*E22, "SIGNIFICANT", "NOT SIGNIFICANT")
```

In this example, the X Coefficient (1.5197) is much greater than 2.5 times the Standard Error of Coefficient (0.012037), so the test passes easily and "SIGNIFICANT" is reported. However, if you are doing multiple or polynomial regression with several coefficients, the test will show how many of the coefficients do not correlate with the predicted Y value. Those "a_i" coefficients that have a S_a that approaches the magnitude of the value of the "a_i" coefficient do not provide a beneficial correlation to the curve fitting equation and should be dropped by setting the coefficients to zero.

For example, suppose you were using the following equation and the Lotus linear regression program to try to fit some experimental data:

$$Y = a_0 + a_1X + a_2X^2 + a_3X^3$$

Further suppose that a_2 had failed the t-test. This information would tell you that the fitting equation does not have a square term in it because the noise at this term makes up a substantial portion of the observation. Based on this error analysis, you would use the following equation to describe your data:

$$Y = a_0 + a_1X + a_3X^3$$

That is, you would eliminate the "SQUARED" column (cells B5..B12) from the template and replace the "SQUARED" data with "CUBED" data (e.g., move the data from cells C5..C12 to cells B5..B12). You would then use the regression program again and repeat the analysis. To be on the safe side, it is often desirable to add the next term to the equation (e.g., a_4X^4) to determine whether the next term becomes significant. Do so by placing the appropriate equation into Column C.

Perhaps a more common use of this analysis is in multiple linear regression, where the following equation is applicable:

$$Y = a_0 + a_1X_1 + a_2X_2 + a_2X_2$$

With this scenario, Standard Error of Coefficient testing will tell you which of the independent variables have influences on the data (and should therefore be included in the equation).

CALCULATING INTERCEPT AND SLOPE ERRORS

The goal of this section is to provide you with information on how to use the error analysis values returned by the Lotus regression program to build tools that can be used to evaluate parameter errors. More specifically, the tools that will be created in this section will evaluate the error in the intercept and slope of a regression line. This evaluation is important to the interpretation of a line because the intercept and X coefficients are only estimates based on sample data, implying the existence of corresponding true values (based on a population). Therefore, predictions based on the estimates will be subject to some finite error.

An error in the intercept of a regression line gives rise to a constant error from any point on the line because a change in the intercept moves the line up or down without changing its slope. Conversely, an error in the slope of a regression line gives rise to an error that is zero at the point $(\bar{x}, \bar{y})$ and increases as x moves away from $\bar{x}$. Any prediction made from the regression equation is therefore also subject to error.

In terms of the intercept, it is beneficial to test whether the estimate for the intercept (a_0) of a regression line differs significantly from some specified value (usually zero). The appropriate analysis for the intercept is called the *Standard Error of the Intercept* and is calculated manually using the following formula

$$s_{a_0}^2 = \frac{S_{xy}^2 \sum x_i^2}{N(\sum x_i^2) - (\sum x_i)^2}$$

where, S_{xy} is the Standard Error of Y Estimate. Thus, the Standard Error of the Intercept can be very conveniently calculated using information returned from the regression program. However, you must first create a new column to perform a summation for x_i^2. Figure 6-15 shows this new column. To create this column type $+A5^{\wedge}2$ into cell I5 and copy the formula to cells I6..I12. To calculate the Standard Error of the Intercept, type the following formula into an open cell of the spreadsheet (e.g., H15):

```
+F16^2*@SUM(I5..I12)/(F18*@SUM(I5..I12)-@SUM(A5..A12)^2)
```

```
--------A--------B--------C--------D--------E--------F--------G--------H--------I----
 1                      FLUORESCENT DNA ASSAY DATA
 2 DATE RUN:31-OCT-88
 3              ****STANDARD CURVE DATA****                      X      REGRESS SQUARE OF
 4 NG DNA/ML ASSAY1   ASSAY2   ASSAY3   ASSAY4   AVERAGE  PREDICT  CURVE  NG DNA/ML
 5      0     93.9    92.9     94.3     94.3      93.9     0.7     92.7       0
 6     25    130.4   134.6    133.4    135.9     133.6    26.9    130.7      625
 7     50    174.5   171.8    173.1    174.1     173.4    53.1    168.7     2500
 8     75    202.9   204.5    204.2    205.2     204.2    73.3    206.7     5625
 9    100    235.0   242.2    242.7    239.9     240.0    96.9    244.7    10000
10    250    473.9   462.0    469.3    473.2     469.6   248.0    472.7    62500
11    500    847.1   854.2    850.4    862.2     853.5   500.6    852.6   250000
12    750   1236.4  1236.4   1226.0   1235.2    1233.5   750.6   1232.6   562500
13         ==============================================
14                     Regression Output:               STD ERROR OF INTERCEPT
15         Constant                    92.74541              2.590111
16         Std Err of Y Est             3.441756         INTERCEPT LIMITS (+/-)
17         R Squared                    0.999939              6.345774
18         No. of Observations                 8         STD ERROR OF SLOPE
19         Degrees of Freedom                  6             -5.5E-06
20                                                       SLOPE LIMITS (+/-)
21         X Coefficient(s)  1.519749                        -1.3E-05
22         Std Err of Coef.  0.004814                    t-TEST
23         ==============================================         315.6281
```

Figure   6-15

In this example, the Standard Error of the Intercept is 2.59. Therefore, the confidence limits (range) for the intercept are

$$A_0 = a_0 +/- t * S_{a_0}$$

$$A_0 = 92.7 +/- 6.35$$

$$86.3 < A_0 < 99.1$$

where, t is determined from a standard table of two-sided t-values for $N - 2 = 6$ Degrees of Freedom and a confidence level of 0.05 (i.e., $t = 2.45$ in this example).

In terms of the slope, the more scatter there is about a regression line, the less precisely the slope (a_1) of the regression line will be known. It is therefore helpful to evaluate the confidence limits of a slope to ensure that they are reasonable. The appropriate analysis for the slope is called the *Standard Error of the Slope* and is calculated manually using the following formula:

$$S_{a_1}^2 = \frac{S_{xy}^2}{(\Sigma x_i^2) - \dfrac{(\Sigma x_i)^2}{N}}$$

where, S_{xy} is the Standard Error of Y Estimate. Thus, the Standard Error of the Slope can also be very conveniently calculated using information returned from the regression program and the new column created to perform the summation of x_i^2 for the calculation of the Standard Error of the Intercept. To calculate the Standard Error of the Slope, type the following formula into an open cell of the spreadsheet:

+F16^2/(@SUM(I5..I12) - (@SUM(A5..A12)^2/F18))

In this example, the Standard Error of the Slope is $-2.0E-06$. Therefore, the confidence limits (range) for the slope are

$$A_1 = a_1 +/- t * S_{a_1}$$
$$A_1 = 1.520 +/- \sim 0$$

where, t is determined from a standard table of two-sided t-values for N-2 = 6 Degrees of Freedom and a confidence level of 0.05 (i.e., $t = 2.45$ in this example).

As stated above, any prediction made from a regression equation will have some finite error due to the uncertainty in the regression coefficients. Publishing the confidence limits of regression coefficients is an informative means to relate the limits of that error.

However, there are other important uses for the confidence limits. Comparing the ranges of the experimental slope and intercept to published, theoretical, or known regression coefficient values can also be useful to ensure that an assay is working within established guidelines and therefore giving accurate predictions. Confidence limit ranges are also helpful when comparing the sensitivity, bias, etc., of two or more methods and can be used to determine whether differences really do exist, or if apparent differences are due just to the precision of the fitted curves.

Another important area where confidence limits are invaluable is when you are comparing the results of samples from one assay to another. For example, suppose you assayed a single set of samples by Assay A and again by Assay B. Further, suppose you found a slope that was different from 1.000 or an intercept that was greater than 0.000. Using ranges for the slope and intercept based on confidence limits, you can quickly view the limits to see if the discrepancy in the two methods was significant (meaning that the methods were truly different), or if the difference was due to normal imprecision in the coefficients.

TRANSFORMATION AND DEPENDENT (PAIRED) SAMPLE T-TESTS

There are two t-tests that can be performed on paired data. The first t-test will make it easier for you to interpret the correlation coefficient. For example, R Squared in Figure 6-10 was 0.999939. In this case it is quite obvious that there is a very good fit of the data by the regression line. Likewise, it is quite obvious that an R Squared value of 0.00001 shows a very poor fit of the data by the regression line. However, what if the correlation coefficient was 0.40?, or 0.80? These questions may not be easy to answer and you may need a tool to help you in the decision-making process.

When variables are at least roughly normally distributed, you can use a technique called *transformation*. With the transformation technique, the correlation coefficient statistic is transformed to a t-statistic prior to evaluation, because the latter has a distribution that is well-defined and routinely published. To accomplish this transformation, you merely need to type one of the following formulae into a cell of the spreadsheet

```
@SQRT(F17)*@SQRT(F19)/@SQRT(1-F17)
```

or

```
@SQRT((F17*F19)/(1-F17))
```

These formulae are the Lotus equivalents of the following:

$$t = |r| \frac{\sqrt{N-2}}{\sqrt{1-r^2}}$$

Using the regression output in Figure 6-15, a *t*-test value of 315.6 is determined using the above equations.

The interpretation of this result is based on the *null hypothesis*. That is, the transformation equation is based on the hypothesis that there is no correlation between *X* and *Y* and that the correlation between the populations is, therefore, zero. In effect, the *t*-test is being applied to the null hypothesis to determine the probability that the assumption was incorrect. Therefore, with this adaptation of the *t*-test, values of *t* that are **higher** than table cutoff values of *t* indicate that the null hypothesis can be rejected and that a significant correlation exists.

For example, with *N*-2 = 6 Degrees of Freedom and a 0.05 level of significance, the critical value for a two-sided *t* from a standard *t*-table is 2.45. Because the experimental *t*-value for this example (315.6) is much greater than 2.45, the linear relationship is therefore real and not due to chance.

Another statistical test that can be conveniently calculated is the *t*-test for paired samples. The *t*-test was introduced in Chapter 5, where it was used to compare unpaired samples. The discussion contained therein applies almost directly to using the *t*-test to compare paired samples. When paired samples are studied, the *t*-test is used to study whether two methods are giving the same average value for a given sample, if two analysts are obtaining the same result on a common set of samples, etc.

Manually, to calculate a *t*-test on paired samples you would need to use the following equation:

$$t = \frac{\overline{d}}{s_d} \sqrt{N}$$

where, *d* is the difference in a pair of values, $\overline{d}$ is the average difference for all pairs of values, *N* is the number of data pairs, and s_d is the standard deviation of the differences between the pairs of observations, defined as follows:

$$s_d = \sqrt{\frac{\sum d^2 - \frac{(\sum d)^2}{N}}{N-1}}$$

With Lotus, it is much easier to perform the paired sample *t*-test. By way of example, let us use this technique to compare the standard curve-predicted *y* results to the experimentally-determined *y* results. To calculate the *t*-test for paired observations, you would create a column in the spreadsheet that contains formulae to find the differences between each data pair. You would then enter a formula (based on the Lotus @AVG, @STD, and @COUNT functions) into a cell to calculate the value of *t*. For example, to add a *t*-test for the data pairs in Columns F and H of Figure 6-15, you would type the formula for finding the differences between $y_{predict}$ and $y_{experimental}$ into cell J5 and copy it to cells J6..J12:

```
+H5-F5
```

The Lotus formula for the paired data *t*-test in this example is:

```
+@AVG(J5..J12)*@SQRT(@COUNT(J5..J12))/@STD(J5..J12)
```

This remarkable simplicity will be readily appreciated by anyone who has performed the calculations manually and is a good demonstration of the power of using a spreadsheet for statistical analyses. The entire process can be easily performed on very large numbers of data pairs in a matter of less than two minutes, as opposed to hours using a calculator.

The determination for significance is exactly the same as presented for the *t*-tests in Chapter 5, except that *N*-1 Degrees of Freedom are used. For this example $t = 1.2E - 14$. For $N - 1 = 7$ Degrees of Freedom and a 0.05 level of significance, the critical value for a two-sided *t* from a standard *t*-table is 2.36. Because the experimental *t*-value for this example (1.2E − 14) is less than 2.36, the *t*-test shows that there are no significant biases in the two sets of data.

THE EFFECT OF REGRESSION ANALYSIS ON THE TEMPLATE

Column G of the Standard Curve section (the one entitled "X PREDICT") relates to the Regression section. In Chapter 3 you placed a Lotus equivalent to the formula $X = (Y - a_0)/a_1$ into each of the cells of this column. This formula is a re-arrangement of the $Y = a_0 + a_1X$ formula, which yields the predicted concentrations, given the fluorescence units. (When you create the template for your application, use an equation that appropriately describes the data for **your** standard curve.)

When a linear regression is performed on the "NG DNA" and "AVERAGE" columns in this example, the X PREDICT column updates using the "Constant" (intercept) and "X Coefficient" (slope) from the linear regression. Using the results, you can quickly check to see how closely each of the points on your standard curve agree with those predicted by the regression analysis.

This equation is also the pertinent one for calculating the unknown values from a simple linear least squares curve. For example, in Chapter 3, you also placed this equation into cells B39..H47. Using regression coefficients and the appropriate formulae used for the regression curve, you can calculate the values for your unknowns from their experimental responses (e.g., fluorescence, absorbance, radioactivity, etc.).

All of the templates in this chapter have been set up using this scenario. However, for assay templates that calculate concentrations from standard curves, it is often more convenient to reverse the roles of the independent and dependent variables. That is, if you use the experimental response as the X variable (independent variable) and concentration as the Y variable (dependent variable), this variable assignment makes it easier to convert data from polynomial, exponential, etc., equations. You can use the regression equation directly to convert unknown responses to concentrations.

VISUALLY CHECKING THE LINEAR LEAST SQUARES FIT

There is no better way to guard against being tricked by regression statistics than to plot the regression curve overlaid onto the experimental data points. This plot will show if there is curvature at the top end, hockey stick effects at the low end, etc. That is, it will show if you are overlooking something in the numerical maze that is associated with all of the error analysis on the regression line. Based on my experience, no matter how many statistics you look at, there is still a possibility that you will miss something if you don't plot the curves.

Figure 6-13 illustrates this point. The figure shows a scatter diagram for some distinctly quadratic data. Overlaid onto the scatter diagram is the least squares straight line. The correlation coefficient of this line is 0.9767. However, note that there is significant curvature of the data points. The next section will show how to fit a quadratic formula to data points, but let us first discuss how to create the graph.

To graphically check the fit of the example data, you need a column of cells that contains values predicted using the regression coefficients. If you have not already created cells H5..H12 in the example template (Figure 6-10), start by typing a formula into cell H5 to convert the independent variable into a predicted y-value. (For your data, you should base the formula on the one used for your regression analysis.) Because the regression was found to be simple-linear, the appropriate equation is

```
+A5*$SLOPE+$INTERCEPT
```

Copy cell H5 to H6..H12. If you are using Lotus 1-2-3, issue the /Graph command to bring up the graphics menu. From that menu, choose Type. Specify an XY graph. Next, choose X (for the X range), specify A5..A12, and press Return. Then, choose A (the first y range), specify F5..F12, and press Return. Now choose B (the second Y range), specify H5..H12, and press Return. Finish the setup by choosing Options Format A Symbols B Lines and Quit. To view the graph, choose View. After viewing the graph, save it for later printing or plotting by choosing the Save menu choice.

If you are using Symphony, issue the /Graph command to bring up the graphics menu. From that menu, choose 1st-Settings and then Type. Specify an XY graph. Next, choose Range and then X (for the X range), specify A5..A12, and press Return. Then, choose A (the first Y range), specify F5..F12 and press Return. Next, choose B (the second y range), specify H5..H12, and press Return. Now, press {ESC}ape. Finish the setup by choosing Format A Symbols B Lines and then choose Quit twice to return to the main Graph menu. To view the graph, choose Preview. After viewing the graph, you can save it for later printing or plotting by choosing the Image-save menu choice.

As stated earlier in this section, Figure 6-13 clearly shows a problem with using a straight line to adequately describe the data. The remainder of this chapter will be devoted to other forms of linear regression that can be used to address this problem.

POLYNOMIAL REGRESSION

When error statistics or a scatter diagram show that a linear characterization of the association between observed values of two variables appears ineffectual, an equation of higher order may be investigated to see if it improves the fit. This equation will be of the general form

$$Y_p(X) = a_0 + a_1 X + a_2 X^2 + \cdots + a_n X^n$$

Polynomial curve fitting is an extension of the fitting technique described for linear regression analysis. For this reason, much of the nomenclature described above is directly applicable to polynomial fitting. Likewise, the same general manual calculation techniques may be applied to the fitting of multiple relationships because additional independent variables are mathematically equivalent to successive powers of a single, independent variable.

Normally, manual polynomial regression is performed by matrix techniques. Setting up the matrices and the techniques necessary to solve them for polynomial regression is rather laborious, particularly if there are a large number of original observations and many powers in the equation. The Lotus regression program uses these matrix techniques. However, the Lotus regression program handles all of the calculation details automatically, making the fitting of data with a 15th degree polynomial almost as easy as fitting a straight line to the data.

Within the Lotus environment, the fitting of a curvilinear relationship by a polynomial is very similar to simple linear regression described previously in this chapter, except that the powers $X^2, X^3, \ldots, X^n$ of the values of the independent variable X are used in addition to observed values of the single independent variable X. Lotus can handle up to a 16th degree polynomial.

Figure 6-16 shows how to implement a cubic fit to the data of Figure 6-13. Notice that two columns have been inserted between the NG DNA/ML column and the columns containing the raw data. These columns hold formulae to calculate X^2 and X^3 from the corresponding X value in Column A. Column I contains Lotus formulae that can be used to graphically evaluate Goodness-of-Fit. These formulae are of the following format:

$$a_0 + a_1 X + a_2 X^2 + a_3 X^3$$

To create this example, start with the template of Figure 6-10 and overtype the old fluorescent units data with the new data. Then, perform the following steps:

1. Issue the /Range Name Create command to assign the name SLOPE to cell E21. (This cell will contain the coefficient for X.)

2. Issue the /Range Name Create command to assign the name SQUARE to cell F21. (This cell will contain the coefficient for X^2.)

3. Issue the /Range Name Create command to assign the name CUBE to cell G21. (This cell will contain the coefficient for X^3.)

4. Issue the /Move command to move cells B5..H12 to cells D5..J12. This action will open Columns B and C for use in the regression analysis and move all of the existing cells two columns to the right.

5. Into cell B5, type the following formula (which calculates the square of X) and copy it to cells B6..B12:

   +A5^2

6. Into cell C5, type the following formula (which calculates the cube of X) and copy it to cells C6..C12:

   +A5^3

7. Into cell I5, type the following formula (which calculates a predicted Y from the X value and the regression coefficients) and copy it to cells I6..I12:

   +$INTERCEPT+$SLOPE*A5+$SQUARE*B5+$CUBE*C5

8. If you are using **Lotus 1-2-3,** issue the /Data Regression command. If you are using **Symphony,** issue the /Range Regress command.

9. Choose the X-range, Y-range, and Output-range commands in turn to specify the following:

Menu Choice	Setting	Range
X-range	Independent Variable	A5..C12
Y-range	Dependent Variable	H5..H12
Output-range	Top-left Corner of Results Range	C14

10. Execute the regression analysis by choosing the Go menu option.

Figure 6-16 shows the outcome of this sequence of actions. Note that the output is very similar to the simple linear regression output, but has some additions. These additions are the X Coefficients and the Standard Error of Coefficients for the square and cube term of the cubic equation. Lotus will return this pair of values for each power of X (up

```
--------A--------B--------C--------D--------E--------F--------G--------H--------I--------J-----
1                            FLUORESCENT DNA ASSAY DATA
2 DATE RUN:31-OCT-88
3                                    ****STANDARD CURVE DATA****                      REGRESS
4 NG DNA/ML SQUARED    CUBED  ASSAY1  ASSAY2  ASSAY3  ASSAY4  AVERAGE   CURVE
5       0        0        0     8.4     9.3     8.6     9.1      8.9     10.2
6      25      625    15625    17.0    18.8    17.4    18.4     17.9     17.7
7      50     2500   125000    28.7    31.7    29.3    31.1     30.2     28.2
8      75     5625   421875    39.9    44.1    40.7    43.2     42.0     41.9
9     100    10000  1000000    55.2    61.0    56.3    59.8     58.1     58.7
10    250    62500 15625000   211.5   233.7   215.9   229.3    222.6    223.2
11    500   250000  1.3E+08   693.8   778.4   716.5   759.7    737.1    736.8
12    750   562500  4.2E+08  1462.2  1618.9  1494.0  1588.2   1540.8   1540.9
13            ===================================================
14              Regression Output:                  |COEFFICIENT ANALYSIS
15            Constant              10.15738         |====================
16            Std Err of Y Est       1.267390        |FIRST: SIGNIFICANT
17            R Squared              0.999996        |SECOND:SIGNIFICANT
18            No. of Observations          8         |THIRD: NOT SIGNIFICANT
19            Degrees of Freedom           4
20
21            X Coefficient(s)  0.237698 0.002484 -1.07E-07
22            Std Err of Coef.  0.016684 0.000057  5.04E-08
23            ===================================================
```

Figure   6-16

to 16). Column I contains the predicted y-values using a cubic equation and the X Coefficients.

After the Lotus regression analysis has been performed, create a graph to evaluate the suitability of the fitted curve. Follow the instructions given in the "Visually Checking the Linear Least Squares Fit" section to create the graph. Specify the following ranges for the graph:

Range	Cells
X	A5..A12
A	H5..H12
B	I5..I12

Look qualitatively at the fitted curve to ensure that the predicted y-values are sufficiently close to the experimental y-values.

To check the significance of the coefficients for square and cubic terms, evaluate the Standard Error of Coefficient for each of the three terms. If you recall (from the "Error Analysis for Linear Regression" section), this evaluation is made with the aid of the following equation:

$$|a_i| = t * S_{a_i}$$

Therefore, into cells I16..I18, type the following formulae:

```
@IF(@ABS(E21)>2.5*E22,"SIGNIFICANT","NOT SIGNIFICANT")
@IF(@ABS(F21)>2.5*F22,"SIGNIFICANT","NOT SIGNIFICANT")
@IF(@ABS(G21)>2.5*G22,"SIGNIFICANT","NOT SIGNIFICANT")
```

In this example, these formulae show that the noise in the coefficient of the cubic term is significantly large in relation to the absolute value of the coefficient to warrant dropping the term from the equation.

MULTIPLE LINEAR REGRESSION

In many experiments, a quantity of interest (such as the yield of a product) will depend on several unrelated independent variables or on more than one property of a sample. In such cases, polynomial curve fitting equations are inappropriate and it is desirable to fit a multiple regression equation to the data, showing the dependence of the data on each of the factors considered.

Linear regression analysis with several independent variables is known as multiple linear regression. Lotus 1-2-3 and Symphony can handle up to 16 independent variables.

The same techniques that were applied to polynomial regression may be applied to the fitting of multiple variables because additional independent variables are mathematically equivalent to successive powers of a single independent variable. The following is the pertinent equation:

$$Y_p(X) = a_0 + a_1X_1 + a_2X_2 + \cdots + a_nX_n$$

As in the case for solving polynomial problems, you could use a matrix technique, but multiple regression matrices are comparably complicated and arduous to set up. The Lotus regression routine is much easier. To implement a multiple regression, follow the protocol given above in the "Polynomial Regression" section of this chapter. However, instead of using the square, cube, etc., formulae, enter the values of the independent variables into sequential columns.

When the regression program is executed, the first X Coefficient will apply to the first column in the X-range, the second X Coefficient will apply to the second column, and so on.

As has become customary, you should verify the accuracy of the regression by graphing a scatter diagram with a regression curve overlaying the experimental data.

MULTIPLE QUADRATIC REGRESSION

A similar generalization of the multiple regression equation, involving powers and products of several independent variables, is also possible. Manually, the required calculations are very difficult because of the large number of complex equations involved, but in Lotus you need only to add columns of the appropriate values or formulae and specify the ranges in a precise and similar manner as shown in the previous two sections. Each power or independent variable entering into the regression equation is treated as if it were a separate, independent variable in setting up the least squares equations for the regression coefficients.

LOGARITHMIC, EXPONENTIAL, ETC., REGRESSION

Linear regressions can also be performed on trigonometric, exponential, logarithmic, log-logit, etc., functions. Linear regressions involving these functions are fitted in a precise and similar manner as shown in the previous three sections.

To implement regression on these functions, follow the protocol given above in the "Polynomial Regression" section. However, instead of using the square, cube, etc., formulae in the new column(s), place formulae containing Lotus @functions into sequential column(s) and use the column(s) containing the formulae as the X-range (omitting the original x-values, if necessary).

For a review of the possible formulae, see the section entitled "Overview of How to Find the Appropriate Linearization Formula" at the beginning of this chapter. For a review of @EXP, @LOG, @LN, @SIN, @COS, @TAN, etc., functions, see Appendix B.

SUMMARY OF WHAT HAS BEEN ACCOMPLISHED

This chapter described methods that you can use to determine a curve that best fits your standard and/or method comparison data. The chapter made extensive use of the Lotus regression program because the program is very efficient and simple to use for even complex relationships. The Lotus regression program examples presented in this chapter used concentration as the independent variable (X) and response as the dependent variable (Y). However, for assay templates the roles of these two variables can be reversed to obtain coefficients that make it more straightforward to calculate concentrations from standard curves.

This chapter also explained the error analysis values returned by the Lotus regression program and described how to create formulae to transform the error analysis values for use in other statistical tests.

Additionally, the chapter stressed the importance of creating graphical overlays to ensure the appropriateness of regression results.

WHAT'S NEXT?

If you cannot find an equation that gives an acceptable least squares fit, then you will need to use non-linear methods. Non-linear methods are described in Chapters 8 and 9. However, before using non-linear methods, you must first develop some simple macro programming skills. Chapter 7 will teach you these skills.

7

Macro Programming

A set of instructions in a format and language that Lotus® can understand is called a Lotus *macro* program. The language that is used is called the *Lotus Command Language (LCL)* and the instructions are called *macro commands*. An overview of the six broad categories of Lotus macro commands is given in Chapter 1. The instructions of the Lotus Command Language relevant to this book are given in Appendix C.

Macro programs are invaluable tools. In addition to exercising your creativity, macro programs can be used to place values in cells rather than assigning the task to active formulae. There are many situations where I use macro programs. Here are some of the situations and the motivation of my preference for using the macro approach.

- As an alternative to active formulae in cells. As mentioned in Chapter 3, the more formulae that you have in cells, the slower the spreadsheet will recalculate. If you have been creating the example templates in the last few chapters, you undoubtedly noticed that the recalculation speed of the spreadsheet is getting slower and slower. The length of time required for recalculation is especially evident if you have a PC or XT personal computer. You could diminish the problem by issuing the /Range Values command to convert the formulae to their values, but then you would have to re-type the formulae into the template each time you wanted to re-use the template with new data. With macro programs, the formulae permanently reside in the spreadsheet, are reusable, and do not affect the recalculation speed of the template. However, because values in cells do not recalculate when they have been placed there by a macro program, you will need to decide which cells of the spreadsheet need to have active formulae and which cells need to have their values placed by macro programs.

- To decrease the chance of user mistakes. In previous chapters, the user was required to issue the /Data Distribute and /Data Regress commands in Lotus 1-2-3 or the /Range Distribute and /Range Regress commands in Symphony to activate Lotus data reduction programs. These command sequences obligated the user to remember to issue the commands and, more importantly, they assumed that the user had issued the proper commands, had properly defined the named ranges, and had correctly specified the ranges when prompted for them. These types of assumptions often lead to incorrect results and in turn, can lead to misinterpretation of the experimental data. If results are automatically placed into a template by a macro program, the results can be reviewed with increased confidence. An even higher degree of safety can be achieved when a macro program places the results of analyses directly into the template as the data are being acquired. The focus of *Laboratory Lotus®: A Complete Guide To Instrument Interfacing* was to perform data reduction as an integral part of data generation. If you have integrated your instrument and spreadsheet using the concepts of that book, then you should think about how to add data reduction capabilities to the existing program. If you are entering your data manually, then you will need to settle for somewhat less of an ideal system. However, by using a custom menu system, you can still prevent a number of user mistakes.

- To improve the "user friendliness" of a spreadsheet. User friendly spreadsheets—spreadsheets that make the user's tasks as easy to discern and accomplish as possible—allow users with little specialized knowledge to use your spreadsheets. As a general rule, the easier it is for a user to learn how to operate a program, the less training that you will need to provide. Spreadsheets can be made user friendly by including programs that have menus, prompts to get information from the user, graphics, status messages, and help messages. This system will make it much less frustrating for the users of your spreadsheet.

- To enhance the overall automation of a data collection and reduction system. As noted above, using macro programs allows data reduction to become part of a larger system that gets the data, then automatically formats the data and performs analyses. This method is not only safer, but minimizes technician time, thereby improving throughput.

- To provide a remedy for situations where formulae in cells or the Lotus menu commands cannot solve a problem. In the next chapter, a nonlinear fit will be made to an "S" shaped line. Because Lotus does not provide menu commands or @functions to accomplish this task, you will need to create a program.

This chapter will give you some macro programming basics. Then, in the next chapter, you will learn how to apply these skills to solving problems that cannot be solved without creating a program.

FOCUS OF THIS CHAPTER

Entire volumes have been written on macro programming. To keep the size of this chapter small enough to allow room for the primary purpose of this book (data reduction), this chapter cannot be exhaustive. Therefore, the topics covered will be treated as they apply to this book. If you want to learn more about macro programming and the Lotus Command Language, you may want review *Laboratory Lotus®: A Complete Guide To Instrument Interfacing* or investigate one of the books listed in the references at the end of Chapter 1.

DESIGNING MACRO PROGRAMS

There are two basic ways to design and develop a macro program that solves a statistical and/or curve fitting problem. The first way is to locate an existing BASIC, C, Pascal, etc., program that solves the problem and then translate it into a Lotus macro program. Optimized and fully evaluated programs that address statistical and/or curve fitting problems are readily available in any of a number of computer statistics and/or numeric methods books.

If you use this method of designing your program, you will want to amend it during translation so that it becomes tailored to your template. This way is often the easiest for a beginner to develop programs and is the fastest way for any programmer to have a program up and running. It is also a good way to learn how to design more sophisticated programs.

However, after you have become more advanced, you will want to use the second method—design your own programs. This way gives a better sense of accomplishment and usually leads to a program that handles your particular applications more efficiently. This way is required for those statistical and/or curve fitting problems for which no text programs exist.

Although often different in format, the Lotus macro programming tools can be used to emulate virtually all BASIC commands. However, Lotus goes beyond BASIC. Lotus has tools that emulate powerful C language operations (such as pointers to variables and pointers to functions). Thus, you can think of Lotus macro programming as a hybrid between BASIC and C.

The next sections in this chapter will give you a brief overview of Lotus macro programming, followed by a comparison of the Lotus Command Language, @functions, and macro programs to BASIC and C. These discussions should get you started (especially if you plan to use the translation method). This chapter also describes how to apply some of the more commonly used macro commands and @functions to create some simple programs. For a quick review of the definitions for Lotus @functions and macro commands, see Appendices B and C, respectively. Chapter 8 contains a relatively large and fully functional program that will illustrate a "real life" implementation of macro programming.

MACRO PROGRAMMING OVERVIEW

The first step in creating any macro program is to create and test a template to hold your data and data reduction summaries. Chapters 3 through 6 gave many pointers regarding template design. After your template is in place, your next step will be to write your macro program.

As stated at the beginning of this chapter, Lotus 1-2-3 and Symphony have a built-in programming language, called the Lotus Command Language. The Lotus Command Language allows you to automate virtually any procedure that you can perform manually through the keyboard. A program that uses the Lotus Command Language to automatically perform Lotus procedures is called a *macro program,* or simply a *macro.*

In their simplest form, macros are time-savers. You can save time and keystrokes by creating a macro that "presses the keys" for you. That is, a simple macro contains the instructions that identify which keystrokes you would use to perform a specific task. These keystrokes are stored as labels (text) within cells of your spreadsheet. When you run a macro, Lotus 1-2-3 or Symphony reads and executes the keystrokes and commands stored in the macro. For example, you can use the following macro to "type" the word "hello" into two adjacent cells (˜ means "press Return" in the Lotus Command Language):

```
\R   hello˜
     {RIGHT}
     hello˜
```

Similarly, you can create a macro to automatically use the menu to perform certain tasks. For example, you would use one of the following macros to erase cells A1 through B7:

For Lotus 1-2-3	For Symphony
\R /REA1..B7˜	\R /EA1..B7˜

The Lotus Command Language has numerous programming commands that allow you to prompt a user for input, perform file operations, calculate formulae, assign values to cells, transfer data from one cell to another, control the display screen, move the cell-pointer, control the path of macro execution, display custom menus, trap errors, and perform a host of other useful functions. In fact, Lotus has more than 40 specialized macro commands available (see Appendix C).

In addition to macro commands, you can create efficient programs by placing @functions within macro commands. This type of programming is called *nesting* and is described more fully below.

Thus, the macro language is like having a built-in BASIC programming language. It is just as easy, or easier, to use and has many advantages over BASIC. These advantages will become readily apparent to you in the comparisons shown later in this chapter and as you proceed through the remainder of this book.

CREATING MACRO PROGRAMS

You create a macro program by typing a list of commands horizontally down a single column of a spreadsheet. These instructions are entered directly into cells just like any other data. Each instruction can be made up of one, or a combination of, the following:

- Keystrokes that you would press if you were entering data.
- Keystrokes that you would use to select an item from a Lotus menu.
- Keystrokes that you would use to respond to a prompt.
- Commands that use one of the Lotus Command Language keywords.
- Commands that are a combination of Lotus Command Language keywords, @functions, and/or formulae.

To name a macro, you would place a name for the macro in the cell to the left of the first cell of the macro and use the /Range Name Label Right command to assign the name to the macro. For example, to create the "hello" example above, you would type '\R (single-quote, backslash R) into a cell of the spreadsheet. To the right of the cell, you would type the program. To "activate" the program, you would move the cell-pointer to the cell containing the '\R, issue the /Range Name Label Right command (/RNLR) and press Return.

Use meaningful names for your macros. You will normally use a backslash (\) and a letter for the first macro in your program. This designation will allow you to start a chain of programs by using the [ALT] key and its corresponding letter. However, that macro will usually call upon other macros, and those macros should have descriptive names.

Choose names that relate to the processes that you are trying to achieve. Try to use names that remind you of the macro's purpose. And, above all, do not use YOUR name as a name for a macro. It can be embarrassing to come back to a program once you have become an accomplished macro writer and find your name as a macro.

Also, observe the following words of caution when you are deciding on a name: Do not use names that look like cell addresses (e.g., N6, AB23); do not use one of the Lotus special keynames, keywords, or macro names (e.g., PUT, QUIT); do not use spaces and symbols within a name; and do not exceed 15 characters. Sometimes you can get away with names that break the rules; but most of the time you cannot. Remember, unconventional names can lead to dangerous programming situations.

You can have as many macros in your spreadsheet as you like. You can also place them anywhere in a spreadsheet.

Although the length of a cell entry is limited to 240 characters, the size of your macro is virtually unlimited because the macro can include as many cells in a column as it takes to accomplish the task that you need it to perform. The only limitation is the amount of available computer memory.

MACRO EXECUTION

To start a macro running, specify it by name. For example, in the simple macros given above, the macros were called by the name \R. To start a macro with this name, you would hold down the [ALT] key, press R, and then release both keys. (That is, you would **not** type the backslash (\)).

If you assign a descriptive macro name in Symphony, you can invoke the macro by pressing the User key (F7), typing in the name, and pressing Return.

Lotus reads program lines starting with the cell to the right of the cell containing the name of the macro and proceeds downward. (That is, Lotus starts with the cell that was specified by the /Range Name Label Right command, which is the cell just to the right of the macro's name.) Lotus performs instructions in sequence. As each instruction is completed, Lotus automatically proceeds to the next cell down.

A macro terminates when it encounters an instruction to stop (the {QUIT} command), a cell entry that contains an instruction that is not recognized, or an empty cell. Most macros in this book terminate with an empty cell.

PROGRAM STRUCTURE

Nearly every program that you write will require the same basic components. The first part of a program is a macro (or set of macros) that is executed at the start-up of the program. Macros of this nature "initialize" the spreadsheet, the main program, and your instrument (if applicable). The purpose of this macro is to combine all of these components into a system that will carry out the tasks specified by the remainder of the program.

As you shall see in the next chapter, initialization macro(s) are important because they set all of the starting values within the system to the prescribed conditions needed by the system to operate. For example, if you recall from Chapter 5, the STAT.app application program had to be attached before it could be used by Symphony. An initialization macro that has been created to auto-execute when a spreadsheet is retrieved can check to see if STAT.app has been attached and, if not, attach it. Initialization macros can also automatically execute menus of your own design, other programs, etc.

The next part of a program is the "main" program. This component is the central control for the remainder of the program and will be returned to whenever the tasks in other modules have been completed. View the "main" program as the central hub from which subroutines are controlled. The "main" program in Figure 7-1 is called \R.

A subroutine is just another macro program that is apart from the main program and can be called upon by the main program. A subroutine usually performs just one very specific task. When a subroutine is called, control is transferred from the main program to the subroutine, the subroutine carries out the task that it was programmed to perform, and then the subroutine returns control back to the main program. For example, some subroutines display custom menus, print, store files, perform statistical calculations and the like. Example subroutines in Figure 7-1 are DO__ROW, DO__COL, CHK__MIN, CHK__MAX, AND WARN.

The last section of a program is known as the "scratch-pad" section. The scratch-pad is a set of cells that temporarily stores data. Cells Q25..Q29 constitute the scratch-pad in Figure 7-1.

View a scratch-pad as a temporary work area . . . the electronic equivalent of a note-pad. Data are placed into the scratch-pad from a variety of sources. For example, a {FOR} command will keep track of how many times its loop has been executed in one of the cells of the scratch-pad. (Programs that are executed in a repetitive manner are referred to as *loops*.) In this context, the scratch-pad cells hold

```
--------P--------Q--------R--------S--------T--------U--------V--------W-----
 1
 2  \R           {GETNUMBER "HOW MANY STANDARDS?",NUM_ROWS}
 3               {GETNUMBER "HOW MANY REPLICATES/STANDARD?",NUM_COLS}
 4               {FOR ROW,0,NUM_ROWS-1,1,DO_ROW}
 5               {CHK_MIN}
 6               {BRANCH CHK_MAX}
 7  FINISH       {CALC}
 8
 9  DO_ROW       {LET ROW_SUM,0}
10               {FOR COL,0,NUM_COLS-1,1,DO_COL}
11               {PUT STDF,0,ROW,ROW_SUM/NUM_COLS}
12
13  DO_COL       {LET ROW_SUM,ROW_SUM+@INDEX(STDSECT,COL,ROW)}
14
15  CHK_MIN      {IF @INDEX(STDF,0,0)<80}{WARN}
16
17  CHK_MAX      {IF @INDEX(STDF,0,NUM_ROWS)>1300}{BRANCH WARN}
18               {BRANCH FINISH}
19
20  WARN         {BEEP}{BEEP}
21
22
23
24
25  NUM_ROWS           8
26  NUM_COLS           4
27  ROW                8
28  COL                4
29  ROW_SUM         4934
30
```

Figure   7-1

numbers that are used for counting in programs that repeat a prescribed number of times. Data that have been input by a user are also stored in the scratch-pad.

For you to fully understand the function of a scratch-pad, you must first know the important concept of a "buffer". A *buffer* is merely a temporary storage area for data and is an excellent example of one of the primary functions of a scratch-pad. A buffer is usually empty at the beginning of a program. When data need to be temporarily stored, they are stored in the buffer. When more data need to be stored, the new data overwrite the current contents of the buffer cell.

As you can deduce from the description, the scratch-pad is usually in a state of constant flux. Information that is placed into cells is overwritten time after time. The scratch-pad is one of the most active portions of the entire spreadsheet and can be informative to watch.

COMPARISON OF LOTUS TO BASIC

With the above groundwork in mind, let us examine macro programming more specifically. Figures 7-1 and 7-2 show two simple programs that accomplish the same tasks. The program in Figure 7-2 is written in BASIC and the program in Figure 7-1 is the translated version using Lotus tools. Both perform the relatively simple task of finding the average fluorescence value for each of the standards in Figure 7-3. Admittedly, the programming in these examples is not very well-designed or efficient, nor

```
 10 DIM A(10,10),B(10)
 20 DATA 93.9,92.9,94.3,94.3
 30 DATA 130.4,134.6,133.4,135.9
 40 DATA 174.5,171.8,173.1,174.1
 50 DATA 202.9,204.5,204.2,205.2
 60 DATA 235.0,242.2,242.7,239.9
 70 DATA 473.9,462.0,469.3,473.2
 80 DATA 847.1,854.2,850.4,862.2
 90 DATA 1236.4,1236.4,1226.0,1235.2
100 INPUT "HOW MANY STANDARDS";NUMROWS
110 INPUT "HOW MANY REPLICATES/STANDARD";NUMCOLS
120 FOR ROW=0 TO NUMROWS-1
130      ROWSUM=0
140      FOR COL=0 TO NUMCOLS-1
150           READ A(ROW,COL)
160           ROWSUM=ROWSUM+A(ROW,COL)
170      NEXT COL
180      B(ROW)=ROWSUM/NUMCOLS
190      PRINT B(ROW)
200 NEXT ROW
210 IF B(0)<80 THEN GOSUB 1000
220 IF B(NUMROWS-1)>1300 GOTO 1000
230 END
1000 BEEP:BEEP
1010 PRINT "ERROR!!!!"
1020 RETURN
```

Figure   7-2

```
--------A--------B--------C--------D--------E--------F--------G--------H-----
 1                      FLUORESCENT DNA ASSAY DATA
 2 DATE RUN:31-OCT-88
 3               ****STANDARD CURVE DATA****                X      REGRESS
 4 NG DNA/ML ASSAY1    ASSAY2    ASSAY3    ASSAY4   AVERAGE PREDICT  CURVE
 5      0      93.9      92.9      94.3      94.3     93.9     0.7    92.7
 6     25     130.4     134.6     133.4     135.9    133.6    26.9   130.7
 7     50     174.5     171.8     173.1     174.1    173.4    53.1   168.7
 8     75     202.9     204.5     204.2     205.2    204.2    73.3   206.7
 9    100     235.0     242.2     242.7     239.9    240.0    96.9   244.7
10    250     473.9     462.0     469.3     473.2    469.6   248.0   472.7
11    500     847.1     854.2     850.4     862.2    853.5   500.6   852.6
12    750    1236.4    1236.4    1226.0    1235.2   1233.5   750.6  1232.6
13              ========================================
14                         Regression Output:
15              Constant                      92.74541
16              Std Err of Y Est               3.441756
17              R Squared                      0.999939
18              No. of Observations                   8
19              Degrees of Freedom                    6
20
21              X Coefficient(s)   1.519749
22              Std Err of Coef.   0.004814
23              ========================================
```

Figure   7-3

are the programs particularly useful. However, the examples fulfill their two primary purposes:

- To quickly and briefly illustrate some of the more important similarities between Lotus and BASIC.

- To give you examples of how to use macro commands and @functions to move data around the spreadsheet.

Before we get into the details of the programs, take a moment to look qualitatively at the two programs and compare them. Note the differences in their format. For example, in BASIC, line numbers are needed to keep track of program flow (e.g., 10, 20, etc.). The program is contiguously aligned. Neither requirements apply to Lotus. Line numbers are not used in Lotus. As stated above, Lotus senses the end of a macro program or subroutine by an open cell. Also note that each subroutine has a name in Lotus, but a line number in BASIC.

Also note the differences in the method of navigating around the program. In BASIC, the GOTO and GOSUB commands are used; in Lotus, the {BRANCH subroutine} and {subroutine} are the equivalent commands.

Finally, in BASIC, all of the variables are either read in from DATA statements or are hidden from routine view. If a matrix is required, a DIMension statement specifies the size of the matrix. When the elements of the matrix are referred to, they are specified using the following formats:

```
A(col,row)
B(row)
```

In Lotus, variables are stored in the cells of a spreadsheet. Temporary storage is often located in the scratch-pad and are buffers that are always available for review. Placing data into individually specified cells is executed via {LET} commands. If a matrix is needed, no DIMension statements are required because the matrix is comprised of cells in the spreadsheet. **To get data into a matrix, you use the {PUT} command. To retrieve data from a matrix, you use the @INDEX function.** Both the {PUT} command and the @INDEX function take a named range, a column offset, and a row offset as specifications for the cell in the matrix being referred to.

Next, take a moment to look at the syntax of some common commands. The following is a side-by-side list of the differences in Figures 7-1 and 7-2:

BASIC	Lotus
INPUT "message"; NUMROWS	{GETNUMBER "message", NUM_ROWS}
FOR ROW=0 TO NUMROWS-1	{FOR ROW, 0, NUM_ROWS-1, 1, DO_ROW}
ROW_SUM=0	{LET ROW_SUM, 0}
ROW_SUM=ROW_SUM+A(COL,ROW)	{LET ROW_SUM, ROW_SUM+@INDEX(STDSECT, COL, ROW)}
IF B(0)<80 THEN GOSUB 1000	{IF @INDEX(STDF, 0, 0)<80}{WARN}
GOSUB 500	{CHK_MIN}
GOTO 600	{BRANCH CHK_MAX}
B(ROW)=ROW_SUM/NUMCOLS	{PUT STDF, 0, ROW, ROW_SUM/NUM_COLS}
BEEP	{BEEP}
No equivalent	{WAIT @NOW+@TIME(0, 0, 5)}

As you can see, although these two languages have slight differences, they are actually very similar. If you are a beginner, this similarity will make it easy for you to "translate" programs in text books into useful programs that manage data in your Lotus templates.

The Lotus program in Figure 7-1 contains the tools and techniques most often used in general programming. For this reason, the remainder of this chapter is dedicated to giving details on the tools and techniques used in Figure 7-1.

OBTAINING INFORMATION FROM THE USER

The main program (\R) of Figure 7-1 illustrates one of the methods that are available for prompting the user for input. In this example, a pair of prompts ask the user to enter the number of standards and the number of replicates/standard.

The command that you use to obtain numbers from a user is the {GETNUMBER} command. For text, use the {GETLABEL} command. The syntax of these two commands is

```
{GETNUMBER prompt, location}
{GETLABEL prompt, location}
```

When either of these two commands is executed in a macro, a prompt is displayed on the second line of the control panel and execution of the macro is paused until input is made. When the user presses a carriage return on the keyboard, the input is placed into the cell specified by "location". In the example program, the number of standards go into the cell named "NUM＿ROWS" and the number of replicates go into the cell named "NUM＿COLS".

User input obtained from a {GETNUMBER} or {GETLABEL} command can be used later by your program. For example, in this program, the "NUM＿ROWS" number will set the limit for the main loop counter and the "NUM＿COLS" number will set the limit for a second loop counter. (A description of how this information is incorporated is given below.)

Another use of the {GETLABEL} command is to pause a program at specific points during execution. That is, by strategically placing a {CALC} command and a {GETLABEL "PAUSE",TXT} command in your program, you can temporarily stop the program to view the current status of the template, scratch-pad, etc. When you press Return, the program will continue until another {GETLABEL} command is encountered. This use makes {GETLABEL} a very valuable learning aid and debugging tool. (See the section entitled "Determining Matrix Offsets" at the end of Chapter 9.)

An alternative programming tool that you can use to get input from a user is the {?} command. The {?} command halts macro execution, thus allowing the user to type any combination of keystrokes, function keys, or arrow keys (to move the cell-pointer). The macro will continue after the user presses Return. This pause allows the user to input text or numbers directly into cells anywhere in the spreadsheet.

Pressing Return has no effect other than to end the {?} command. That is, pressing Return with the {?} command will NOT permanently enter data into a cell. To permanently enter user input, you must include a tilde (˜) in the cell, after the {?}. A tilde means "Press Return." Without a tilde after the {?} command, the data that were input may be lost.

DATA REDUCTION USING LOOPS

There are two nested {FOR} loops in the program shown in Figure 7-1. Nested {FOR} loops are perhaps the most commonly used methods for data reduction calculations, especially if you are dealing with data matrices. You will probably use them often for your data reduction chores, so we will take a closer look at how they work.

The outer {FOR} loop is in the main program and simply executes the inner loop NUM_ROWS−1 times. One is subtracted because the loop executed starts at zero, not one. The zero starting point will become more obvious later when we discuss the offsets for @INDEX functions and {PUT} commands.

The format of the outer loop is as follows:

```
{FOR ROW, 0, NUM_ROWS-1, 1, DO_ROW}
```

where,
DO_ROW is the name of the macro that you want the spreadsheet to execute repeatedly, and the arguments ROW, 0, NUM_ROWS, and 1 represent the scratch-pad cell that will contain the current loop number, the starting value, the scratch-pad cell containing the stop value, and the step size, respectively.

The inner {FOR} loop is in the DO_ROW subroutine. After the ROW_SUM is reset to zero, the {FOR} loop executes the DO_COL subroutine NUM_COLS-1 times. DO_COL performs a summation on each of the fluorescence values in a standard's row.

By constructing this combination of commands for the standard section in Figure 7-3, you can get an average of the four replicates readings for each of the eight standards in the standards section. The results are then placed into sequential rows of the STDF range of the spreadsheet template (based on the ROW counter of the first {FOR} loop).

GETTING DATA FROM A TEMPLATE

The @INDEX function in the {LET} command of DO_COL retrieves data from the template matrix.

@INDEX is one of the most useful @functions in a class of @functions that Lotus calls *lookup functions*. In Chapter 2, we encountered two other lookup @functions (@HLOOKUP and @VLOOKUP). The @functions in this class take a table and "look up" one of the pieces of data in that table based on its position. Thus, these @functions require a definition of the table AND the offset(s) that pinpoint the piece of data that you want.

The @INDEX function is a very handy programming tool. The tables that the @INDEX function uses are named ranges in the spreadsheet. The syntax of @INDEX is

```
@INDEX(range,column_offset,row_offset)
```

This @function pinpoints a cell located within a range based on the specified column and row offsets and returns the value of the cell. Using @INDEX, you can access any cell within a two-dimensional array (matrix) and get its contents. In this example, STDSECT has been defined as cells B5..E12. The column and row offsets in an @INDEX function begin at (zero,zero). This starting position is one of the reasons for starting the {FOR} loop counters at zero and not one. The @INDEX in Figure 7-1 has the following format:

```
@INDEX (STDSECT, COL, ROW)
```

Thus, as the COL and ROW counters in the two {FOR} loops increment, they specify the column and row offsets for @INDEX. Using COL and ROW as the column and row offsets allows you to increment the offsets and thereby change the data being utilized in the calculation. For example, when COL and ROW are both zero, @INDEX takes on the value of cell B5 (column-offset = zero, row-offset = zero). In Figure 7-3, the @INDEX value is 93.9. When COL increments to 1, @INDEX takes on the value of cell C5 (column-offset = one, row-offset = zero). In Figure 7-3, that value is 92.9. This process of advancing through the matrix using loop counters continues for the duration of both loops.

MOVING DATA INTO A TEMPLATE

After a full row's summation has been completed, the program uses a {PUT} command to transfer the data into the appropriate cell of the template. The format of the {PUT} command is

```
{PUT range, col_ offset, row_ offset, value}
```

where,

range defines the block of cells in which the {PUT} could place the result; col—offset specifies the column (left/right) offset within the range; row—offset specifies the row (up/down) offset within the range; and value is the result that is to be placed into the cell specified by the column and row offsets. Value can be a cell address or the name of a cell containing data, or it can be a formula, @function, etc. As in the @INDEX case, the first row and column of the range have the offset of (zero, zero).

Thus, the {PUT} macro command pinpoints a cell located within a range based on the specified column and row offsets and places a value into the cell. Using {PUT}, you can change the value of any cell within a two-dimensional array (matrix). In the example program of Figure 7-1, STDF has been defined as cells F5..F12. Because the column offset in a {PUT} command begins at zero, the column offset has been permanently set to zero. That is, it is a one-dimensional array (i.e., a column).

The row offset in this example program comes from the outer loop counter. The row offset for a {PUT} command begins at zero. This starting position is the second reason for starting the {FOR} loop counters at zero. The {PUT} command in Figure 7-1 has the following format:

```
{PUT STDF, 0, ROW, ROW_ SUM/NUM_ COLS}
```

Thus, as the ROW counter in the outer {FOR} loop increments, it specifies the row offset for the {PUT} command. Using COL and ROW as the column and row offsets thereby allows you to increment offsets and change the cell being modified. For example, when COL and ROW are both zero, {PUT} places the value of ROW__SUM/NUM__COLS into cell F5 (column-offset = zero, row-offset = zero); when ROW increments to 1, {PUT} places the value of ROW__SUM/NUM__COLS into cell F6 (column-offset = zero, row-offset = one); and so on.

The {PUT} command has a built-in feature that can add a measure of safety to your program. Notice that I used a single column (F) as the range of acceptable cells in the {PUT} command. This specification provides a measure of safety because it does not allow the program to place data in sections of the template or program where it will overwrite formulae and/or macro commands. This feature is an important one that you will want to exploit in your programs.

{PUT} commands provide a very desirable purpose and using them in place of template @functions should be considered often. As stressed at the beginning of this chapter, @functions and formulae slow the recalculation speed of spreadsheets. If @functions are "nested" within a macro {PUT} command, the calculation would only be performed at the time that the {PUT} command is executed. This method can therefore improve the recalculation speed of your spreadsheet.

CALLING AND BRANCHING TO MACRO PROGRAMS

From a macro program, you can execute a second macro program in one of two ways. The first way is to call the second macro as a subroutine. If the second macro is called as a subroutine, control of the program is temporarily shifted to the subroutine, the subroutine carries out the tasks it was programmed to perform, and then control is returned to the original, calling program.

The main program in Figure 7-1 calls CHK__MAX as a subroutine. Notice the method that is used to call the subroutine. **Whenever you call a subroutine from either a macro or menu, place the name of the subroutine within { } brackets.** For example, to call CHK__MIN, type the following text into your program:

```
{CHK_MIN}
```

If your macro has a backslash-letter name (e.g., \R), then type the following text into your program to call the macro subroutine:

```
{\R}
```

In the example, after the CHK__MIN program has completed its execution, program control returns to the main program.

The second way to execute a macro is to branch to it using the {BRANCH} command. The {BRANCH} command permanently transfers control of the program to the cell named in the {BRANCH} command. That is, when a macro is executed with the {BRANCH} command, the program "jumps" to the new macro. Execution terminates in the original macro and total control transfers to the new macro. Then, the new macro executes. When the last instruction in the new macro is encountered, program execution will terminate (unless another branching command is encountered).

Branching is illustrated in the main program of Figure 7-1. The example is

{BRANCH CHK_MAX}

If you note the CHK__MAX macro has a {BRANCH FINISH} command. This {BRANCH} command is required to return to the main program. Without it, the program would terminate. This termination is illustrated by the branching to WARN when the {IF} test in CHK__MAX is true. Because WARN is not called as a subroutine and does not have a {BRANCH} command, the program will terminate via this pathway.

There is a subtle difference in these two methods of executing macros. On the surface, the main difference is that a subroutine call will automatically return to the calling program, while a {BRANCH} will not. However, the ramifications of this subtle difference can be very important to the reliability of a program's execution because Lotus limits the number of subroutine levels that you can call while within a program.

If a macro calls another macro as a subroutine, Lotus is supporting two levels of subroutines. Likewise, if a second subroutine calls a third subroutine, Lotus is supporting three levels, and so on. The number of subroutine levels that either Lotus 1-2-3 or Symphony can support (without returning to the first macro level) is 31.

If you exceed the subroutine limit by calling a macro from itself and/or by trying to nest too many subroutines, the macro aborts and an error message appears.

Whenever possible, I prefer {BRANCH}ing as opposed to subroutine calls because {BRANCH}ing allows more levels of macro nesting, thereby allowing programs to be longer and more efficient.

{IF} TESTS

The {IF} tests in the CHK__MIN and CHK__MAX macros check to ensure that the fluorescence units are within acceptable limits. This programming technique may be required by an assay to ensure that the "blank" is not too high, that there is sufficient sensitivity, etc. However, the real intention of the example is to illustrate the {IF} command.

An {IF} command allows a macro program to perform a specific task based on the result of a TRUE/FALSE test. If the result of the expression within the {IF} command does not have a numeric value of zero, Lotus considers the expression to be TRUE and the macro executes the instructions immediately after the {IF} command (**in the same cell**). If the result of the expression within the {IF} command has the numeric value zero, Lotus considers the expression to be FALSE and the macro continues **in the cell below** the one containing the {IF} command.

In this example, the two {IF} tests call the subroutine {WARN} if the argument specified in the command is TRUE. If the argument to the {IF} test is FALSE, the program just skips the call to the WARN subroutine and proceeds down to the next command line. You can place any Lotus macro command, subroutine call, menu command, branch, etc., after the {IF} command.

FORCING SPREADSHEET RECALCULATION

The {CALC} command at the end of the program forces a recalculation of all formulae in the spreadsheet and updates, or re-displays, the current data. When a macro program executes, it does not necessarily recalculate the formulae in a template. **Placing a {CALC} command at the end of a macro is very important to avoid user confusion**.

More importantly, if a user forgets to recalculate a spreadsheet before printing it, the results may not be correct because they may not reflect new values that have been added to the template by the macro program. To prevent the possibility of catastrophe, get into the habit of placing a {CALC} command at the end of all of your macro programs and, more importantly, before any automated printing is performed by your program.

THE "SCRATCH-PAD"

Earlier in this chapter, we looked briefly at the concept of a scratch-pad. A scratch-pad is a group of cells that temporarily stores data and holds values for loop counters. Your program relies very heavily on the data in your scratch-pad. This relationship makes the scratch-pad an important part of your program and makes it an area where you can greatly enhance the efficiency of your program.

For example, in Figure 7-1, COL and ROW serve three purposes. Their primary function is to serve as loop counters. When the {FOR} loops in this example begin execution, they set the value of their loop counters to zero. Each time a loop is repeated, the {FOR} command increments the counter by one. Therefore, a {FOR} loop needs a cell in the spreadsheet to hold the current loop value. COL and ROW fulfill this requirement.

The second function of the loop counters is that they serve as the column and row offsets for an @INDEX function. As COL and ROW increment with the loops, the appropriate cell in the range is designated and the @INDEX takes on the value of the cell.

The ROW loop counter has a third function. This function is to specify the row offset values in the {PUT} command. Thus, as the ROW counter in the outer {FOR} loop increments, it automatically increments the row coordinates within the STDF range and the average data are placed in the specified cells. This system of {FOR} loops, @INDEX functions, and {PUT} commands is a very handy way to position data and you will find that you will be using it often.

The COL and ROW cells thereby make neat programming packages, tying everything together. This kind of package programming is very efficient and decreases the chance of error.

Another important feature of the scratch-pad is the cell called ROW__SUM. We discussed the concept of a buffer earlier in this chapter, but it seems appropriate to repeat some of the discussion as it applies to the scratch-pad. If you recall, a buffer is a temporary storage area for data. It is usually empty at the beginning of a program. When data come in from a program, they are stored in the buffer. When more data come in, they overwrite what was previously in the buffer.

In the example program, ROW__SUM is initialized to zero each time the outer {FOR} loop increments and calls DO__ROW. Thereafter, each time DO__COL is called by the inner {FOR} loop, the existing data in ROW__SUM are queried to find the existing summation, the new value is added to it, and the old value is overwritten with the new value.

User input frequently requires the use of a buffer. For example, the {GETNUM-BER} and {GETLABEL} commands require a single cell specification (e.g., NUM__ROWS). This requirement means that a buffer cell is needed to receive the data. Therefore, if you use one of these commands to input data for template cells, you need to set up a buffer to receive the data. Once received, you can transfer the data to any cell of the spreadsheet with a {PUT} command. When subsequent user input is made, the old data are overwritten with new, processed, and so on.

Another common use of buffers is to obtain data from instruments. In my previous book, **Laboratory Lotus®: A Complete Guide To Instrument Interfacing,** buffers were required whenever data were acquired from an instrument. In this context, the buffer served as a staging area for {READLN} commands. {READLN} needs a specific location to place the data that are read. {READLN} cannot pinpoint data with column- and row-offsets the way that {PUT} can. Once data have been placed in a buffer, they are parsed with @functions and then moved to another cell in the spreadsheet with a {PUT} command.

GETTING ACQUAINTED WITH CREATING MACROS

Even though you will not be using the example application in "real life" situations, type the program in Figure 7-1. Use the template spreadsheet that you created in the previous chapters. **Make certain that the macro is to the right of the last column of cells in the template**. Also, make sure that you leave at least one blank row between each macro; Lotus 1-2-3 and Symphony look for these blank rows to determine the end of a macro.

It is easy to prepare a section of the spreadsheet for your scratch-pad: Just type the names of the variables into the same column as the macro names.

The next task in creating any program is to "activate" the program and scratch-pad. To accomplish this task, assign label names to the macros and the scratch-pad cells. Do so by issuing the /Range Name Label Right (/RNLR) command. When the spreadsheet prompts you for the range of labels, press {ESC}ape, move the cell-pointer to the cell containing the first macro name, type a period (.), use the down arrow key to highlight all of the macro names AND the scratch-pad names, and press Return. This activity defines the cells to the right of the column as having the names specified in the name column. You should now be able to appreciate the convenience of aligning range names into a single column.

Finally, give cells B5..E12 the name STDSECT. Do so by issuing the /Range Name Create (/RNC) command. When the spreadsheet prompts you for the name, type STDSECT, and press Return. When prompted for the range of cells, type B5..E12 and press Return. Repeat this process to give cells F5..F12 the name STDF. Now, save the spreadsheet. To execute the program, press [ALT]-R.

TEST IT!

As with a template, it is better to test macro programs as you create each section rather than to wait until the entire program has been completed. If you wait until the end of the programming process, the many subroutines and program lines will make it more difficult to isolate the errors. However, if you test as you build, you can verify that each section works correctly before you add another level of complexity.

Test your macro program with actual experimental data. This real-life testing often uncovers problems that cannot be found by just testing the system with contrived data.

After you have completed the entire program, perform a final, exhaustive test of the complete system. A template and macro program that have not been thoroughly tested with actual experimental data should not be used to report results. Important decisions may be based on a report that looks formal, but contains erroneous results. For this reason, you should compare the results of the macro program calculations in your template with the results that you obtain from a calculator, etc.

BACK UP SEVERAL COPIES

Once you have added all of your enhancements and completed final testing, your spreadsheet will represent a considerable time investment. You must protect this investment.

Save the spreadsheet (without data) on several, clearly labeled floppy disks. Also, print out the program for hard copy documentation. Save all of the items in several different, safe places.

MAKING YOUR JOB A LITTLE EASIER

The remainder of this chapter covers topics that will help you create more professional programs. This section will cover the topic of using named ranges to achieve program flexibility.

Whenever you prepare a program, make the programming as flexible as possible to allow for future modification. That is, if you need to modify a program in the future, you will want to be able to implement the update with a minimum of changes. As a general rule, the fewer changes that are required to implement an update, the less chance of forgetting to edit something that will affect the program's performance. One very good programming technique that can be of help achieving this goal can be seen in the {PUT} commands and @INDEX functions that appear in the DO__ROW and DO__COL macros of Figure 7-1.

If you recall, earlier in this chapter, named ranges were placed into a {PUT} command and @INDEX functions. Rather than using STDSECT and STDF in the program, you could have specified B5..E12 and F5..F12, respectively.

However, if you change a template (by moving cells, inserting or deleting rows or columns, etc.) and if the {PUT} and {LET} commands and/or nested @functions of your program contain cell addresses, all the addresses would need to be edited. In contrast, if you change the cell address specifications of named ranges and if named ranges have been specified in macro commands or @functions nested into macro commands, then you would not have to change a thing. For example, {PUT B5..E12, . . . } would need to be edited if you moved the range over by just one column, but {PUT DATASECT, . . . } would not. If you have a large program with many macro commands, editing is not only a massive amount of work, but a prime target for bugs in your program.

There is another reason why you should get into the habit of using named ranges instead of cell addresses for your {PUT} and {LET} commands and @functions: Using named ranges will make your program easier to read and to follow. STDSECT is certainly more informative than B5..E12. Names become very important in large spreadsheets and macro programs.

A FEW COMMENTS ON COMMENTING

You should add comments to your programs for one VERY important reason: Comments make it easier for someone (including yourself) to read and understand your program. Just try to return to a program after several weeks and attempt to remember what you were trying to accomplish in a certain section of the program! Or, try to remember a pitfall that you thought about while you were programming. If you fully document a program section at the time that you write it, the program will be at its freshest in your mind and the comments will make the best sense. In other words, comments will be more "readable" and it will be easier to correct or modify a program, if necessary.

Commenting your program also helps to clarify your own concept of what the program does and often exposes design flaws. Commenting is such an essential part of every program that for every line of the program that you write, you should try to write a line of comments.

What is the best way to comment your program? No best way exists because there is so much flexibility in what you can do. So, implement a system that seems the most comfortable to yourself and stay with the system. Recall that when Lotus reads a macro, it reads straight down a column. If a comment is placed just one column over, it will be ignored. The /*comment*/ format that you see in this book is the standard way to comment in C. Figure 7-4 shows an example of how to comment using this convention.

You can use any commenting format that you like, as long as the comments are not in a cell within the program. Because macro execution begins at the cell that

```
--------P--------Q--------R--------S--------T--------U--------V--------W-----
 1 /*PROGRAM TO FIND AVERAGE FLUORESCENCE UNITS FOR STANDARD CURVE*/
 2 \R        {GETNUMBER "HOW MANY STANDARDS?",NUM_ROWS}    /*REV: 12/27/88*/
 3           {GETNUMBER "HOW MANY REPLICATES/STANDARD?",NUM_COLS}   /*LMM*/
 4           {FOR ROW,0,NUM_ROWS-1,1,DO_ROW}       /*OUTER LOOP*/
 5           {CHK_MIN}          /*SEE IF BLANK VALUE LOW ENOUGH*/
 6           {BRANCH CHK_MAX}   /*SEE IF ENOUGH SENSITIVITY*/
 7 FINISH    {CALC}             /*FORCE RECALCULATION OF SPREADSHEET*/
 8
 9 DO_ROW    {LET ROW_SUM,0}    /*INITIALIZE BUFFER TO ZERO*/
10           {FOR COL,0,NUM_COLS-1,1,DO_COL}      /*FIND THE ROW'S SUM*/
11           {PUT STDF,0,ROW,ROW_SUM/NUM_COLS} /*PUT AVERAGE INTO TEMPLATE*/
12                             /*ADD CURRENT CELL TO SUMMATION*/
13 DO_COL    {LET ROW_SUM,ROW_SUM+@INDEX(STDSECT,COL,ROW)}
14
15 CHK_MIN   {IF @INDEX(STDF,0,0)<80}{WARN}
16
17 CHK_MAX   {IF @INDEX(STDF,0,NUM_ROWS)>1300}{BRANCH WARN}
18           {BRANCH FINISH}
19
20 WARN      {BEEP}{BEEP}       /*SOUND SPEAKER*/
21
22
23           /*SCRATCH PAD*/
24
25 NUM_ROWS       8             /*USER DEFINED NUMBER OF STANDARDS*/
26 NUM_COLS       4             /*USER DEFINED REPLICATES/STANDARD*/
27 ROW            8             /*ROW COUNTER FOR OUTER LOOP*/
28 COL            4             /*COLUMN COUNTER FOR INNER LOOP*/
29 ROW_SUM     4934             /*SUMMATION ACCUMULATOR*/
30
```

Figure   7-4

corresponds to the macro's name, any text above the cell will be ignored. Just make sure that you leave at least one row between the synopsis and the macro above it. Otherwise, Lotus will think that the comment is part of the previous macro.

Another commenting practice to follow is to write a brief synopsis of the function of a program at the beginning of the program. A good synopsis contains the objectives of the program, type of testing performed, textbook references, name of the curve fitting algorithm, revision dates, programmer's initials, etc.

HANDLING SCIENTIFIC FORMULAE

Scientific data often require the use of logarithms, natural logs, exponentials, sines, cosines, etc. You can handle these cases by adding the appropriate Lotus mathematical @function(s) into the {PUT} command that places the data into cells. The following is a list of common @functions:

@LOG (number)	Log of number, base 10
@LN (number)	Log of number, base e
@EXP (number)	The number e raised to the number power
@SIN (number)	Sine of number
@COS (number)	Cosine of number
@TAN (number)	Tangent of number

@ASIN (number)	Arc sine of number
@ACOS (number)	Arc cosine of number
@ATAN (number)	Two-quadrant arc tangent of number
@SQRT (number)	Positive square root of number

For example, to transform the value of a BUFFER cell using the @LOG function and place the transformed value into a cell in the range called ABSDATA of a template, use the following {PUT} command:

{PUT ABSDATA, COL, ROW, @LOG (BUFFER) }

NOT FOR C PROGRAMMERS ONLY

As noted at the beginning of this chapter, Lotus has tools that emulate pointers. The @@(CELL—REF) function is the redirection operator for cell contents. The redirection @function returns the contents of the cell referenced by CELL—REF. That is, the range name or cell location (e.g., F3) that appears in the CELL—REF cell is used as the address of a cell. The value returned by the @@ function will be the contents of the cell at this address. For example, if the following formula was typed into cell Q30 of Figure 7-1:

@@ (CELL—REF)

and if the contents of the cell called CELL—REF was the label "ROW—SUM", then the value of the ROW—SUM cell would appear in Q30.

The {DISPATCH macro—spec} macro command is the Lotus equivalent of a pointer to a function (subroutine) in C. When encountered in a macro program, the {DISPATCH} command will cause a branch to the macro that begins at the macro's name specified in the MACRO—SPEC cell, and the new macro will begin executing. For example, suppose that a cell named MACRO—SPEC was placed in Figure 7-1 and that the contents of MACRO—SPEC was the label "WARN". Further suppose that the following command was issued in the \R macro:

{DISPATCH MACRO_ SPEC}

When the {DISPATCH} command was encountered, {DISPATCH} would branch to the macro named WARN and begin execution.

SUMMARY OF WHAT HAS BEEN ACCOMPLISHED

This chapter described two ways of creating macro programs. The first way was to translate a BASIC, C, etc., program from a textbook into Lotus, modifying the program as needed. To this end, the common BASIC, C, and Lotus commands were compared and contrasted.

The second way was to create your own program from scratch. To assist you in this endeavor, this chapter presented an overview of program design techniques that accomplish some of the more common tasks required by nearly every data reduction macro program. To this end, a number of fundamental macro programming concepts were given:

- How to determine when to use macro programs vs. active formulae in cells.
- What to do if Lotus menu commands, formulae, @functions, etc., cannot solve a statistical or curve fitting problem.
- What a macro program is and what it is comprised of.
- How to design and create a macro program.
- How to execute macro programs.
- How to temporarily and permanently transfer program control from one macro to another.
- How to implement a scratch-pad.
- What a buffer is, its importance, and how it is used.
- How to get information from a template using @INDEX.
- How to move data into a template using {PUT} and {LET} commands.
- The importance of testing macro programs and templates.
- The importance of using named ranges in macro commands.
- The importance of commenting macro programs.

WHAT'S NEXT?

Now that you have learned some basic programming skills, the next chapter will show you a fully functional curve fitting program. Besides being a very useful program in its own right, the program presented in the next chapter should give you a good example of programming techniques that you can use when you create your own programs.

8

Fitting Nonlinear Data

Chapter 6 dealt with the subject of estimating linear, multiple linear, and curvilinear relationships between variables. The chapter made extensive use of the Lotus® regression program. In cases where these straightforward relationships failed to give a good fit for the data, an attempt was made to improve the fit by using complex functions (exponential, logarithmic, etc.) to transform the data and then re-executing the Lotus regression program. By using this scheme, seemingly complicated equations could often be reduced to a form that was **linear in the coefficients**. It is important for you to recognize that after this transformation, the data were actually being fit to a straight line using the Lotus linear regression program. That is, all of the equations had the following format

$$Y = a_0 + a_1 F_1 + a_2 F_2 + \cdots + a_p F_k + \epsilon$$

$$Y = a_0 + \sum_{j=1}^{k} a_j F_j$$

where F_j represents functions of the independent variables $X_1, X_2, \ldots, X_k$ and ϵ is an error term.

In many cases, these approaches work quite well. However, not all real-world scientific data can be fitted by transformation to a straight line, especially data from the areas of immunology, cell biology, pharmacology, etc. In cases where expressions are *nonlinear functions of unknown parameters,* linear curve fitting procedures are inadequate and it is necessary to use nonlinear fitting techniques. (A common example of data that require application of nonlinear fitting techniques are data that follow a sigmoidal curve, as illustrated in Figure 8-1.)

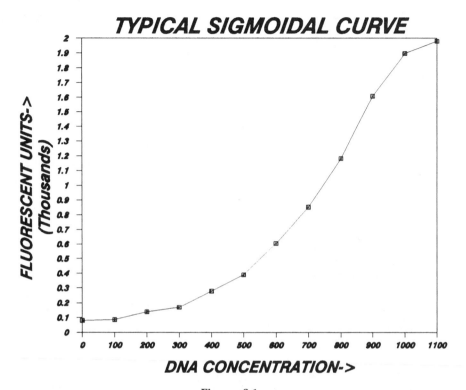

Figure   8-1

To clarify the difference between linear and nonlinear data, consider applying the methods discussed in Chapter 6 to the use of natural logarithms for transforming a set of exponential data into linear form. The method would involve the addition of a column to hold formulae that would transform the independent variable data points to their natural logarithms. The formulae would be based on logarithmic (@LN) functions. For example, data that follow the equation

$$Y = a_0 e^{a_1 x}$$

can be linearized by taking the natural logarithm,

$$\text{Ln } (Y) = \text{Ln } (a_0) + a_1 X$$

and determining a_0 and a_1. The important point here is that the equation is now linear **in its parameters** a_0 and a_1. If values for x are specified in the equation, Ln(y) will vary linearly. The following equation is also linear in the parameters:

$$\ln Y = a_0 + a_1 t^2$$

Both of these equations are linear because they are of the form

$$Y = a_0 + a_1 F_1 + a_2 F_2 + \cdots + a_n F_n + \epsilon$$

However, there are many situations where this tactic cannot be used. For example, suppose the data follow a double exponential:

$$Y = a_0 e^{a_1 X} + a_2 e^{a_3 X}$$

or a single exponential with two parameters:

$$Y = a_0 e^{(-a_1 X - a_2 X)}$$

or a Gaussian peak plus a quadratic background:

$$Y = a_0 \exp -\frac{1}{2}\frac{X - a_2^2}{a_1} + a_4 + a_5 X + a_6 X^2$$

or the following transcendental function:

$$Y = a_0 \sin (a_1 X)$$

or any other function that contains parameters that cannot be separated into different components of a summation. Then you cannot linearize the data unless you already know some of the parameters.

In addition to the above, there are many situations in which the form of an equation is either not known or doesn't adequately describe the data. For example, in Chapter 6 it was stressed that it was important to plot the original data and the fitted line. This plot may make it apparent that a linearly fitted line does not describe the data accurately.

In all of these situations you will need to use nonlinear data fitting techniques. The term *nonlinear regression* should be avoided for nonlinear curve fitting (estimation) procedures because nonlinear regression refers specifically to fitting an equation that is linear in unknown parameters. In nonlinear curve fitting, Least Squares equations are no longer linear in terms of the unknown parameters. For this reason, alternative techniques for their solution are required. There are several methods available for nonlinear curve fitting. Successive approximation (iterative) methods will be presented in this chapter and interpolation methods (point-to-point, polynomial, and spline) will be presented in Chapter 9.

This chapter has the following main objectives:

- To show you how to choose and implement an iterative technique to fit nonlinear curves.
- To show you how to apply the macro tools and concepts that you learned in Chapter 7 to the design and creation of programs that will address your own statistical and/or curve fitting applications.
- To show you how to obtain statistical information on how well a curve fits the data.

Before discussing these objectives, let us first examine some differences in linear and nonlinear data fitting.

STANDARD NONLINEAR NOMENCLATURE AND ERROR ANALYSES

In most text books, the guidelines for standard nonlinear nomenclature are different from those of linear Least Squares. The following is a comparison of the notation used in this book. This notation follows the dominant system in current use.

Category	Linear	Nonlinear
Response	Y	Y
Observation subscripts	$i = 1, 2, \ldots, N$	$u = 1, 2, \ldots, N$
Predictor variables	$X_1, X_2, \ldots, X_k$	$\xi_1, \xi_2, \ldots, \xi_k$
Predictor subscripts	$j = 1, 2, \ldots, k$	$j = 1, 2, \ldots, k$
Parameters	$a_0, a_1, \ldots, a_p$	$\theta_1, \theta_2, \ldots, \theta_p$
Parameter subscripts	$m = 0, 1, \ldots, p$	$m = 1, 2, \ldots, p$

Based on the new notations, a postulated model would be of the following form:

$$Y = f(\xi_1, \xi_2, \ldots, \xi_k; \theta_1, \theta_2, \ldots, \theta_p) + \epsilon$$

where ϵ is an error term.

Because it is common to have multiple predictor variables, parameters, etc., in nonlinear data fitting, the use of matrices to simplify notation is prevalent. For example, we can use the following matrices

$$\{\xi\} = (\xi_1, \xi_2, \ldots, \xi_k)^T$$

$$\{\theta\} = (\theta_1, \theta_2, \ldots, \theta_p)^T$$

to simplify the model. The superscript "T" in these matrix designations means *transpose*. The transpose matrix is obtained by writing all rows as columns in the order in which they occur in the original matrix so that the columns all become rows. For example, the transpose of the $\{\xi_1, \xi_2, \cdots, \xi_k\}$ matrix is

$$\{\xi\} = \begin{vmatrix} \xi_1 \\ \xi_2 \\ \cdot \\ \cdot \\ \xi_k \end{vmatrix}$$

These transposed matrices simplify the model to the following:

$$Y = f(\{\xi\}, \{\theta\}) + \epsilon$$

When N observations are made, the original equation takes on the following form:

$$Y_u = f(\xi_{1u}, \xi_{2u}, \ldots, \xi_{ku}; \theta_1, \theta_2, \ldots, \theta_p) + \epsilon_u$$

where ϵ_u is the error in the uth observation and $u = 1, 2, \ldots, N$. This equation can be shortened to

$$Y_u = f(\{\xi_u\}, \{\theta\}) + \epsilon_u$$

where, $\{\xi_u\} = (\xi_{1u}, \xi_{2u}, \cdots, \xi_{ku})^T$.

Using the above notation, a sum of squares function, $S(\{\theta\})$, can be defined to represent the error of the model. The following is the equation:

$$S(\{\theta\}) = \sum_{u=1}^{N} \{Y_u - f(\{\xi_u\}, \{\theta\})\}^2$$

In this equation, Y_u and $\{\xi_u\}$ have fixed values. Therefore, the sum of squares is a function of $\{\theta\}$ alone. The notation $\{\hat{\theta}\}$ in this equation is used to represent the matrix containing least-squares estimates of $\{\theta\}$ and is thereby a value of $\{\theta\}$ that minimizes $S(\{\theta\})$. Thus, to determine $\{\hat{\theta}\}$, $S(\{\theta\})$ is minimized by differentiating the sum of squares equation with respect to $\{\theta\}$ and equating the differential to zero. This minimization leads to the following set of p equations (one for each parameter), which must be solved for $\{\hat{\theta}\}$:

$$\sum_{u=1}^{N} w_u \{Y_u - f(\{\xi_u\}, \{\hat{\theta}\})\} \left[\frac{\partial f(\{\xi_u\}, \{\theta\})}{\partial \theta_m} \right]_{\theta = \hat{\theta}} = 0$$

where, $\hat{\theta}$ is the current estimate of θ.

That is, there is an equation that must be solved for each combination of $m = 1, 2, \ldots, p$ parameters and $u = 1, 2, \ldots, N$ observations. When solving these sum of squares equations, all of the θs in the derivative are replaced by the corresponding $\hat{\theta}$'s with the same subscript.

In the equation, w_u is a weighting factor. The weighting factor is included so that the same formula may be applied to data from a variety of statistical distributions. There are three main categories of weighting factors:

- Instrumental weight, with $w_u = 1/\sigma_u^2$.
- Equal weight, with $w_u = 1$.
- Statistical weight, with $w_u = 1/y_u$.

The instrumental weighting system is the most generic way to apply weighting. It pertains to the situation in which each data point in the analysis is itself comprised of the average of several measurements. In this case, the standard deviation is known and can therefore be used as the weighting factor. When this information is not available, the equal weight factor is the best choice. When the variance depends on the square root of the response, statistical weighting is appropriate. For example, statistical weighting should be used for data that follow a Poisson distribution that approximates a normal distribution with the variance increasing with the square root of the response. In this chapter, the equal weighting scheme will be used exclusively.

The nonlinear sum of squares formula is very similar to the one used in linear regression. If you recall from Chapter 6, in linear regression the best-fit straight line was found by minimizing the following sum of squares function:

$$\text{Sum of Squares} = \sum w_i (y_i - y_i(x_i))^2$$

CHOOSING A NONLINEAR CURVE FITTING TECHNIQUE

As you may have been able to deduce from the previous few paragraphs, as more parameters become involved, solving nonlinear curve fitting problems can become complicated and it can be difficult to ascertain the final parameterized equation. The technique that you use to solve a particular nonlinear application will depend heavily upon whether an approximate solution is acceptable, how disheveled the data are, if a theoretical equation is available, and how much statistical characterization of the fitted curve is required. Ideally, you want a program to provide you with all of the following features:

- The equation that the program yields should allow easy calculation of paired values in both directions. That is, if the final equation will be used only for plotting smooth graphs, then you only need the final equation to allow the x_i values to be incremented sequentially and yield the y_i values. If, however, you also want to use the final equation as a standard curve, then you will want an equation that also conveniently yields x_i values, given y_i values. The requirement for dual purpose fitted curves is the rule, rather than the exception, when experimental data are being reduced.

- A corollary to the above is that it is usually better to have a single function than to fit data piecemeal. That is, it is better to have a single, continuous equation to describe the entire relationship between independent and dependent variables than to cut the range into a series of subranges and interpolate between known points on the curve to find the corresponding values.

- The program should provide a high degree of efficiency when converging to the solution (i.e., the number of iterations it takes to define an equation that gives an acceptable error). A program that requires a large number of iterations to locate the parameters for the best fit curve can take an inordinate amount of time, especially if the program takes a long time for each iteration. Some iterative approaches are easy to program, but take 20 minutes to hours to converge. These approaches may be useful for infrequent or one-time only usage, but should be avoided for routine use.

- If the program finds parameters by successive approximations, the program should have a feature that automatically adjusts the "correction factors." These adjustments to the parameters should be large at the beginning of the search and should automatically diminish in size as the search approaches the final answer. A program that uses this approach will give the greatest performance per iteration and will, in turn, use the least amount of time to complete the function.

- Similarly, you will want a program that can automatically switch between iterative algorithms (depending on the speed and accuracy needs at the particular point in the iterative process). Some algorithms are very efficient at determining approximate values, but are not particularly accurate. Other algorithms are slow, but very accurate. If your program can start with a fast, crude algorithm to start and then switch to a fine-tuning algorithm to finish, you can have the best of both worlds.

- You will want a program that can operate as independently of the user as possible. Many iterative programs require the user to make initial guesses at the parameters and the adjustment values to be applied. The program then proceeds to apply corrections to fine-tune the initial guesses. However, for some programs, if the initial guesses are even slightly amiss, the programs will run sub-optimally (taking several thousand iterations) and take an extraordinarily long time to complete. For this reason, a program that automatically sets starting guesses is very desirable.

- A program that provides an analysis of the error associated with Goodness-of-Fit is desirable. As with all of the other analyses in this book, it is difficult to make sound judgments about the results of an experiment without knowing the magnitude of the error associated with the conclusions.

- The curve fitting program should converge reliably and should not "ring" or oscillate around the solution. Be aware that some algorithms use higher order polynomials to get a better fit to the data. However, if the order is too high, the curve will often oscillate. You should also be cognizant of the fact that although a high order curve may hit all of the data points, it may not be a good predictor of points between them.

- The program should be as convenient for the user to operate as possible.

- The program should contain features that provide as much freedom from user mistakes as possible.

However by their very nature, even the fastest nonlinear curve fitting techniques tend to be slow. This drawback is especially true for disheveled curves. Therefore, depending on how disheveled your curves are, you may need to balance the above wish list against the length of time that you are willing to wait for your answers.

OVERVIEW OF NONLINEAR LEAST SQUARES METHODS

As mentioned above, most algorithms that have been developed for the determination of nonlinear parameters have been based on finding a set of coefficients (θ_0, θ_1, . . . , θ_i) that provide the best fit line for the data. These determinations invariably focus on using iterative techniques for minimizing the sum of squares. This minimization was outlined above and the following minimizing equation was presented:

$$\sum_{u=1}^{N} w_u \{Y_u - f(\{\xi_u\}, \{\hat{\theta}\})\} \left[\frac{\partial f(\{\xi_u\}, \{\theta\})}{\partial \theta_m} \right]_{\theta = \hat{\theta}} = 0$$

The three most common iterative techniques that utilize this methodology are

- Linearization
- The Method of Steepest Descent
- Marquardt's Compromise.

The following is a very brief overview of these three techniques. For a more complete discussion on the three methods, see the References listed at the end of this chapter.

However, it is very important to point out at the onset that **a firm understanding of the theoretical discussions on the methods in this book or in the References is not a requirement for you to be able to ascertain how the macro programs that implement the methods work.** And in fact, it is more likely for you to find the opposite to be true. By spending more time on the structure of the program and the flow involved than on the theory behind the mathematics, you should have a much better comprehension of the three methods. So, do not get too caught up in the complicated theoretical details of these three procedures. Spend the bulk of your valuable time studying the program.

The Linearization method is an offshoot of the Gauss-Newton method. The Gauss-Newton method is a method for generating equations with candidate coefficients for sum of squares evaluation and is based on expanding the model equation as a Taylor series. Optimal values for corrections to the parameters are determined using the method of Least Squares. The parameter increments calculated at each iteration (based on the assumption of local linearity) are then applied to the parameters, the predicted dependent variables are calculated, and the predicted dependent values are compared to the experimental values using the sum of squares equation. The Taylor series is expanded again using the new estimates for the parameters. This process is repeated until the solution converges. Convergence is usually defined as that point in the successive iterations where the ratio of the change in each θ_m is small (e.g., <0.0000001) compared to the corresponding value of θ_m.

The general form of the Taylor series is

$$f(x) = f(a) + \frac{f'(a)}{1!} (x - a) + \frac{f''(a)}{2!} (x - a)^2 + \frac{f'''(a)}{3!} (x - a)^3$$

$$+ \cdots + \frac{f^n(a)}{n!} (x - a)^n$$

The following is the form of the Taylor series expansion using the nomenclature defined at the beginning of this chapter and limiting the expansion to the first derivative:

$$f(\{\xi_u\}, \{\theta\}) = f(\{\xi_u\}, \{\theta_0\}) + \sum_{m=1}^{p} \left[\frac{\partial f(\{\xi_u\}, \{\theta\})}{\partial \theta_m} \right]_{\theta = \hat{\theta}} (\theta_m - \theta_{m0})$$

Usually, only the first two or three terms of the Taylor series are implemented when reducing experimental data because they typically give adequate accuracy for the fit in an acceptable amount of time. If the Taylor series is truncated at the first derivative when the values for the $\{\theta_0\}$ matrix are close to the values in the $\{\theta\}$ matrix, the form is linear and can be solved using linear Least Squares.

The *Method of Steepest Descent* (or *Gradient Search*) also uses an iterative approach to minimizing the sum of squares function. It was originally proposed by A. Cauchy in 1845. Basically, the Method of Steepest Descent involves moving from an initial point $\{\theta_0\}$ along a single directional matrix (a vector) constructed from the following components (the values of which change continuously as the path is followed):

$$\left| -\frac{\partial S(\{\theta\})}{\partial \theta_1}, \; -\frac{\partial S(\{\theta\})}{\partial \theta_2}, \; \ldots, \; -\frac{\partial S(\{\theta\})}{\partial \theta_p} \right|^T$$

A vector matrix comprised of first partial derivatives is called a *gradient*. The negative direction of this matrix is "downhill" in the sense that the sum of squares function $S(\{\theta\})$ will decrease as long as the distance moved along the gradient is small. The gradient search provides a **single** correction factor, which is used to prepare **all** of the θs for the next iteration. The process is repeated (re-computing the gradient after each change in the parameters) until the method converges.

Although in theory the Linearization method will always converge, it is quite slow. The Method of Steepest Descent is quite fast initially, but becomes agonizingly slow after the first few iterations. The sluggishness in the Method of Steepest Descent is due to the fact that as the search approaches small sum of squares values, the derivatives consist of taking differences between nearly equal numbers. For this reason, the Method of Steepest Descent is rarely used alone and a second method is usually combined with the Method of Steepest Descent to locate the convergence values once the Method of Steepest Descent has closely approached the point of convergence.

After considerable search and evaluation, I have found Marquardt's Compromise (also called the Levenberg-Marquardt method) to be the best choice to fulfill the criteria set forth in the previous section. Marquardt's Compromise is a combination of the Linearization method and the Method of Steepest Descent. In practice, this combination works very well. An overview of Marquardt's Compromise is given in the next section, followed by a number of sections describing a program that implements Marquardt's Compromise.

MARQUARDT'S COMPROMISE

Marquardt's Compromise employs the Method of Steepest Descent and the Linearization method. With the implementation presented in this chapter, the Method of Steepest Descent is used when the model curve is far from the "true" curve. Marquardt's Compromise switches smoothly and continuously from the Method of Steepest Descent to the Linearization method as the "true" curve is approached. In this way, Marquardt's Compromise combines the best features of the Steepest Descent and Linearization methods, while avoiding their most serious limitations. Marquardt's Compromise works very well in practice (i.e., it converges relatively quickly and reliably) and has become the standard of nonlinear routines.

More specifically, Marquardt's Compromise takes a series of N experimental data point pairs (ξ_i, y_i) and finds the parameter set $\{\theta\}$ that will minimize the sum of squares function

$$S(\{\theta\}) = \sum w_i(y_i - \hat{y}(\xi_i, \{\theta\}))^2$$

where w_i is the weighting factor (discussed at the beginning of this chapter), y_i is the experimental value, and $\hat{y}$ is the value of the parameterized equation at ξ_i.

The Marquardt minimization process takes place in a rather clever matrix that simultaneously contains calculations for both constituent methods. The matrix has the following elements:

$$\alpha_{jk} = \begin{cases} (1 + \lambda)w_i \dfrac{\partial \hat{y}}{\partial \theta_j} \cdot \dfrac{\partial \hat{y}}{\partial \theta_k} & \text{for } j = k \\[12pt] w_i \dfrac{\partial \hat{y}}{\partial \theta_j} \cdot \dfrac{\partial \hat{y}}{\partial \theta_k} & \text{for } j \neq k \end{cases}$$

where $j, k = 1, 2, \ldots , N$. These equations lead to a matrix of the following format (assuming four parameters):

$$A = \begin{vmatrix} (1 + \lambda)\alpha_{11} & \alpha_{12} & \alpha_{13} & \alpha_{14} \\ \alpha_{21} & (1 + \lambda)\alpha_{22} & \alpha_{23} & \alpha_{24} \\ \alpha_{31} & \alpha_{32} & (1 + \lambda)\alpha_{33} & \alpha_{34} \\ \alpha_{41} & \alpha_{42} & \alpha_{43} & (1 + \lambda)\alpha_{44} \end{vmatrix}$$

The λ parameter in the above matrix is adjusted as part of the minimization process, assuring efficient and reliable movement towards the minimum of $S(\theta)$. For safety reasons, the initial value of λ is set to a small value and $S(\theta)$ is evaluated. If it is determined that $S(\theta)$ is far from the minimum, λ will be increased to be much greater than unity. In this way, the diagonal elements of the matrix (Steepest Descent) are used in the calculation of the increments for the parameters θ_m to move the system towards a minimum. As the minimum is approached, the value of λ will be decreased to a point where the complete matrix is used (Linearization method).

The solution of Marquardt's Compromise is found by inverting matrix A and multiplying A^{-1} by a vector matrix containing gradient equations. This multiplication yields a vector matrix of correction factors for the parameters. The following will illustrate these other two matrices:

$$\delta_a = \begin{vmatrix} \delta_{a1} \\ \delta_{a2} \\ \delta_{a3} \\ \delta_{a4} \end{vmatrix}$$

$$g = \begin{vmatrix} g_1 \\ g_2 \\ g_3 \\ g_4 \end{vmatrix}$$

where, g is the gradient matrix of the first partial derivatives of $S(\theta)$ with respect to θ_i and δ_a is a vector matrix containing the corrections for the parameters. The partial derivatives in the gradient matrix are defined as

$$\frac{\partial S(\{\theta\})}{\partial \theta_u} = \sum_{u=1}^{N} w_u(y_u - \hat{y}(\xi_u, \{\theta\})) \frac{\partial \hat{y}(\xi_u, \{\theta\}))}{\partial \theta_u}$$

where, $\hat{y}(\xi_u, \{\theta\})$ indicates the value of the model equation comprised of the current estimates of $\{\theta\}$.

The process of solving the matrices, applying the δ_{a_u} corrections, and then re-evaluating continues until the values within the δ_a vector are sufficiently small.

Figures 8-2 and 8-3 show a template containing nonlinear data and a plot of the data, respectively.

Figures 8-4 through 8-11 illustrate an implementation of Marquardt's Compromise to fit the data using Rodbard's Four-Parameter Logistic equation (described below).

In the figures, initial guesses are provided for two parameters based on some characteristics of the equation. Initial guesses for the other two parameters are made with the help of the Lotus linear regression program. Given the initial guesses for the set of fitted parameters (θ_i), the protocol is:

1. Compute the error as a sum of squares, $S(\{\theta\})$.

2. Pick a modest value for lambda (e.g., 0.001).

3. Solve the linear equations for δ_a and evaluate $S(\{\theta\} + \delta_a^*\{\theta\})$.

4. If $S(\{\theta\} + \delta_a^*\{\theta\}) >\, = S(\{\theta\})$, increase lambda by a factor of 10 and go back to Step 3.

5. If $S(\{\theta\} + \delta_a^*\{\theta\}) < S(\{\theta\})$, decrease lambda by a factor of 10, update the trial solution, and go back to Step 3.

6. Stop when δ_a is very small.

These features will become more apparent as you proceed through the sections that describe the program. However, before examining the program, the next two sections describe an example template, some example data, and Rodbard's Four-Parameter Logistic equation.

```
--------A--------B--------C--------D--------E--------F--------G--------H-----
 1                           FLUORESCENT DNA ASSAY DATA
 2 DATE RUN:31-OCT-88
 3                   ****STANDARD CURVE DATA****        AVERAGE
 4 NG DNA/ML ASSAY1      ASSAY2      ASSAY3      ASSAY4  RESPONSE PRED_VAR LOG_CONC
 5      31.2     81.0     86.0      81.8      85.1      83.5 0.000269 3.440418
 6      62.5    139.0    171.0     140.4     169.3     154.9 0.000264 4.135166
 7       125    323.0    309.0     326.2     305.9     316.0 0.000252 4.828313
 8       250    604.0    616.0     610.0     609.8     610.0 0.000230 5.521460
 9       500   1182.0   1106.0    1193.8    1094.9    1144.2 0.000193 6.214608
10      1000   1897.0   1839.0    1916.0    1820.6    1868.1 0.000140 6.907755
11
12
13
14
15
16
17
18
19
20 TABLE OF VALUES                  STATISTICS                RODBARD PARAMETERS
21 ===============                  ==========                ==================
22 A_PARAM   -1.4E+07               NUM_STDS        6         A
23 B_PARAM   3656.029               ERROR     67.29376        B
24 C_PARAM   3820.159               HIGHEST       1000        C
25 D_PARAM   1.179422               SD_EST    5.800593        D
26                                  R_SQUARE  0.999971
27
28
```

Figure 8-2

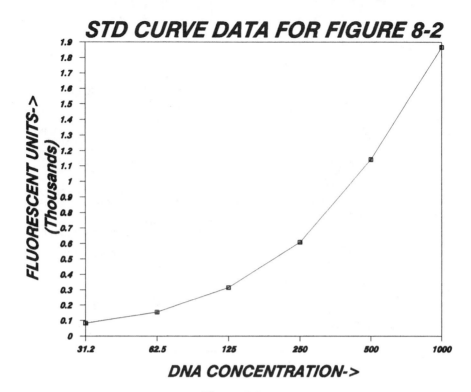

STD CURVE DATA FOR FIGURE 8-2

FLUORESCENT UNITS-> (Thousands)

DNA CONCENTRATION->

Figure 8-3

```
--------P--------Q---------R--------S--------T--------U--------V--------W-----
 1                      /*NON-LINEAR DATA FITTING PROGRAM*/
 2 /*PERFORMS RODBARD'S 4-P, SPLINE, LINEAR REGRESSION, AND LAGRANGE INT*/
 3                      /*REVISED 6JAN89; LOUIS M. MEZEI*/
 4
 5           /*AUTO-EXECUTING MACRO FOR LOTUS 1-2-3*/
 6 \0        {LET VERSION,0}    /*IDENDIFY SPREADSHEET AS 123*/
 7           {LET INIT_REG,"/DRXPRED_VAR~"}        /*SELF-MODIFY REGRESSION*/
 8           {LET REG_CMD,"/DRG"}                  /*COMMANDS*/
 9           {MENUBRANCH MENU1}         /*BRING UP THE MAIN MENU*/
10
11           /*AUTO-EXECUTING MACRO FOR SYMPHONY*/
12 AUTOEXEC  {IF @ISERR(@APP("STAT",""))}{ADD_STAT}/*SEE IF STAT ATTACHED*/
13           {LET VERSION,1}    /*IDENDIFY SPREADSHEET AS SYMPHONY*/
14           {LET INIT_REG,"/RRXPRED_VAR~"}        /*SELF-MODIFY REGRESSION*/
15           {LET REG_CMD,"/RRG"}                  /*COMMANDS*/
16           {MENUBRANCH MENU1}         /*BRING UP THE MAIN MENU*/
17
18 ADD_STAT  {S}AASTAT~Q        /*ADD STAT.APP PROGRAM TO SYMPHONY*/
19
20 \M        {MENUBRANCH MENU1}         /*BRING UP THE MAIN MENU*/
21
```

Figure 8-4

```
--------AB----AC---------------AD--------------AE--------------AF-------------AG-----------------AH-------AI----
 1                              /*MAIN MENU FOR NON-LINEAR CURVE FITTING*/
 2 MENU1 4P              SPLINE          REGRESS          POINT           CUBIC              QUIT
 3     FOUR-PARAMETER    CUBIC SPLINE    LINEAR REGRESSION POINT-TO-POINT  CUBIC INTERPOLATION QUIT THE PROGRAM
 4     {MAIN}            {SPLINE}        {LIN RGRS}       {P_TO_P}        {CUBIC}            {QUIT}
 5     {MENUBRANCH MENU1}{MENUBRANCH MENU1}{MENUBRANCH MENU1} {MENUBRANCH MENU1} {MENUBRANCH MENU1}
 6
```

Figure 8-5

155

```
--------AN-------AO-------AP-------AQ-------AR-------AS-------AT-------AU----
15              /*MAIN PROGRAM FOR MARQUARDT'S FIT TO 4-P EQUATION*/
16 MAIN         {HOME}              /*MOVE CELL-POINTER TO CELL A1*/
17              {SET_STDS}          /*SET UP THE STANDARDS SECTION IN MATRIX*/
18              {INIT_4P}           /*SET UP INITIAL 4P PARAMTERS*/
19              {SET_RNGS}          /*RESET RANGES USED IN REGRESSION*/
20              {INIT_REG}          /*INITIALIZE REGRESSION RANGES*/
21 FULL_ITER{LET SUM_SQRS_0,ERROR}      /*UPDATE SUM OF SQUARES VALUE*/
22              {MAKE_MTX}          /*SET UP THE MARQUARDT MATRIX*/
23 PART_ITER{SOLVE_MTX}         /*SOLVE THE MARQUARDT MATRIX*/
24              {MOD_PARAMS}        /*UPDATE B AND D USING DELTA VALUES/
25              {REGRESS}           /*FIND VALUES FOR LINEAR PARAMETERS A AND C*/
26              {LET SUM_SQRS_1,ERROR}   /*UPDATE THE SUM OF SQUARES*/
27              {LET TEMP,@ABS(DELTA_B/MID_0)}
28              {LET TEMP,TEMP+@ABS(DELTA_D/EXPO_0)}
29              {IF TEMP<0.002}{CALC}{RETURN} /*TEST FOR CONVERGENCE*/
30              {IF SUM_SQRS_1<SUM_SQRS_0}{BRANCH BIG_INC}
31              {LET LAMBDA_1,LAMBDA_1*10} /*INCREASE LAMBDA BY FACTOR OF 10*/
32              {BRANCH PART_ITER}       /*ITERATE*/
33
34              /*DECREASE MATRIX DIAGONAL MULTIPLIER AND RESET B AND D TERMS*/
35 BIG_INC      {LET LAMBDA_1,LAMBDA_1/10} /*DECREASE LAMBDA BY FACTOR OF 10*/
36              {LET MID_0,MID_1}         /*UPDATE ARCHIVED MIDPOINT*/
37              {LET EXPO_0,EXPO_1}       /*UPDATE ARCHIVED EXPONENTIAL TERM*/
38              {BRANCH FULL_ITER}        /*ITERATE*/
39
```

Figure   8-6

```
--------AZ-------BA-------BB-------BC-------BD-------BE-------BF-------BG-------BH-------BI----
 1                      /*INITIALIZATION MACROS FOR MARQUARDT*/
 2
 3                      /*CONVERT THE CONCENTRATIONS TO THEIR LOGS*/
 4 SET_STDS {LET NUM_STDS,@COUNT(CONC)}       /*DETERMINE NUMBER OF STDS*/
 5          {FOR INC1,0,NUM_STDS-1,1,LOG_LOOP} /*CONVERT ALL STDS TO NATURAL LOGARITHMS*/
 6          {CALC}                       /*FORCE RECALC OF SPREADSHEET*/
 7
 8                 /*CONVERT ONE OF THE STD'S CONC TO ITS NATURAL LOG*/
 9              /*(IF A CONCENTRATION IS ZERO, THEN ASSIGN IT A LOG OF -2*/
10 LOG_LOOP {LET TMP_CONC,@INDEX(CONC,0,INC1)} /*GET CURRENT CONCENTRATION*/
11          {LET TMP_CONC,@IF(TMP_CONC=0,-4.606,@LN(TMP_CONC))}  /*CONVERT TO LN*/
12          {PUT LOG_CONC,0,INC1,TMP_CONC}   /*PLACE LN INTO THE TEMPLATE*/
13
14                 /*SET INITIAL VALUES FOR THE 4-P PARAMETERS*/
15 INIT_4P  {LET HIGHEST,@MAX(CONC)}   /*FIND HIGHEST CONCENTRATION*/
16          {LET MID_0,HIGHEST/2}      /*INITIALIZE MIDPOINT PARAMETER (B)*/
17          {LET EXPO_0,1}             /*SET INITIAL EXPONENTIAL TERM (D) TO 1*/
18          {LET LAMBDA_1,0.001}       /*SET INITIAL DIAGONAL MULTIPLIER TO A SMALL VALUE*/
19          {LET B_PARAM,MID_0}        /*SET B TO HALF THE HIGHEST CONCENTRATION*/
20          {LET D_PARAM,EXPO_0}       /*SET D (CURVE FACTOR) TO 1*/
21
22          /*ADJUST RANGE SIZES FOR NUMBER OF STANDARDS*/
23 SET_RNGS /RNCRESPONSE~{ESC}.{DOWN NUM_STDS-1}~ /*RESET FLUORESCENT UNITS (Y) RANGE*/
24          /RNCPRED_VAR~{ESC}.{DOWN NUM_STDS-1}~
25                      /*RESET PREDICTOR VARIABLE RANGE FOR (B+EXP(D+LN X)) VALUES*/
26
27          /*SET UP REGRESSION RANGES*/
28 INIT_REG /RRXPRED_VAR~         /*SET UP REGRESSION X RANGE TO PREDICTOR VARIABLES*/
29          YRESPONSE~            /*SET UP REGRESSION Y RANGE TO FLUORESCENT RESPONSE*/
30          OREG_DATA~            /*SET UP REGRESSION OUTPUT RANGE*/
31          Q                     /*QUIT THE REGRESSION MENU*/
32          {REGRESS}             /*PERFORM A REGRESSION TO GET INIT VALUES FOR A AND C*/
33
```

Figure   8-7

```
--------BN-------BO-------BP-------BQ------BR-------BS-------BT------BU-------BV---
 1 /*PERFORM REGRESSION TO FIND LINEAR PARAMS (A AND C) AND UPDATE THE TEMPLATE*/
 2 REGRESS   {FOR INC1,0,NUM_STDS-1,1,UPDATE}    /*UPDATE PREDICTED VALUES*/
 3 REG_CMD   /RRG                  /*PERFORM REGRESSION ON CURRENT LINE*/
 4           {LET A_PARAM,SLOPE}       /*UPDATE THE TEMPLATE WITH NEW VALUES*/
 5           {LET C_PARAM,INTERCEPT}
 6           {LET ERROR,STDERR^2*DEG_FRED}    /*UPDATE ERROR CELL*/
 7           {LET R_SQUARE,R_VALUE}    /*UPDATE R SQUARED CELL IN TEMPLATE*/
 8           {LET SD_EST,@SQRT(ERROR/(DEG_FRED-2))}}
 9                                      /*UPDATE STANDARD ERROR OF ESTIMATE CELL*/
10
11           /*RECALCULATE THE NONLINEAR PREDICTOR VALUES*/
12 UPDATE    {LET TMP_CONC,1/B_PARAM}  /*USE 1/MIDPOINT IF CONCENTRATION IS ZERO*/
13           {IF @INDEX(CONC,0,INC1)<>0}{CALC_X} /*CALC NEW PREDICTOR VALUE*/
14           {PUT PRED_VAR,0,INC1,TMP_CONC}    /*PUT VALUE IN TEMPLATE*/
15
16           /*CALCULATE THE NONLINEAR PORTION OF THE EQUATION*/
17 CALC_X    {LET TMP_CONC,1/(B_PARAM+@EXP(D_PARAM*@INDEX(LOG_CONC,0,INC1)))}}
18
19 REG_DATA        Regression Output:
20           Constant                 3820.159
21           Std Err of Y Est         4.101638
22           R Squared                0.999971
23           No. of Observations             6
24           Degrees of Freedom              4
25
26           X Coefficient(s)   -1.4E+07
27           Std Err of Coef.   36778.54
28
```

Figure   8-8

```
--------CA---------CB-------CC-------CD-------CE-------CF-------CG-------CH-------CI----CJ----
 1
 2                    /*SET UP THE MARQUARDT MATRIX*/
 3 MAKE_MTX  {BLANK A_VALS}            /*INITIALIZE ALL MATRIX SUMMATIONS TO ZERO*/
 4           {BLANK GRAD_3}            /*INITIALIZE 3rd TERM OF THE GRADIENT MATRIX*/
 5           {BLANK GRAD_4}            /*INITIALIZE 4th TERM OF THE GRADIENT MATRIX*/
 6           {FOR INC1,0,NUM_STDS-1,1,SET_UP_MTX}/*CALC PARTIAL DERIVS AND ALPHAS*/
 7           {LET ALPHA_22,NUM_STDS}   /*SET MATRIX TERM ALPHA_22 TO NUMBER STDS*/
 8
 9           /*ADD TO PARTIAL DERIVATIVE SUMATIONS BASED ON CURRENT X,Y DATA PAIR*/
10 SET_UP_MTX {LET X_TO_D,0}   /*SET A VALUE FOR X TO THE Dth POWER TERM*/
11           {IF @INDEX(CONC,0,INC1)<>0}{DO_X_TO_D}   /*CALCULATE Xi^D */
12           {LET DERIV_A,1/(X_TO_D+B_PARAM}}   /*PARTIAL DERIV WRTO A*/
13           {LET DERIV_B,-A_PARAM*DERIV_A^2}   /*PARTIAL DERIV WRTO B*/
14           {LET DERIV_D,DERIV_B*X_TO_D*@INDEX(LOG_CONC,0,INC1)} /*PARTIAL DERIV WRTO D*/
15           {LET ALPHA_11,ALPHA_11+DERIV_A^2}   /* SUMMATION FOR dF/dA*dF/dA */
16           {LET ALPHA_12,ALPHA_12+DERIV_A}     /* SUMMATION FOR dF/dA*dF/dC = dF/dA*1 */
17           {LET ALPHA_13,ALPHA_13+DERIV_A*DERIV_B}    /* SUMMATION FOR dF/dA*dF/dB */
18           {LET ALPHA_14,ALPHA_14+DERIV_A*DERIV_D}    /* SUMMATION FOR dF/dA*dF/dD */
19           {LET ALPHA_23,ALPHA_23+DERIV_B}     /* SUMMATION FOR dF/dB*dF/dC = dF/dB*1 */
20           {LET ALPHA_24,ALPHA_24+DERIV_D}     /* SUMMATION FOR dF/dD*dF/dC = dF/dD*1 */
21           {LET ALPHA_33,ALPHA_33+DERIV_B^2}   /* SUMMATION FOR dF/dB*dF/dB */
22           {LET ALPHA_34,ALPHA_34+DERIV_B*DERIV_D}    /* SUMMATION FOR dF/dB*dF/dD */
23           {LET ALPHA_44,ALPHA_44+DERIV_D^2}   /* SUMMATION FOR dF/dD*dF/dD */
24           {LET DIF_TRM,@INDEX(RESPONSE,0,INC1)-A_PARAM*DERIV_A-C_PARAM}
25           {LET GRAD_3,GRAD_3+DIF_TRM*DERIV_B} /*UPDATE GRADIENT MATRIX THIRD TERM*/
26           {LET GRAD_4,GRAD_4+DIF_TRM*DERIV_D} /*UPDATE GRADIENT MATRIX FOURTH TERM*/
27
28           /*CALCULATE X TO THE POWER OF D_PARAM (X^D)*/
29 DO_X_TO_D {LET X_TO_D,@EXP(D_PARAM*@INDEX(LOG_CONC,0,INC1))}}
30
```

Figure   8-9

```
--------CN---------CO-------CP-------CQ------CR-------CS------CT------CU-------
 1
 2 /*SOLVE THE MATRIX BY INVERTING MARQUARDT AND MULTIPLYING BY GRADIENT*/
 3 SOLVE_MTX {LET LAMBDA_0,1+LAMBDA_1}  /*UPDATE MATRIX DIAGONAL MULTIPLIER*/
 4           {LET LAMBDA_2,LAMBDA_0^2}  /*FIND SQUARE OF LAMBDA*/
 5           {LET X_TO_D,ALPHA_12^2-ALPHA_11*LAMBDA_2*ALPHA_22}
 6           {LET X_TWO,(ALPHA_11+LAMBDA_0*ALPHA_23-ALPHA_12*ALPHA_13)/X_TO_D}
 7           {LET Y_TWO,(ALPHA_11+LAMBDA_0*ALPHA_24-ALPHA_12*ALPHA_14)/X_TO_D}
 8           {LET X_ONE,(ALPHA_22+LAMBDA_0*ALPHA_13-ALPHA_12*ALPHA_23)/X_TO_D}
 9           {LET Y_ONE,(ALPHA_22+LAMBDA_0*ALPHA_14-ALPHA_12*ALPHA_24)/X_TO_D}
10           {LET BFR_R,ALPHA_13*X_ONE+ALPHA_23*X_TWO+ALPHA_33*LAMBDA_0}
11           {LET BFR_S,ALPHA_13*Y_ONE+ALPHA_23*Y_TWO+ALPHA_34}
12           {LET BFR_U,ALPHA_14*Y_ONE+ALPHA_24*Y_TWO+ALPHA_44*LAMBDA_0}
13           {LET BFR_T,ALPHA_14*X_ONE+ALPHA_24*X_TWO+ALPHA_34}
14           {LET X_TO_D,BFR_S*BFR_T-BFR_R*BFR_U}
15           {LET DELTA_B,(GRAD_4*BFR_S-GRAD_3*BFR_U)/X_TO_D}
16           {LET DELTA_D,(GRAD_3*BFR_T-GRAD_4*BFR_R)/X_TO_D}
17
18           /*UPDATE B AND D USING DELTA VALUES FROM MATRIX SOLN*/
19 MOD_PARAMS {LET MID_1,MID_0+DELTA_B}       /*UPDATE B PARAMETER*/
20           {LET EXPO_1,EXPO_0+DELTA_D}      /*UPDATE D PARAMETER*/
21           {LET B_PARAM,MID_1}     /*UPDATE B IN TEMPLATE*/
22           {LET D_PARAM,EXPO_1}    /*UPDATE D IN TEMPLATE*/
23
24
```

Figure   8-10

```
--------CZ---------DA---------DB--------DC-------DD---------DE-------DF------
 1
 2                              SCRATCHPAD
 3                              ==========
 4          MID_0      3654.0561       DERIV_A    0.000140
 5          MID_1      3656.0290       DERIV_B    0.274702
 6          EXPO_0     1.1793279       DERIV_D    6549.128
 7          EXPO_1     1.1794222
 8          DELTA_B    1.9728801       LAMBDA_0   1.000000
 9          DELTA_D    0.0000943       LAMBDA_1   1.0E-10
10                                     LAMBDA_2   1.000000
11          ALPHA_11   0.0000003
12          ALPHA_12   0.0013516       X_TO_D     -56.4874
13          ALPHA_13   0.0010614
14          ALPHA_14   3.0190881       X_ONE      -5724.04
15          ALPHA_22           6       X_TWO      0.556889
16          ALPHA_23   4.3955637       Y_ONE      50635901
17          ALPHA_24   16200.005       Y_TWO      -14106.9
18          ALPHA_33   3.6309723       BFR_R      0.002921
19          ALPHA_34   8213.7286       BFR_S      -46.0560
20          ALPHA_44   76404554.       BFR_T      -46.0560
21                                     BFR_U      745408.6
22          GRAD_3     0.0014188       DIF_TRM    -0.38460
23          GRAD_4     -20.54337
24                                     TEMP       0.000619
25          SUM_SQRS_0 67.294630       INC1             6
26          SUM_SQRS_1 67.293765       TMP_CONC   0.000140
27                                     VERSION          1
```

Figure   8-11

BRIEF DESCRIPTION OF THE EXAMPLE DATA

One of the most common nonlinear curves is sigmoidal in shape. Figure 8-1 illustrates this type of curve. Sigmoidal curves are exceedingly common in many disciplines. Among the assays that produce sigmoidal curves are: radioimmunoassays (RIA), immunoradiometric assays (IRMA), Enzyme ImmunoAssays (EIA), Enzyme Linked ImmunoSorbent Assays (ELISA), Enzyme Multiplied Immunological Techniques (EMIT), bioassays, pharmaceutical dose-response studies, CytoPathic Effect assays (CPE), and cell proliferation assays.

Nonlinear curve fitting procedures are required to fit sigmoidal curves. The Marquardt procedure can usually be modified and applied to many other shapes of curves. For this reason, a sigmoidal curve will be used as the example for this chapter. For discussions on curve fitting procedures and translatable programs for other curve shapes, see the "**References**" Section at the end of this chapter.

Figure 8-2 is the template that we have been using throughout this book. If you will note, it has been modified to contain data representative of a portion of a sigmoidal curve. The remainder of this chapter will use this template to illustrate Marquardt's nonlinear curve fitting method.

FOUR-PARAMETER LOGISTIC CURVE FITTING

One of the most commonly used nonlinear equation models in immunoassay is the four-parameter logistic model of Rodbard (1974). The *4-P* equation can be written in a variety of forms; a common form being

$$y = \frac{E - G}{1 + (x/F)^H} + G$$

In this equation,

- y is the analytical response (e.g., fluorescence).
- x is the concentration.
- E is the response at $x = 0$.
- F is the 50% intercept, (i.e., the response halfway between the upper and lower plateaus of the dose response curve).
- G is the response at infinite dose of unlabeled ligand.
- H is the slope factor (exponential term).

This equation is popular because it offers many advantages. The following is a list of some of the advantages of this equation:

- A wide variety of response variables can be used (e.g., B/T, B/F, B/B_0, bound count, etc.), meaning that an analyst can observe a natural response without making special transformations.

- The equation is applicable to RIA, IRMA, and EIA.
- The equation provides a fairly satisfactory uniformity of variance for the dose response curve compared to logit transformation.
- The equation also estimates "F" (ED_{50}), and ED_{50} standard errors. This information is useful in characterizing the cross reactivity of several closely related ligands. The "F" parameter is also useful in cytopathic effect (CPE) assays and lethal dose (LD50) studies.
- The equation permits a comparison of methods using the "H" (slope) values together with the standard errors of the slopes.

However, the 4-P method requires that a curve be fairly symmetrical around its midpoint. This assumption is usually valid, even for curves containing data that incompletely span the entire range (e.g., see Figure 8-3). By using five, six, or seven parameters, it is possible to solve some very unusual problems. Nonetheless, the use of more than four parameters is quite uncommon. For this reason, the 4-P equation will be used for the Marquardt example. If you have previously determined that another equation can be iteratively used to fit your data, then you should use the equation and modify the Marquardt example given below for your equation. If you do not know which equation to use, refer to Bevington for a BASIC version or Press for a C version of a generic Marquardt method.

The Rodbard equation can be transformed to a simpler form. The transformation process is referred to as *re-parameterization* and involves a redefinition for each of the four parameters as follows:

$$E = \frac{A}{B} + C$$

$$F = B^D$$

$$G = C$$

$$H = D$$

Using the re-parameterized definitions yields the following form of the equation:

$$y = \frac{A}{(x_i^D + B)} + C$$

This form of the equation makes programming easier because it allows A and C to be calculated using linear Least Squares. More importantly, the equation is in a form that can be used by the Lotus linear regression program, thereby improving the speed of program execution. That is, it is in the following form:

$$y_i = \theta_0 + \theta_1 \, \xi_i$$

with

$$\xi_i = 1/(x_i^D + B)$$

This form can be used directly by Lotus. Another nice feature of the re-parameterized form of an equation is that starting values are required only for B and D, and crude estimates will suffice. D can be set equal to one and B can be set equal to one-half, the highest concentration. Then, on each iteration, a new value will be determined for the nonlinear parameters (B and D) and the linear parameters (A and C) can be determined by the Lotus regression program. This setup provides efficient program execution, an important consideration when using iterative procedures (which all tend to be slow).

To apply Marquardt's Compromise to the re-parameterized Rodbard's equation, notice that there are four parameters. These parameters correspond to the following θ values:

$$\begin{array}{lll} \text{Linear:} & A = \theta_1 \\ & C = \theta_2 \\[6pt] \text{Nonlinear:} & B = \theta_3 \\ & D = \theta_4 \end{array}$$

A 4×4 Marquardt matrix is therefore required. Because the matrix contains the partial derivatives with respect to each of the parameters, a good starting point would be to find the partial derivatives of the equation with respect to each of the parameters. To simplify the partial derivative equations, define ξ as follows:

$$\xi = x^D = \text{EXP}\ (D * L(K))$$

where $L(K)$ is the natural logarithm of x_k. Using ξ, the partial derivative equations are

$$\frac{\partial F}{\partial \theta_1} = \frac{\partial F}{\partial A} = 1/(\xi + B)$$

$$\frac{\partial F}{\partial \theta_2} = \frac{\partial F}{\partial C} = 1$$

$$\frac{\partial F}{\partial \theta_3} = \frac{\partial F}{\partial B} = -A/(\xi + B)\hat{\ }2$$

$$\frac{\partial F}{\partial \theta_4} = \frac{\partial F}{\partial D} = (-A * \xi * L(k))/(\xi + B)\hat{\ }2$$

where, $k = 1, 2, \ldots, N$ data points.

For the gradient matrix, you will need the partial derivatives of the sum of squares function $S(\{\theta\})$ with respect to each of the parameters. If you recall, the sum of squares function and its derivative were defined in the section entitled "Standard Nonlinear Nomenclature and Error Analyses" as

$$S(\{\theta\}) = \sum_{u=1}^{N} \{Y_u - f(\{\xi_u\}, \{\theta\})\}^2$$

$$\frac{\partial S(\{\theta\})}{\partial \theta} = \sum_{u=1}^{N} w_u \{Y_u - f(\{\xi_u\}, \{\hat{\theta}\})\} \left[\frac{\partial f(\{\xi_u\}, \{\theta\})}{\partial \theta_m} \right]_{\theta = \hat{\theta}}$$

Applying these formulae to the re-parameterized 4-P equation yields the following:

$$S(\{\theta\}) = \sum_{u=1}^{N} \left\{ Y_u - \frac{A}{(x_i^D + B)} + C \right\}^2$$

$$\frac{\partial S(\{\theta\})}{\partial \theta} = 2 \sum_{u=1}^{N} \left\{ Y_u - \frac{A}{(x_i^D + B)} + C \right\} \cdot \left[\frac{\partial f(\{\xi_u\}, \{\theta\})}{\partial \theta_m} \right]_{\theta=\hat{\theta}}$$

More specifically,

$$\frac{\partial S(\{\theta\})}{\partial A} = 2 \sum_{u=1}^{N} \left\{ Y_u - \frac{A}{(x_i^D + B)} + C \right\} \cdot \left[\frac{\partial f(\{\xi_u\}, \{\theta\})}{\partial A} \right]_{\theta=\hat{A}}$$

$$\frac{\partial S(\{\theta\})}{\partial C} = 2 \sum_{u=1}^{N} \left\{ Y_u - \frac{A}{(x_i^D + B)} + C \right\} \cdot \left[\frac{\partial f(\{\xi_u\}, \{\theta\})}{\partial C} \right]_{\theta=\hat{C}}$$

$$\frac{\partial S(\{\theta\})}{\partial B} = 2 \sum_{u=1}^{N} \left\{ Y_u - \frac{A}{(x_i^D + B)} + C \right\} \cdot \left[\frac{\partial f(\{\xi_u\}, \{\theta\})}{\partial B} \right]_{\theta=\hat{B}}$$

$$\frac{\partial S(\{\theta\})}{\partial D} = 2 \sum_{u=1}^{N} \left\{ Y_u - \frac{A}{(x_i^D + B)} + C \right\} \cdot \left[\frac{\partial f(\{\xi_u\}, \{\theta\})}{\partial D} \right]_{\theta=\hat{D}}$$

where the caps ($\wedge$) above the parameters indicate that they are the current estimates of the parameters. That is, the gradient equations are based on a comparison of the y value predicted from the parameters and the experimental y value.

Now, note from the discussion immediately above on the derivatives for the Marquardt matrix

$$\frac{\partial F}{\partial A} = 1/(\xi + B)$$

That is, $\partial F/\partial A$ can be used in place of the denominator of the gradient derivatives. Substituting this information into the gradient derivatives gives a form that has a nice programming feature: Once the partial derivative with respect to A has been calculated, it can be re-used several times. Performing fewer calculations makes program execution faster. Implementing this substitution and simplifying the nomenclature a bit, the third and fourth gradient equations take on the following form:

$$G_3 = \sum_{u=1}^{N} \left(y_i - A \cdot \frac{\partial F}{\partial A} - C \right) \frac{\partial F}{\partial B}$$

$$G_4 = \sum_{u=1}^{N} \left(y_i - A \cdot \frac{\partial F}{\partial A} - C \right) \frac{\partial F}{\partial D}$$

where, $\partial F/\partial B$ and $\partial F/\partial D$ were also calculated as part of the Marquardt matrix. Re-using the values of these two derivatives provides another program execution time savings.

Now, let us begin to discuss the macros in Figure 8-4. Although these macros are not actually part of Marquardt's Compromise, they are required to initialize an environment for optimal performance of Marquardt's Compromise. These macros will also teach you some important programming concepts that you will want to use for your applications.

AUTO-EXECUTING MACROS

You can program Lotus 1-2-3® and Symphony® to automatically start a macro running when the worksheet that contains the macro is retrieved. This special macro is the one that you would normally create whenever you start programming a new application.

A macro program that automatically executes when a spreadsheet is retrieved is called an *auto-executing macro*. Auto-executing macros are very commonly used to perform tasks that **MUST** be accomplished **BEFORE** the template or other macros in the spreadsheet can be used successfully. For example, auto-executing macros are frequently used to place starting values into key cells of the template.

However, the most common and important use of auto-executing macros is to coordinate the process of creating an environment that can be used by your program. For example, before a Symphony add-in application program can be used, it must be attached. This task is a prime candidate for an auto-executing macro. Other prime candidates are tasks that are required to start up user menus, start up the main program, etc.

These processes are referred to as *initialization processes* and will be explained in detail in the next few sections. The example auto-executing macro programs shown in this chapter are auto-executing macros that aid in configuring a program to run under both Lotus 1-2-3 and Symphony spreadsheets. The sections will also give you a brief review of some of the concepts of Chapter 7.

DIFFERENCES IN LOTUS 1-2-3 AND SYMPHONY AUTO-EXECUTING MACROS

Lotus 1-2-3 and Symphony handle most macros identically. However, auto-executing macros are handled in slightly different manners by the two spreadsheets. For example, Lotus 1-2-3 requires that an auto-executing macro be given a special, unique name (\0). Symphony allows any legitimate macro name for the auto-executing macro as long as you tell Symphony the name.

You can include both macros in the same worksheet if you like and the two macros will not interfere with each other. That is, if you have Symphony and have specified the name AUTOEXEC as the auto-executing macro in the spreadsheet's Settings, Symphony will ignore the \0; while Lotus 1-2-3 will only recognize \0.

If you are writing programs that are designed to run within both Lotus 1-2-3 and Symphony spreadsheets, including two different auto-executing macros in the same spreadsheet is a good way to ensure portability.

As you can tell from the previous discussions in this book, the two Lotus spreadsheets have minor differences in the way that some of the menu commands are issued; but macro commands and @functions are identical. Therefore, if you can detect which Lotus spreadsheet is using a macro program, you can issue the appropriate menu commands. In other words, you can exploit the differences in Lotus 1-2-3 and Symphony auto-executing macros to implement a system that will allow a spreadsheet to be independent of whether it is working under Lotus 1-2-3 or Symphony. This system, in turn, allows you to share your program with colleagues or upgrade from Lotus 1-2-3 to Symphony without performing an extensive rewrite. This flexibility will allow the spreadsheet to be moved back and forth from one Lotus program to the other.

With this system, you may need to re-define a few graphics or print settings, re-specify the name of the auto-executing macro (if stored in Lotus 1-2-3 and used in Symphony), rename the file extension (e.g., from ".wk1" to ".wr1" when you go from Lotus 1-2-3 to Symphony), etc., but you will not have to change the program.

To utilize this portability, you must first make certain that the auto-executing macro for Lotus 1-2-3 is named '\0 and the auto-executing macro for Symphony has another, different name. Once you have done so, you can use either of two different methods to customize subsequent macro programs. These methods are shown in Figure 8-4 and are described in more detail below.

Because of the slight differences in the two Lotus programs, auto-executing macros are separated by section. The next section describes the auto-executing macro for Lotus 1-2-3 and the following section describes the auto-executing macro for Symphony. Again, the difference in auto-executing macros is a special case. With the "portability" outlined above, most of the macros that you create will translate directly back and forth between the two spreadsheets.

LOTUS 1-2-3 AUTO-EXECUTING MACROS

To make an auto-executing macro for Lotus 1-2-3, just give the macro the special name \0 (backslash zero). Assigning this name to a macro causes Lotus 1-2-3 to automatically execute it as soon as it retrieves the worksheet.

The auto-executing macro for this example is shown in Figure 8-4. A line in the auto-executing macro places a number into the cell called VERSION. The command that accomplishes this placement is the {LET VERSION,0} command. This information will be used later to determine which set of menu commands to issue, thereby achieving program portability.

The other {LET} commands in \0 change the program directly. That is, they change the contents of the cells named INIT__REG and REG__CMD. These cells are in the line of flow of the INIT__REG and REGRESS programs of Figures 8-7 and 8-8. This method of direct modification is another way to achieve program portability and will be discussed in more detail later.

To create the example macro, start by typing the text as you see it in Figure 8-4. Before you can use any macro, you must assign it a formal name and tell Lotus 1-2-3 where the macro begins. Assigning a formal name to the macro will activate the macro. To activate the macro, move the cell-pointer to cell P6, issue the /Range Name Label Right (/RNLR) command, and press Return. Store the spreadsheet.

SYMPHONY AUTO-EXECUTING MACROS

The format of a Symphony auto-executing macro is shown in Figure 8-4. This macro is a bit more elaborate than its Lotus 1-2-3 equivalent because it must check if the Lotus STAT.app application program has already been attached before it can invoke it. As you know, although regression analysis is a permanent part of Lotus 1-2-3, the STAT.app program must be attached to Symphony before it can be used.

The two @functions in the auto-executing macro (@ISERR and @APP) work together to determine whether the STAT application program is attached. The @APP requires two pieces of information. The first is the application program's name (STAT, within quotation marks); the second is a text string (which could be used for an error message to the user). @APP normally returns the text string if the application has already been attached; if not, it returns an "ERR". Because the second argument is irrelevant, the example uses empty quotation marks to satisfy the @function's requirements.

The @APP is nested within @ISERR. @ISERR will translate an ERR returned from @APP into a "true" (that is, a one). If an ERR is not returned by @APP, then @ISERR will return a "false" (zero). Thus, the value returned from @APP is translated into something that an {IF} command can understand.

If you recall from Chapter 7, an {IF} command allows a macro program to perform a specific task based on the result of a TRUE/FALSE test. If the result of the expression within the {IF} command does not have the numeric value of zero, Lotus considers the expression to be TRUE and the macro will execute the instructions in the same cell, immediately after the {IF} command. If the result of the expression within the {IF} command has the numeric value zero, Lotus considers the expression to be FALSE and the macro will continue down to the cell below the {IF} command.

Whenever you want to call a second macro from an {IF} test, enclose the name of the second macro within brackets ({ }). For example, to call ADD__STAT, type {ADD__STAT} after the {IF} test. The {IF} command in this example calls upon the ADD__STAT macro if the argument specified in the command is TRUE. By doing so, the STAT application program is attached by the ADD__STAT macro. If the argument to the {IF} test is FALSE, then STAT has already been attached and the program just skips down to the next command line.

The three {LET} commands in this macro achieve the same goals as their corresponding commands in the Lotus 1-2-3 \0 auto-executing macro. However, they customize the cells with Symphony-specific information. The next section will discuss these commands in more detail.

You can create a similar macro program for your application by modifying the example macro for your particular needs. To create the example macro, start by typing the text as you see it in Figure 8-4. It is very important that there be at least one blank row between two macros because a blank row signals the end of a macro to Symphony.

When you create your own auto-executing macro, the name can be a single quote-backslash ('\) followed by a letter or a zero (\0) can be used for an auto-executing macro only). Alternatively, the name can be a string of text characters (without the backslash). When choosing a name for a macro, be as descriptive as possible. For this reason, I chose the name AUTOEXEC. To use it, you would type AUTOEXEC into cell P12. (Naming the macro AUTOEXEC would still allow you to run the macro with the USER key (F7) anytime during your session.)

Before you can use the AUTOEXEC and ADD__ STAT macros, you must assign them formal names (the names in Column P). These names tell Symphony where the macros begin. Assigning a formal name to a macro will activate the macro. Name a macro by issuing the / Range Name Label Right (/RNLR) command. When prompted for the range of labels, press ESCape, use the arrow keys to move to the cell containing the AUTOEXEC, type a period (.), and arrow down to highlight both the AUTOEXEC and the ADD__ STAT cells. Press Return.

Symphony needs to know more about the AUTOEXEC macro before it can use AUTOEXEC for auto-execution. Specifically, Symphony needs to know that the macro called AUTOEXEC is the one to run when the spreadsheet is retrieved. To achieve the flexibility that allows a user to specify names other than \0 for auto-executing macros, Lotus requires you to place the name of an auto-executing macro into the spreadsheet's settings. Do so by issuing the {Services} Settings Auto-Execute Set command and typing AUTOEXEC. Store the spreadsheet. (NOTE: If the spreadsheet had been stored as a Lotus 1-2-3 file and is now being used as a Symphony spreadsheet, the Auto-Execute specification would be lost and you would need to re-set the specification.)

PROGRAM PORTABILITY

As stated previously, the two spreadsheets have minor differences in the way that some of the menu commands are issued; but the macro commands and @functions are identical. Therefore, if you can detect which Lotus spreadsheet is using a macro program, you can issue the appropriate menu commands. The {LET} commands in each of the two auto-executing macros of Figure 8-4 help you to do just that.

At this point, Lotus 1-2-3 uses '\0 as an auto-executing macro and Symphony is configured to use AUTOEXEC as its auto-executing macro.

As explained above, depending on the auto-executing macro that is activated, a {LET VERSION,0} or {LET VERSION,1} command will place a 0 or 1 into the cell named VERSION, indicating Lotus 1-2-3 and Symphony, respectively. By testing for a zero or a one in the VERSION cell with {IF} commands, the program can decide which macro subroutines to execute or which sets of menu commands to issue. For example, you could use the following two lines in a program to issue the menu command to begin execution of the regression program:

```
        :
{IF VERSION=0}/DRG
{IF VERSION=1}/RRG
        :
```

The other {LET} commands in these macros achieve the same goals. However, they change the program directly. That is, the appropriate auto-executing macro will customize cells that lie in the path of macro execution. After the initialization macro completes execution, these cells will contain spreadsheet-specific commands that will execute when encountered by the program. In this example, they change the contents of the cells named INIT__REG and REG__CMD. These cells are in the line of flow of the INIT__REG and REGRESS programs of Figures 8-7 and 8-8. As you can see, /RRXX__PREDICT⁻ and /RRG have been placed into these cells. These commands indicate that the program was started in Symphony and that the correct commands will be issued when appropriate.

STARTING MENUS FROM AUTO-EXECUTING MACROS

A feature that you will want to include in your program is one that allows users to retrieve a spreadsheet and be automatically presented with a menu of all functional options available in a program. A program that automatically takes care of all details for the user without the user having knowledge of them is extremely user friendly.

The following discussion describes how this system works:

After VERSION has been updated and the program lines have been modified, the {MENUBRANCH MENU1} command displays a custom menu called MENU1 in the control panel. A custom menu looks and works just like the ones that Lotus 1-2-3 or Symphony presents to you when you press the / key.

Each menu option has its own capsule description (help line) that is displayed when the menu option is highlighted with the menu-pointer. Each menu option is also tied to a macro routine. When a user selects a menu option, the corresponding macro routine executes.

You can execute a menu in one of two different ways. The first way is to use the {MENUCALL menu__name} command; the second is to use the {MENUBRANCH menu__name} command. These two commands are similar. They both pause macro execution, go to a specified location, and display the menu at the location.

However, each command treats a menu differently. {MENUCALL} treats the menu like a subroutine. That is, it returns to the original macro when the subroutine instructions in the menu have been completed. With {MENUBRANCH}, the macro "jumps" to the menu. Execution terminates in the original macro and total control transfers to the menu. Then, the menu executes. When the last instruction in the menu is encountered, macro execution will terminate (unless another branching command is encountered).

As you can see, a significant difference exists between the two menu commands. This dissimilarity can be important because Lotus limits the number of subroutine levels that you can call from within a program. This concept was introduced for macros in Chapter 7. Let us review this concept and apply the concept to menus.

If a macro calls another macro as a subroutine, Lotus is supporting two levels of subroutines. Likewise, if a second subroutine calls a third subroutine, Lotus is supporting three levels, and so on. The number of subroutine levels that either Lotus 1-2-3 or Symphony can support (without returning to the first macro level) is 31.

If you exceed the subroutine limit by calling a macro from itself and/or by trying to nest too many subroutines, the macro aborts and an error message appears. Because a {MENUCALL} acts just like a subroutine call, the same problem would occur if you tried to repeatedly re-use a menu from a {MENUCALL}.

For the reason outlined above, it is advisable to execute the {BRANCH} command whenever using a macro and the {MENUBRANCH} command whenever using a menu.

MENUS

Figure 8-5 is an example of a menu. Note that this figure has an expansion of the column width to expose all of the text. MENU1 is a simple menu. Take a moment to examine this menu so that you will know how to construct a menu for your own applications.

All custom menus are structured with the same basic format. The first row of cells defines the options that will appear when the menu is displayed. The text for the first option must be in the cell specified in the {MENUBRANCH} or {MENUCALL} command. The text for the other options in the menu must be in the same row and in the columns immediately to the right of the column that contains the first option. If you want to be able to select a menu option by pressing the first letter of a prompt, you must designate the first letter of each option as a unique upper-case letter, digit, or symbol from the keyboard.

Option prompts can be of variable length. Symphony and Lotus 1-2-3 automatically place two empty spaces between prompts. The length of the prompts plus spaces must not exceed 80 characters. If the total length exceeds 80 characters, the spreadsheet truncates all entries longer than a specific length; the length depending on the number of prompts used.

Your menu can have a maximum of eight selections. If you try to use more than eight selections, the spreadsheet will ignore them. If you have fewer than eight selections, the cell to the right of the last menu choice must be empty. An empty cell signals the end of a menu list to Lotus.

The row immediately below the options contains the prompt's help message. The menu displays a help message when an option is highlighted. The longest possible help message is 75 characters. The spreadsheet will ignore all additional characters when a message is displayed.

The cells below a help message contain the instructions that the menu choice should execute if a user makes that particular choice. These cells may contain macro commands, a subroutine call, a branch to a macro, a branch to another menu, or a branch to the same menu.

CALLING SUBROUTINES FROM MENUS

As was just stated, the cells below help messages of a menu may contain subroutine calls. A subroutine call from a menu is handled in the same fashion as a subroutine call from a macro program: Control of the program is temporarily shifted to the subroutine, the subroutine carries out the tasks it was programmed to perform, and then control is returned to the menu.

The "4P" menu choice in MENU1 is an example of this concept. This menu choice calls the MAIN macro program as a subroutine. Notice the method that is used to call the subroutine. In the example, after the MAIN program has completed its execution, program control returns to the menu. The {MENUBRANCH} command at the end of this menu choice is required for continuation of the program. Without it, the program would terminate. When the program executes the {MENUBRANCH} it branches back to the menu and displays the menu again so that the user can make another selection.

BRANCHING FROM MENUS

Some of the other menu choices show you another method that you can use to program a menu. The SPLINE menu choice shows you another way to execute a macro from within a menu. SPLINE uses the {BRANCH} command.

If you use the {BRANCH} command to execute a macro, you must have a {MENUBRANCH} command at the end of the macro. This command is needed to return to MENU1 because if {MENUBRANCH} were omitted, the program would terminate after the macro was completed.

QUITTING MENUS

Always include a Quit menu choice at the end of every menu. This choice allows a user to abort the menu's execution. One of the most important reasons for using a spreadsheet is that it allows a user to explore, manipulate, graph, perform statistics, and perform unique operations on data. A Quit menu choice allows a user to utilize these capabilities.

The {QUIT} command that appears under the example menu choice terminates all macro and menu execution and returns the user to the spreadsheet.

THE MAIN PROGRAM

Now, let us examine the program that performs Marquardt's Compromise. Program structure was discussed in Chapter 7. As part of that discussion, the concept of a *main program* was introduced and described as the central control for the remainder of a program. The main program will be returned to whenever the tasks in other modules have been completed. The MAIN program in Figure 8-6 is a good example of this concept.

If you note, the MAIN program calls upon several subroutines, branches, issues commands directly, etc. It also contains a mechanism to return program control to MENU1 (c.f., cell AO29). The style of using small *modules* from which to build a program is a very important one to follow because it is more efficient, reliable, and easier to troubleshoot if the need arises.

Let us begin by examining the flow within the main program in more detail because it blueprints Marquardt's Compromise. After moving the cell-pointer to cell A1, the program calls upon SET__STDS and INIT__4P. These two macros are shown in Figure 8-7 and are described in more detail below.

The SET__STDS macro begins by completing an important function. It determines the number of data points in the curve (NUM__STDS) using the @COUNT function. This determination provides the specification for the number of times each of the many {FOR} loops throughout the program will be repeated. The determination thereby allows re-use of the template for standard curves (or experiments) of varying sizes. The SET__STDS macro then takes the concentrations of the standards and converts them to their natural logarithms.

The INIT__4P macro initializes the B and D parameters and the lambda multiplication factor for the matrix diagonal. As noted earlier, because Marquardt's Compromise uses a Steepest Descent algorithm to provide quick, front-end narrowing towards the convergence point, the initial values of B and D can be crude. Therefore, the starting values for B and D are chosen as one-half, the highest concentration, and one (i.e., hyperbolic), respectively. The subscripts "0" for MID__0 and EXPO__0 are used because archived values for MID and EXPO will be required for comparisons later in the program. The value for the matrix diagonal multiplier (LAMBDA) is set to the conservatively small value of 0.001. This value setup means that the entire matrix will have nearly equal weighting for the first iteration.

After these initialization procedures have been done, the SET__RNGS macro of Figure 8-7 customizes the sizes of the RESPONSE and X__PREDICT named ranges, based on NUM__STDS. These customizations are important because the ranges are used by the Lotus linearization program and errors will occur if the sizes are set incorrectly.

The final initialization tasks specify these named ranges in the Lotus regression program and perform an initial regression to get the starting values for A and C (using the INIT__REG macro of Figure 8-7). The process begins at the cell named INIT__REG (one of the cells modified by the auto-executing macros) and proceeds to redefine the X, Y, and Output ranges. These three ranges need to be redefined only once during the program's execution, so the process is performed at the beginning of the main program and not in the REGRESS macro. This method speeds macro execution and saves time.

To perform the initial regression analysis, REGRESS is called. This initial regression provides starting values for the linear parameters (A and C). The REGRESS macro (Figure 8-8) contains the REG__CMD cell with the Symphony-specific Go command (/RRG) customized by the auto-executing macro. This command is equivalent to /Range Regress Go.

Take a moment to examine the method used to provide portability in INIT__REG and REG__CMD. If you recall, portability could have been achieved using a pair of {IF} tests and the information in the VERSION cell to issue the appropriate GO command. However, this alternate scheme is much slower than customizing the cell once and issuing the command directly.

After the current sum of squares error value is archived, the Marquardt matrix is prepared (MAKE__ MTX), and solved (SOLVE__ MTX). (See Figures 8-9 and 8-10, respectively.) The solution provides correction factors for *B* and *D* (called DELTA__ B and DELTA__ D). After these corrections are applied to *B* and *D* using the MOD__ PARAMS macro, regression is performed again using REGRESS to find the new values of *A* and *C*.

After a check is made to see if the combined corrections to *B* and *D* are small enough to quit the iteration process, the current sum of squares is compared to the previously archived sum of squares. If the current sum of squares is smaller, lambda is decreased by a factor of ten (giving less importance to the diagonal of the matrix), MID__ 0 and EXPO__ 0 are adjusted, and the program returns to FULL __ ITER to prepare a new Marquardt matrix for another iteration. If the current sum of squares is larger than the previous one (i.e., error is increasing), then lambda is made larger by a factor of ten (giving more importance to the diagonal) and the iteration continues at PART__ ITER using the previous Marquardt matrix. This second category of iteration, therefore, differs from the previous one because the multiplication factor for the diagonal is **increased** to give the diagonal increased importance in the solution and the matrix is **not** recalculated.

Now that you have an overview of the program's flow, let us consider the inner workings of some macro subroutines.

INITIALIZING VARIABLES

Figure 8-7 shows macros that perform preliminary initializations. Let us look at each in turn.

The SET__ STDS macro uses the @COUNT function to determine the number of standards in the curve. The @COUNT function counts the number of non-blank cells in a range. The @COUNT function ignores blank cells, but includes text and error values. The @COUNT function, therefore, is a very handy way to determine how many elements there are in a range. The number returned by @COUNT can be subsequently used to specify the number of repetitions to be performed by {FOR} loops. For example, cells A5..A19 comprise the range called CONC. However, only cells A5..A16 are filled. Therefore, 12 would be returned. If the next curve only had cells A5..A10 filled, a 6 would be returned, and so on.

The second program line in SET__ STDS illustrates this customization. Due to a quirk in the way that the {FOR} command increments the loop counter and tests it, one must be subtracted from @COUNT. Based on this information, the {FOR} command repeatedly calls LOG__ LOOP to convert the concentrations of the standards to their natural logarithms and uses a {PUT} command to place the Ln value into the template (cells H5..H16). The test in the LOG__ LOOP macro checks to see if the standard value is zero. If it is, the Ln is set to a very small value (because the Ln of zero will produce an error).

The INIT__ 4P macro initializes *B*, *D*, lambda, and an archive system for *B* and *D*. The @MAX function determines the maximum concentration in the CONC range (A5..A19). In this way, the standard curve does not need to be entered in any particular order. For example, the 50 in cell A15 could have just as easily been placed in cell A10.

For the reasons explained in the last section, the archive systems for *B, D* and LAMBDA are initialized to half the highest concentration, 1, and 0.001, respectively. The archive values are then copied to *B* and *D*.

The SET__RNGS macro requires some preliminary setup in the template. That is, before you can adjust a range, it must have been previously created. Note the {ESC} command in the two program lines. If RESPONSE and PRED__VAR had not been previously created, the {ESC} command would back up one menu and prompt for the range name again. However, if the ranges are given preliminary definitions (e.g., F5..F19 and G5..G19, respectively) the {ESC} command would anchor the first cell of the range and continue with the {DOWN} command to highlight the appropriate area. Note the appearance of the NUM__STDS in the {DOWN} command. Also note the tilde (~). This character means "Press Return" in Lotus Command Language.

The INIT__REG macro uses these two ranges as the *X* and *Y* ranges of a linear regression. If you recall from the section on "Four-Parameter Logistic Curve Fitting" the re-parameterized Rodbard equation allows the *A* and *C* parameters to be calculated using linear Least Squares. That is, the re-parameterized equation is in the following form:

$$y_i = \theta_0 + \theta_1 \xi_i$$

with

$$\xi_i = (x_i^D + B)$$

The values of ξ_i will later be placed in the predictor variable range (PRED__VAR) by the REGRESS macro (Figure 8-8). However, at this point, the INIT__REG macro merely defines PRED__VAR and RESPONSE as the *X* and *Y* ranges for the Lotus regression program, respectively. INIT__REG also defines the Output range as REG__DATA. This range is shown in Figure 8-8.

The INIT__REG macro then quits the Lotus regression program. The REGRESS macro that is called by INIT__REG is used by other macros in the program. REGRESS assumes that the regression menu has not been previously accessed. The Quit in INIT__REG therefore returns the status of the regression menu to the same point that is anticipated by REGRESS.

The call to REGRESS at the end of INIT__REG performs a preliminary regression on the PRED__VAR and RESPONSE ranges. This preliminary regression yields first guesses at the linear parameters (*A* and *C*).

LINEAR REGRESSION BASED ON PREDICTOR VARIABLES

The REGRESS macro begins by updating the predictor variables. As noted in the previous section, the non-linear parameters (*B* and *D*) and x_i are combined to form a set of predictor variables. (For a discussion of the difference between linear x_i and nonlinear ξ_i, see "Standard Nonlinear Nomenclature and Error Analysis" section at the beginning of this chapter.) The following equation is used for the conversion:

$$\xi_i = (x_i^D + B)$$

or

$$y_i = C + A\xi_i$$

As noted in the previous section, the predictor variables are used as the X range and the RESPONSE (observed fluorescent values) are used as the Y range for the Lotus regression program. The Lotus regression program returns several useful pieces of information. The form of the line was defined in the previous section as

$$y_i = \theta_0 + \theta_1 \xi_i$$

which means that C is the same as the "Constant" returned by the Lotus regression program and A is the same as the "X Coefficient". Cells BR20 and BQ26 have been given the names INTERCEPT and SLOPE, respectively.

The Lotus regression program also returns some very useful error analysis information. The "Std Err of Y EST" (STDERR) is converted to a sum of squares and displayed in the template of Figure 8-2 as ERROR. STDERR is cell BR21. ERROR, in turn, is used at various points in the MAIN program to assign values to the previous and current sum of squares cells. These cells are later used in the detection of the error trend, which in turn sets the next value for the lambda diagonal multiplier.

The R Squared value (R__VALUE) is also displayed in the template of Figure 8-2 and is another useful error statistic. R__VALUE is cell BR22.

PREPARING THE MATRICES

Figures 8-9 and 8-10 prepare and solve Marquardt matrices. The process begins at MAKE__MTX. MAKE__MTX starts by using {BLANK location} commands to erase the contents of A__VALS (DB11..DB20, Figure 8-11), GRAD__3, and GRAD__4. In effect, this erasure is equivalent to placing zeros into each of the alpha positions of the Marquardt matrix and Rows three and four of the gradient matrix. This process is required because it initializes the partial derivative summations. If you recall, summations are performed on the partial derivatives with respect to each parameter for each (x_i, y_i) data pair.

An alternative way to accomplish this zeroing process is to use the following set of commands:

```
MAKE_MTX    {FOR INC1,0,9,1,INIT_MTX}
            {LET GRAD_3,0}
            {LET GRAD_4,0}
INIT_MTX    {PUT A_VALS,0,INC1,0}
```

However, this second approach is much slower and does not improve reliability.

After the matrix terms have been zeroed out, the matrix is filled. At the onset, it should be pointed out that although the matrix is 4 x 4, only 10 of the elements will actually be used. In the interest of speed, the other six elements will not be calculated. It should also be pointed out that because the partial derivative of F with respect to C ($\partial F / \partial C$) is one, the multiplication of the partial derivatives at alpha positions (1,2), (2,3), and (2,4) reduce to the partial derivatives with respect to A, B, and D, respectively.

MAKE__MTX forms the partial derivative summations using a {FOR} command to repeatedly call SET__UP__MTX. For each increment of the {FOR} command, the current value of $Ln(x_i)$ is used to calculate X^D using the DO__X__TO__D macro. This value is then used to solve the partial derivative equations for parameters A, B, and D. (Recall, $\partial F / \partial C = 1$.) For a review of the equations used in SET__UP__MTX, see the previous section entitled "Four-Parameter Logistic Curve Fitting."

After the partial derivatives are calculated, the pertinent alpha positions can be computed using the equations given in the section entitled "Marquardt's Compromise" (above). If you recall, the calculations are based on the following equation:

$$\alpha_{jk} = \frac{\partial \hat{y}}{\partial \theta_j} \cdot \frac{\partial \hat{y}}{\partial v_j}$$

Note that the lambda factor is not applied within this loop. It will be applied when the matrix is solved in SOLVE__MTX. Also note that subscripts for the alpha positions are:

$$
\begin{array}{lll}
\text{Linear:} & 1 = A \\
& 2 = C \\
\text{Nonlinear:} & 3 = B \\
& 4 = D
\end{array}
$$

For example, ALPHA__23 is a multiplication of the partial derivatives with respect to C and B and would therefore be

$$1 * (-A * (B + x_i{}^D)^{\wedge} - 2$$

Only two elements of the gradient matrix are actually required for the solution of the problem, the ones at row positions three and four. (Recall that corrections are not needed for A and C, because their new values are determined from linear regression.) Calculating only GRAD__3 and GRAD__4 saves time.

The GRAD__3 and GRAD__4 gradient elements are the partial derivatives of $S(\theta)$ with respect to B and D, respectively, and were derived in the previous section entitled "Four-Parameter Logistic Curve Fitting." The equations are (recalling the calculation time savings that can be achieved by substituting $\partial F / \partial A$ for the denominator, because they are equal)

$$\text{GRAD\_3} = \sum_{u=1}^{N} (y_i - A \cdot \frac{\partial F}{\partial A} - C) \frac{\partial F}{\partial B}$$

$$GRAD\_4 = \sum_{u=1}^{N} (y_i - A \cdot \frac{\partial F}{\partial A} - C) \frac{\partial F}{\partial D}$$

where, $\partial F/\partial B$ and $\partial F/\partial D$ were calculated as part of the Marquardt matrix. The DIF__TRM variable in SET__UP__MTX corresponds to the difference term in the parentheses of these two equations and is prepared by obtaining the current experimental response from the template (using the @INDEX) and subtracting the predicted response, (which, in turn, is based on the current set of parameters). Re-using DIF__TRM and the values of the derivatives provides program execution time savings.

MAKE__MTX ends by specifying the value of the Marquardt matrix at position (2,2). Because $\partial F/\partial C = 1$, the value at alpha position (2,2) is just the sum of the number of observations. That is, ALPHA__22 is equal to NUM__STDS.

SOLVING THE MATRIX

The SOLVE__MTX macro solves the Marquardt-Gradient matrices using algebraic techniques. Alternatively, you can use the Lotus matrix inversion and multiplication programs (i.e., the /DM (/Data Matrix) macro command for Lotus 1-2-3 or the /RM (/Range Matrix) macro command for Symphony). Instructions on how to implement a macro program that uses the Lotus matrix programs is given in Chapter 9.

Note that the SOLVE__MTX macro applies the lambda multiplication factor to scale the diagonal of the matrix. This application provides a progressive and smooth switch from the Method of Steepest Descent to the Linearization method, based on the trend of the error.

The two important outputs of SOLVE__MTX are the values for DELTA__B and DELTA__D. DELTA__B and DELTA__D are the correction factors for the current values of the midpoint and exponential terms (B and D, respectively). However, the corrections are not made within this macro. They are applied when MOD__PARAMS is called from the MAIN program. After B and D have been updated, REGRESS is called to update the linear parameters, A and C. After all four parameters have been updated, a test is made to see if convergence has been achieved. If the solution has converged, the iteration process terminates.

THE SCRATCH-PAD

The scratch-pad in this example program is divided into two parts. The first part contains the data that need to be presented to the user and appears as part of the template of Figure 8-2. The remainder of the scratch-pad is shown in Figure 8-11.

Because all of the named cells have been defined in previous sections, they will not be explained again. However, an explanation of the "__0" and "__1" subscripts are in order. For some of the variables, the distinction between a subscript "__0" and "__1" lies in whether the value refers to the current value of the term or the previous (archived) value. An example will explain this concept.

In the MAIN program, the current sum of squares is compared to the previous sum of squares to determine how to treat the lambda correction. SUM__SQRS__0 holds the previous sum of squares. SUM__SQRS__0 is archived by the {LET} command in the FULL__ITER cell of MAIN. The {LET} command at FULL__ITER transfers the value of ERROR just after it has been used. After the matrix is solved, etc., the new sum of squares is assigned to SUM__SQRS__1. This assignment allows the previous value (SUM__SQRS__0) to be compared with the current value (SUM__SQRS__1) and the trend detected accordingly. MID__0 vs. MID__1 and EXPO__0 vs. EXPO__1 are analogous situations.

Conversely, GRAD__3 and GRAD__4 refer to the third and fourth row of the gradient matrix, respectively. LAMBDA__1 is the archived value for lambda, LAMBDA__0 is the (1+lambda) term, and LAMBDA__2 is the square of LAMBDA__0. The latter two terms are therefore calculated from LAMBDA__1. LAMBDA__0 and LAMBDA__2 are used in SOLVE__MTX and have no archival significance.

CREATING THE PROGRAM

To create the program, type the text into your spreadsheet as you see it in Figures 8-2 and 8-4 through 8-10. Activate the program using the /Range Name Label Right command. In addition to the named ranges for the macros and scratch-pad, there are some other named ranges that need to be created. They are

Named range	Cells
A__VALS	DB11..DB20
INTERCEPT	BR20
CONC	A5..A19
DEG__FRED	BR24
LOG__CONC	H5..H19
PRED__VAR	G5..G16
RESPONSE	F5..F16
R__VALUE	BR22
SLOPE	BQ26
STDERR	BR21

Figure 8-12 shows a table that lists all of the required named ranges. It was created using the /Range Name Table command. You should issue this command, specify cell DL4 as the starting cell for the table, and compare your table with the one in Figure 8-12. This comparison will show if you have missed or incorrectly identified any named ranges. This check is important because accuracy in naming ranges is an absolute requirement for a program to execute properly. For example, if you were to have missed creating a named range that was used in a calculation within a {PUT} or {LET} command, the command would place the name into the specified cell, not the value of the named range cell. This erroneous placement, in turn, would lead to a program that would not converge because a cell that has a label in it would always evaluate to zero.

```
--------DL---------DM-------DN----DO---------DP-------D----DR---------DS----
 1                        TABLE OF NAMED RANGES
 2                        =====================
 3
 4  ADD_STAT    Q18           DERIV_D     DE6        PRED_VAR    G5..G16
 5  ALPHA_11    DB11          DIF_TRM     DE22       REGRESS     BO2
 6  ALPHA_12    DB12          DO_X_TO_D   CB29       REG_CMD     BO3
 7  ALPHA_13    DB13          D_PARAM     B25        REG_DATA    BO19
 8  ALPHA_14    DB14          ERROR       E23        RESPONSE    F5..F16
 9  ALPHA_22    DB15          EXPO_0      DB6        R_SQUARE    E26
10  ALPHA_23    DB16          EXPO_1      DB7        R_VALUE     BR22
11  ALPHA_24    DB17          FULL_ITER   AO21       SD_EST      E25
12  ALPHA_33    DB18          GRAD_3      DB22       SET_RNGS    BA23
13  ALPHA_34    DB19          GRAD_4      DB23       SET_STDS    BA4
14  ALPHA_44    DB20          HIGHEST     E24        SET_UP_MTX  CB10
15  AUTOEXEC    Q12           INC1        DE25       SLOPE       BQ26
16  A_PARAM     B22           INIT_4P     BA15       SOLVE_MTX   CO3
17  A_VALS      DB11..DB20    INIT_REG    BA28       STDERR      BR21
18  BFR_R       DE18          INTERCEPT   BR20       SUM_SQRS_0  DB25
19  BFR_S       DE19          LAMBDA_0    DE8        SUM_SQRS_1  DB26
20  BFR_T       DE20          LAMBDA_1    DE9        TEMP        DE24
21  BFR_U       DE21          LAMBDA_2    DE10       TMP_CONC    DE26
22  BIG_INC     AO35          LOG_CONC    H5..H19    UPDATE      BO12
23  B_PARAM     B23           LOG_LOOP    BA10       VERSION     DE27
24  CALC_X      BO17          MAIN        AO16       X_ONE       DE14
25  CONC        A5..A19       MAKE_MTX    CB3        X_TO_D      DE12
26  C_PARAM     B24           MENU1       AC2        X_TWO       DE15
27  DEG_FRED    BR24          MID_0       DB4        Y_ONE       DE16
28  DELTA_B     DB8           MID_1       DB5        Y_TWO       DE17
29  DELTA_D     DB9           MOD_PARAMS  CO19       \0          Q6
30  DERIV_A     DE4           NUM_STDS    E22        \M          Q20
31  DERIV_B     DE5           PART_ITER   AO23
32
```

Figure   8-12

FINDING CONCENTRATIONS FROM RESPONSE DATA IN THE TEMPLATE

There are basically two methods that you can use to convert observed data into concentrations. The first method is to just place active formulae into cells. With this method, the template is constructed in the same fashion as was shown in Chapter 3 for linear models. The following is the appropriate formula:

```
@EXP(@LN($A_PARAM/(B36-$C_PARAM)-$B_PARAM)/$D_PARAM)
```

As discussed in the section entitled "Creating the Template for this Example" in Chapter 3, a dollar sign will allow you to fix the column and row coordinates for the parameters so that a formula can be conveniently copied with no changes in the cell designations for the four parameters. The only required modification would then be a customization of the reference to the dependent variable cell.

However, cells with active formulae slow the recalculation of the spreadsheet. As discussed in Chapter 7, you can avoid this decline in speed by using a macro program to place values into the template. The program in Figure 8-13 and its associated template in Figure 8-14 will illustrate this concept. Note that the program in Figure 8-13 is similar to the program in Chapter 7 in that it has a pair of nested {FOR} loops and a {PUT} command.

```
--------DX-------DY-------DZ-------EA-------EB-------EC-------ED-------EE-------EF-------EG-------EH-------EI----
1
2             /*PROGRAM TO UPDATE THE DATA SECTION BASED ON 4-P*/
3                  /*PARAMETERS AND FLUORESCENT UNITS*/
4
5  \R        {DO_ROD} /*UPDATE THE RODBARD PARAMETERS*/
6            {LET NUM_COLS,@COUNT(FLUOR_SECT)/9} /*DETERMINE HOW MANY COLUMNS OF DATA*/
7            {FOR COL,0,NUM_COLS-1,1,DO_COL}    /*UPDATE COLUMNS*/
8            {CALC}                             /*FORCE SPREADSHEET RECALCULATION*/
9
10            /*CONVERT THE RE-PARAMETERIZED VALUES BACK TO RODBARD*/
11 DO_ROD    {LET A,A_PARAM/B_PARAM+C_PARAM}
12           {LET B,B_PARAM^(1/D_PARAM)}/*RODBARD EQUATION IS:*/
13           {LET C,C_PARAM}                       A-C
14           {LET D,D_PARAM}             Y = ------------ + C
15                                            1+(X/B)^D
16
17                 /*UPDATE A COLUMN, ONE ROW AT A TIME*/
18 DO_COL    {FOR ROW,0,8,1,DO_CELL}
19
20                 /*UPDATE A SINGLE CELL*/
21 DO_CELL   {PUT DATA_SECT,COL,ROW,@EXP(@LN(A_PARAM/(@INDEX(FLUOR_SECT,COL,ROW)-C_PARAM)-B_PARAM)/D_PARAM)}
22
23                 /*SCRATCH PAD*/
24 NUM_COLS     7 /*NUMBER OF COLUMNS IN THE FLUORESCENT DATA RANGE*/
25 ROW          9 /*BUFFER FOR ROW COUNTER*/
26 COL          7 /*BUFFER FOR COLUMN COUNTER*/
27
```

Figure   8-13

The program begins by calling DO— ROD. DO— ROD changes the re-parameterized coefficients back to Rodbard parameters and displays them in the template. These representations of the parameters are more conventional and are often easier for the user to interpret.

Next, the pair of {FOR} loops cycle through the fluorescent data range, convert the fluorescent data points to concentrations, and place the concentrations into the data section of the template. In this example FLOUR— SECT and DATA— SECT have been defined as B33..H41 and B44..H52, respectively.

Let us look a little closer at some of the features of this program. The @COUNT function in the {LET} command of \R shows one method (of many) to determine how many data points are within the FLOUR— SECT range. Determining the number of data points within a range at run time allows you to fill the range with varying numbers of data points. In this example, the information from the @COUNT is divided by 9 (the number of data points in a single column) and the result is stored in NUM— COLS for use by the column {FOR} loop.

A pair of {FOR} loops and a {PUT} command place the data into the spreadsheet. The key to understanding the interaction between these three components can be found in the {PUT} command:

```
{PUT DATA_SECT, COL, ROW, @EXP (@LN (A_PARAM/
            (@INDEX (FLUOR_SECT, COL, ROW)
               -C_PARAM) -B_ PARAM) /D_PARAM) }
```

```
--------A--------B--------C--------D--------E--------F--------G--------H----
 1                       FLUORESCENT DNA ASSAY DATA
 2 DATE RUN:31-OCT-88
 3                  ****STANDARD CURVE DATA****        AVERAGE
 4 NG DNA/ML ASSAY1     ASSAY2     ASSAY3     ASSAY4  RESPONSE PRED_VAR LOG_CONC
 5    31.2     81.0      86.0       81.8       85.1     83.5 0.000269 3.440418
 6    62.5    139.0     171.0      140.4      169.3    154.9 0.000264 4.135166
 7     125    323.0     309.0      326.2      305.9    316.0 0.000252 4.828313
 8     250    604.0     616.0      610.0      609.8    610.0 0.000230 5.521460
 9     500   1182.0    1106.0     1193.8     1094.9   1144.2 0.000193 6.214608
10    1000   1897.0    1839.0     1916.0     1820.6   1868.1 0.000140 6.907755
11
12
13
14
15
16
17
18
19
20 TABLE OF VALUES            STATISTICS              RODBARD PARAMETERS
21 ===============            ==========              ==================
22 A_PARAM   -1.4E+07         NUM_STDS         6      A          25.00179
23 B_PARAM   3656.029         ERROR      67.29376     B          1049.491
24 C_PARAM   3820.159         HIGHEST        1000     C          3820.159
25 D_PARAM   1.179422         SD_EST     5.800593     D          1.179422
26                            R_SQUARE   0.999971
27
28
29                  ****SAMPLE FLUORESCENCE DATA****
30 SAMPLE IDXX102-7  XX102-7   XX102-7    XX102-7    XX102-7  XX102-7  XX102-7
31 COMMENTS                    SAMPLE PRECISION STUDY
32
33 SAMPLE A1   428.8     410.1      413.6      409.3     428.8    433.5    423.7
34        2    433.5     407.2      394.3      418.1     410.1    407.2    405.7
35        3    423.7     405.7      416.5      410.5     413.6    394.3    416.5
36 SAMPLE B1   410.1     414.0      410.2      408.4     409.3    418.1    410.5
37        2    415.8     405.5      405.8      421.3     410.1    415.8    416.9
38        3    416.9     400.5      411.6      402.3     414.0    405.5    400.5
39 SAMPLE C1   401.9     418.7      410.1      424.5     410.2    405.8    411.6
40        2    406.4     415.8      412.2      416.0     408.4    421.3    402.3
41        3    408.9     420.4      408.9      418.9     408.9    412.2    418.9
42
43                  ****SAMPLE CONCENTRATION DATA****
44 SAMPLE A1   172.7     165.2      166.6      164.8     172.7    174.6    170.7
45        2    174.6     164.0      158.8      168.4     165.2    164.0    163.4
46        3    170.7     163.4      167.7      165.3     166.6    158.8    167.7
47 SAMPLE B1   165.2     166.7      165.2      164.5     164.8    168.4    165.3
48        2    167.5     163.3      163.4      169.7     165.2    167.5    167.9
49        3    167.9     161.3      165.8      162.0     166.7    163.3    161.3
50 SAMPLE C1   161.8     168.6      165.2      171.0     165.2    163.4    165.8
51        2    163.7     167.5      166.0      167.5     164.5    169.7    162.0
52        3    164.7     169.3      164.7      168.7     164.7    166.0    168.7
53
```

Figure   8-14

Note that the COL and ROW loop counters each appear twice in this command. They are first used to specify the column and row offsets for the @INDEX function. The @INDEX function locates and returns the fluorescence value based on these offsets. The COL and ROW loop counters are then used as offsets by the {PUT} command to place the returned value into the DATA__SECT range. In this way, a one-to-one correlation is maintained between the data in the two sections.

USING OTHER EQUATIONS

The program shown in Figures 8-4 through 8-11 can easily be modified to use other nonlinear curve fitting equations as long as they can be re-parameterized in such a way that they have two linear and two nonlinear parameters. Most nonlinear equations fit this category. For example, double exponential equations can be usually re-parameterized to the following form:

$$Ae^{-Bx} + Ce^{-Dx}$$

This form has two linear parameters (A and C) and two nonlinear parameters (B and D). This example equation can therefore be used in the program.

   If your equation fits into this category, you will need to adapt the program for your application. For example, you will need to

- re-parameterize the equation to yield two linear and two nonlinear parameters.
- determine the partial derivatives of the equation with respect to each of the parameters.
- modify, if necessary, the equation in LOG__LOOP to convert the concentrations to a form appropriate to your equation.
- replace the equation in CALC__X with your equation.
- replace the DERIV__ # equations in SET__UP__MTX with the partial derivatives for your equation.
- replace DIF__TRM in SET__UP__MTX with your equation.
- replace the equation in DO__X__TO__D with your equation.
- update the equations in MOD__PARAMS to coincide with your equation.
- add weighting factors (w_u), as needed.

If your equation cannot be re-parameterized to two linear and two nonlinear parameters, then you will need to make more extensive changes to the program. In addition to the above changes, you will also need to adjust the sizes of the Marquardt and gradient matrices, the SOLVE__MTX macro, etc. These modifications may seem like major undertakings, but really are not. The hard part in designing nonlinear programs is getting the program flow correct and Figures 8-4 through 8-11 already illustrate the correct flow.

   If you have no idea which equation to use, then you should refer to Bevington for a BASIC version or Press for a C version of a generic Marquardt method.

HOW FAST IS FAST?

All iterative nonlinear procedures are by no means instantaneous. Although Marquardt's Compromise is one of the most efficient iterative procedures and although this program makes use of many tricks to speed execution, the program will still take a longer time to complete than most other Lotus programs that you use. For example, it took 34 seconds for the data in Figure 8-2 to converge on my XT with a PCturbo 286e coprocessor board. (For a more complete discussion of execution time vs. computer configuration and version of Lotus, see Chapter 9.) These data are fairly well-behaved. Although many curves converge faster (16 seconds), longer times should be anticipated for

- more data points.
- "S" shaped curves that have extended "plateaus" at one or both ends of the range (i.e., large areas of "flatness" at the top or bottom).
- abrupt "bending" in the "S" just before or after the plateau.

With **very** nasty curves, Marquardt's Compromise can take several minutes to converge. So, just be patient when using this program. Rest assured that the program will converge to an accurate solution and that you will get your curve fitted.

LINEAR REGRESSION MACROS

You should create a macro program (called LIN__RGRS) to perform a linear regression on the CONC and RESPONSE data ranges of Figure 8-14. You already have most of the commands that you will need. Your tool box can be found in the AUTOEXEC, \0, SET__STDS, SET__RNGS, and INIT__REG macros of Figures 8-4 and 8-7. You should use the auto-executing macros to place the appropriate regression menu commands into the cell(s) of the macro. The customized commands in the INIT__REG and REG__CMD cells are examples of this concept.

You should modify MENU1 (Figure 8-5) to execute this macro program as the third menu option.

SUMMARY OF WHAT HAS BEEN ACCOMPLISHED

This chapter presented an overview of nonlinear curve fitting, its nomenclature, and how to choose a data fitting method. More importantly, the chapter presented a method that I feel to be the best iterative approach to curve fitting: Marquardt's Compromise. In doing so, the chapter introduced a number of programming techniques that you will find useful for your statistical or curve fitting applications. For example, the issues of program flow, auto-executing macros, attaching Symphony add-in programs, program portability, menus, solving matrices, and linear regression were discussed.

WHAT'S NEXT?

The next chapter is actually an extension of this chapter. It presents methods that you can use to estimate the values at points intermediate to known values. These methods are collectively known as Interpolation methods.

REFERENCES

NONLINEAR DATA FITTING

BARHAM, R. H., and DRANE, WANZER, "An Algorithm for Least Squares Estimation of Nonlinear Parameters When Some of the Parameters Are Linear," *Technometrics, 14* (1972), 757.

BEVINGTON, PHILIP R., *Data Reduction and Error Analysis for the Physical Sciences*, New York, NY.: McGraw-Hill Book Company, Inc., 1969.

HAMMING, R. W., *Numerical Methods for Scientists and Engineers*, 2nd ed., New York, NY.: Dover Publications, Inc., 1973.

KAHANER, DAVID, MOLER, CLEVE, and NASH, STEVEN, *Numerical Methods and Software*, Englewood Cliffs, NJ.: Prentice Hall, Inc., 1989.

LAWTON, WILLIAM H., and SYLVESTRE, EDWARD A., "Estimation of Linear Parameters in Nonlinear Regression," *Technometrics, 13* (1971), 461.

MARQUARDT, DONALD W., "An Algorithm for Least-squares Estimation of Nonlinear Parameters," *J. Soc. Indust. Appl. Math., 11* (1963), 431.

PRESS, WILLIAM H., FLANNERY, BRIAN P., TEUKOLSKY, SAUL A., and VETTERLING, WILLIAM T., *Numerical Recipes In C: The Art of Scientific Computing*, New York, NY.: The Press Syndicate of the University of Cambridge, 1988.

SEDGEWICK, ROBERT, *Algorithms*, 2nd ed., Menlo Park, CA.: Addison-Wesley Publishing Company, 1988.

FOUR-PARAMETER LOGISTIC EQUATION

DUDDLEY, R., EDWARDS, R., FINNEY, D., McKENZIE, I., RAAB, G., RODBARD, D., and ROGERS, R., "Guidelines for Data Processing," *Clin. Chem., 31* (1985), 1264.

FINNEY, D. J., "Radioligand Assay," *Biometrics, 32* (1976), 721.

HOOTON, J., GIBBS, C., and VERNER, P., "Interaction of Interleukin 2 with Cells: Quantitative Analysis of Effects," *The Journal of Immunology, 135* (1985), 2464.

RODBARD, DAVID, "Statistical Quality Control and Routine Data Processing for Radioimmunoassays and Immunoradiometric Assays," *Clin. Chem., 20* (1974), 1255.

RODBARD, DAVID, and McCLEAN, S. W., "Automated Computer Analysis for Enzyme-Multiplied Immunological Techniques," *Clin. Chem., 23* (1977), 112.

WALKER, W., "An Approach to Immunoassay," *Clin. Chem., 23* (1977), 384.

9

Interpolation

Approximation of data by polynomials is one of the oldest numerical analysis methods and remains one of the most heavily used as well. Approximation methods, collectively called *interpolation methods,* literally "read between the lines" of a table of data. That is, a table of data is divided into segments, each segment being the range between sequential, known data pairs. For each segment, a line passing through adjacent set(s) of known data points is determined and used to predict ("interpolate") values falling within the segment. Each segment's line will be a smooth function and will exactly match the (x,y) values used for the calculation of the line.

One common use of interpolation is to determine values from scientific or engineering tables. In this case, you have independent variable values (x) and you interpolate the values for the corresponding dependent variables (y). Conversely, interpolation can be used to determine x values (given y responses) from standard curves that cannot be conveniently or accurately described using regression or iterative techniques (as long as the standard curves always increase in value). Interpolation techniques are particularly useful for data curves that are asymmetric or "camel hump" shaped. In these cases, no reasonable function will fit very well and interpolation is the best approach.

This chapter describes the use of the three most common interpolation techniques: Linear, Cubic, and Spline. All three of these interpolation methods can be used in either direction (given x, find y; or given y, find x). Switching directions is quite easy and involves merely changing named range specifications in the macro program. A discussion of this process is provided at the end of this chapter.

Because linear equations are actually used in interpolation and because there is a bidirectionality to these equations, the variable names used in this chapter will be v

and z. Hopefully, this designation will help you to differentiate interpolation variables from the nonlinear variables of the previous chapter and the dependent/independent variable directionality of Chapter 7.

POINT-TO-POINT LINEAR INTERPOLATION

The *Point-to-Point Linear Interpolation* method is the "connect-the-dots" approach and is one of the easiest methods for fitting data. With the Linear Interpolation method, linear equations are determined for each segment. A test is performed to determine which segment is to be used and the corresponding equation is used to calculate the interpolated value.

The classical Lagrangian form of the equation used for Linear Interpolation is

$$v = \frac{(z - z_1)}{(z_0 - z_1)} v_0 + \frac{(z - z_0)}{(z_1 - z_0)} v_1$$

where, z_0, z_1, v_0, and v_1 are data points from a table and v is the value being interpolated from the segment between v_0 and v_1.

The advantage of the Point-to-Point Linear Interpolation method is that it is relatively simple to program. The disadvantages to the Point-to-Point method are that it

- requires replicates (e.g., triplicates) at each level to provide stability and protect against the disastrous affects of outliers.
- provides no statistics for the properties of the curve.
- makes it difficult to compare methods.

Figure 9-1 shows a template that will be used for the remainder of this chapter. Figure 9-1 is the same template as we have been using throughout this book, except that it has an enlarged standard curve section to accommodate more data points.

Figure 9-2 shows the menu that was prepared in Chapter 8. This menu can be used to provide a user friendly way to switch between curve fitting techniques.

Figure 9-3 shows a macro program for updating the data section using Linear Interpolation. In this example, fluorescent units correspond to z values and DNA concentration corresponds to v values. For a discussion of how to reverse the definitions of variables, see the section entitled "Switching the Independent and Dependent Variable Definitions" at the end of this chapter. To understand the template and macro program, begin by creating them.

1. Type the template column and row text headings into the appropriate cells. Into cell F67, type the following formula and copy it to cells F5..F28:

```
@IF(@ISERR(@AVG(B5..E5)),0,@AVG(B5..E5))
```

```
--------A--------B--------C--------D--------E--------F--------G--------H----
```

	A	B	C	D	E	F	G
1				FLUORESCENT DNA ASSAY DATA			
2	DATE RUN:31-OCT-88						
3		****STANDARD CURVE DATA****				AVERAGE	
4	NG DNA/ML	ASSAY1	ASSAY2	ASSAY3	ASSAY4	RESPONSE	OFFSET
5	0	81.1	83.4	78.6	81.8	81.2	0
6	100	86.1	88.6	83.4	86.9	86.3	1
7	200	139.1	143.2	134.8	140.4	139.4	2
8	300	171.1	176.1	165.9	172.7	171.5	3
9	400	279.3	287.4	270.6	281.8	279.8	4
10	500	389.4	400.7	377.3	392.9	390.1	5
11	600	604.5	622.1	585.9	610.0	605.6	6
12	700	852.8	877.6	826.4	860.5	854.3	7
13	800	1183.1	1217.5	1146.5	1193.8	1185.2	8
14	900	1610.5	1657.3	1560.7	1625.1	1613.4	9
15	1000	1859.1	1802.1	1991.9	1934.9	1897.0	10
16	1100	1881.6	1824.0	2016.0	1958.4	1920.0	11
17						0.0	12
18						0.0	13
19						0.0	14
20						0.0	15
21						0.0	16
22						0.0	17
23						0.0	18
24						0.0	19
25						0.0	20
26						0.0	21
27						0.0	22
28						0.0	23
29		****SAMPLE FLUORESCENCE DATA****					
30	SAMPLE ID						
31	COMMENTS						
32							

	A	B	C	D	E	F	G	H
33	1	60	393	726	1059	1392	1725	2058
34	2	97	430	763	1096	1429	1762	2095
35	3	134	467	800	1133	1466	1799	2132
36	4	171	504	837	1170	1503	1836	2169
37	5	208	541	874	1207	1540	1873	2206
38	6	245	578	911	1244	1577	1910	2243
39	7	282	615	948	1281	1614	1947	2280
40	8	319	652	985	1318	1651	1984	2317
41	9	356	689	1022	1355	1688	2021	2354
42								
43		****SAMPLE CONCENTRATION DATA****						
44	1	NA	503	646	754	919	792	NA
45	2	235	536	663	765	933	776	NA
46	3	209	558	679	777	943	785	NA
47	4	298	573	694	793	946	828	NA
48	5	393	584	707	811	940	916	NA
49	6	407	593	718	833	925	1055	NA
50	7	400	603	727	855	899	NA	NA
51	8	420	615	736	878	863	NA	NA
52	9	460	630	745	900	825	NA	NA
53								

Figure 9-1

```
--------AB----AC-------------------AD----------------AE----------------AF-----------------AG------------------AH-------AI---
```

	AB	AC	AD	AE	AF	AG	AH	AI
1				/*MAIN MENU FOR NON-LINEAR CURVE FITTING*/				
2	MENU1 4P		SPLINE	REGRESS	POINT	CUBIC	QUIT	
3		FOUR-PARAMETER	CUBIC SPLINE	LINEAR REGRESSION	POINT-TO-POINT	CUBIC INTERPOLATION	QUIT THE PROGRAM	
4		{MAIN}	{SPLINE}	{LIN_RGRS}	{P_TO_P}	{CUBIC}	{QUIT}	
5		{MENUBRANCH MENU1}	{MENUBRANCH MENU1}	{MENUBRANCH MENU1}	{MENUBRANCH MENU1}	{MENUBRANCH MENU1}		
6								

Figure 9-2

```
--------EK----------EL-------EM-------EN-------EO-------EP-------EQ-------ER-------ES-------ET----
1                        /*PROGRAM TO UPDATE THE DATA SECTION USING LINEAR INTERPOLATION*/
2
3  \M          {MENUBRANCH MENU1}          /*CALL UP THE CURVE FITTING MENU*/
4
5                        /*PERFORM LINEAR INTERPOLATION (POINT TO POINT)*/
6  P_TO_P     {LET NUM_COLS,@COUNT(RAW_DATA)/9}   /*DETERMINE HOW MANY COLUMNS OF DATA*/
7             {BLANK CONC_DATA}            /*REMOVE OLD DATA FROM CONCENTRATION SECTION*/
8             {FOR COL,0,NUM_COLS-1,1,DO_INT_COL} /*UPDATE COLUMNS*/
9             {CALC}                       /*FORCE SPREADSHEET RECALCULATION*/
10
11                       /*UPDATE A COLUMN, ONE ROW AT A TIME*/
12 DO_INT_COL {FOR ROW,0,8,1,DO_INT_CELL}
13
14                       /*UPDATE A SINGLE CELL*/
15 DO_INT_CELL {LET FL_UNITS,@INDEX(RAW_DATA,COL,ROW)}     /*GET CURRENT DATA FROM TEMPLATE*/
16             {IF FL_UNITS>@MAX(FL_U)#OR#FL_UNITS<LOWEST}{PUT CONC_DATA,COL,ROW,@NA}{RETURN}
17             {LET OFST,@VLOOKUP(FL_UNITS,FL_U,2)}        /*FIND OFFSET TO USE FOR EQN*/
18             {LET BUFFERA,(FL_UNITS-@INDEX(FL_U,0,OFST+1))*@INDEX(DNA,0,OFST)}
19             {LET BUFFERA,BUFFERA/(@INDEX(FL_U,0,OFST)-@INDEX(FL_U,0,OFST+1))}
20             {LET BUFFERB,(FL_UNITS-@INDEX(FL_U,0,OFST))*@INDEX(DNA,0,OFST+1)}
21             {LET BUFFERB,BUFFERB/(@INDEX(FL_U,0,OFST+1)-@INDEX(FL_U,0,OFST))}
22             {PUT CONC_DATA,COL,ROW,BUFFERA+BUFFERB}
23
24                       /*SCRATCH PAD*/
25 NUM_COLS        7 /*NUMBER OF COLUMNS IN THE FLUORESCENT DATA RANGE*/
26 ROW             9 /*BUFFER FOR ROW COUNTER*/
27 COL             7 /*BUFFER FOR COLUMN COUNTER*/
28 OFST           10 /*INTERPOLATION OFFSET FOR THE LOOKUP TABLE*/
29 FL_UNITS     2354 /*FLUORESCENT UNIT VALUE FOR CURRENT CELL*/
30 BUFFERA  24.40931 /*BUFFER FOR FIRST HALF OF THE EQUATION*/
31 BUFFERB  1073.149 /*BUFFER FOR THE SECOND HALF OF THE EQUATION*/
```

Figure   9-3

These formulae will allow for a partially filled template. If @AVG had been used alone, then ERRs would appear in all rows with no raw data. These ERRs would adversely affect the macro program. Therefore, using the @IF test can make a template more generic.

2. Type the standard curve and sample fluorescence data into the template. Make certain that you also include the OFFSET numbers in cells H5..H25. These data will be used as table lookup offsets by the macro program.

3. Type the macro program and menu into the spreadsheet.

4. Issue the /Range Name Create command to create the following ranges:

Range name	Cells
RAW_ DATA	B33..H41
FL _ U	F5..F28
DNA	A5..A28
CONC_ DATA	B44..H52
LOWEST	F5

5. Issue the /Range Name Label Right command and highlight all of the labels in Column EK to activate the program. Repeat for Column AB (if appropriate).

The template and macro program are now ready for use. Let us examine the program more closely because it demonstrates some techniques that you will find useful for automating other applications.

The program begins by determining the number of columns containing raw data. This technique is the same one used in Chapter 8, Figure 8-13. You may want to revise the technique or prompt the user for the number of columns and rows (using the {GETNUMBER} command) to customize the application for your use.

The two {FOR} loops work identically to their counterparts in Figure 8-13. The DO__INT__CELL macro fills the CONC__DATA cells with concentration data, based on the corresponding fluorescence unit data in the RAW__DATA section.

The DO__INT__CELL macro begins by placing a copy of the raw data from the appropriate RAW__DATA cell into the cell called FL__UNITS (in the scratch-pad). This technique is used because by using an @INDEX function only once, you can substantially increase the speed of your macro program. Next, a test is performed to ensure that the value of FL__UNITS is between the fluorescence units of the lowest and highest standards. This test is important because the Lagrangian formula is undefined outside of these ranges and would yield inaccurate values that would be placed into the template. If the fluorescence units are outside of the required range, an @NA is placed into the cell to alert the user of a potential problem. You can revise the {PUT} command to conform to your policy or skip the {PUT} command altogether, leaving the cell blank.

The cell called LOWEST is used in the {IF} test because in a partially filled template, some of the average cells will have zeros in them and zero will become the minimum value in the range. (There are many other ways around this problem, depending on your application. For example, you could re-set the range using the /RNC menu command and a {DOWN NUM} command, where NUM is based on an @COUNT of the number of data points in DNA.)

Next, the interpolation segment is determined using an @VLOOKUP function. This function returns the value of the corresponding cell in Column H and specifies the row offsets for the @INDEX functions that return the values for z_0, z_1, v_0, and v_1. The "2" in the @VLOOKUP specifies Column H. Because this macro program is built on an @VLOOKUP function and @VLOOKUP requires that identifiers be in increasing order (see Chapter 2), the table z_i values **MUST** therefore be increasing in magnitude (although the v_i values need not be). This requirement is common to most (if not all) interpolation programs.

The series of {LET} commands in DO__INT__CELL use these offsets to perform the Lagrangian calculation. Two buffers are used. BUFFERA corresponds to the v_0 portion of the equation and BUFFERB corresponds to the v_1 portion. The {PUT} command at the end of DO__INT__CELL adds BUFFERA and BUFFERB and places the value into the appropriate cell of the template.

When all rows are filled, a {CALC} command is executed to force a recalculation of the spreadsheet.

CUBIC INTERPOLATION

The *Cubic Interpolation* method takes the Linear Interpolation method two steps further. It fits a cubic (third-order) polynomial to four consecutive points, with interpolation between the two center points. Cubic interpolation is usually more accurate than quadratic interpolation. This increase in accuracy is not only due to the fact that it is a higher order, but because it is more symmetrical about the range being interpolated. The following is the Lagrangian form of the equation used for Cubic Interpolation:

$$v = \frac{(z - z_1)(z - z_2)(z - z_3)}{(z_0 - z_1)(z_0 - z_2)(z_0 - z_3)} v_0$$

$$+ \frac{(z - z_0)(z - z_2)(z - z_3)}{(z_1 - z_0)(z_1 - z_2)(z_1 - z_3)} v_1$$

$$+ \frac{(z - z_0)(z - z_1)(z - z_3)}{(z_2 - z_0)(z_2 - z_1)(z_2 - z_3)} v_2$$

$$+ \frac{(z - z_0)(z - z_1)(z - z_2)}{(z_3 - z_0)(z_3 - z_1)(z_3 - z_2)} v_3$$

where, z_0, z_1, z_2, z_3, v_0, v_1, v_2, and v_3 are consecutive data points from a table and v is the value being interpolated from the segment between v_1 and v_2.

Figure 9-4 shows a macro program for updating the data section using Cubic Interpolation. This program uses the same template as the Linear Interpolation macro. To understand the macro program, begin by adding it to the spreadsheet. Type the text into the spreadsheet as you see it in Figure 9-4. Issue the /Range Name Label Right command and specify the labels in Column EK to activate the macro. Then, issue the /Range Name Create command to create a one-column table called TERMS. The appropriate cells to specify are EM39..EM42. As we shall see in just a moment, because there are now four parts to the equation, an increase in programming efficiency can be achieved by using a {PUT} command and a table to hold each of the equation parts.

As you can see, the program in Figure 9-4 is very similar to the Linear Interpolation program in Figure 9-3. Let us skip over the steps that are identical and proceed to the nuances. These nuances begin with the following command:

{LET C_OFST, @VLOOKUP (FL_DATA, FL_U, 2) -1}

One is subtracted from the lookup offset. One is subtracted because if you recall, the interpolation is between v_1 and v_2, not v_0 and v_1. @VLOOKUP will return the correct offset value for v_1. However, a streamlining of the program's {FOR} loops, @INDEXs, and {PUT} commands can be achieved if they start at zero. Therefore, to find v_0 a one is subtracted from the v_1 offset. As a by-product of this action, the program will also be easier to read and interpret and the subscripts will correspond directly to the equations in this text. Next, the following command executes:

{FOR TRM,0,3,1,ASSEMBLE}

```
--------EK----------EL--------EM-------EN-------EO-------EP-------EQ-------ER-------ES-------ET----
 1                    /*PROGRAM TO UPDATE THE DATA SECTION USING CUBIC INTERPOLATION*/
 2
 3 CUBIC       {LET NUM_C_COLS,@COUNT(RAW_DATA)/9} /*DETERMINE HOW MANY COLUMNS OF DATA*/
 4             {BLANK CONC_DATA}              /*REMOVE OLD DATA FROM CONCENTRATION SECTION*/
 5             {FOR C_COL,0,NUM_C_COLS-1,1,DO_CUB_COL}       /*UPDATE COLUMNS*/
 6             {CALC}                              /*FORCE SPREADSHEET RECALCULATION*/
 7
 8                       /*UPDATE A COLUMN, ONE ROW AT A TIME*/
 9 DO_CUB_COL  {FOR C_ROW,0,8,1,DO_CUB_CELL}
10
11                       /*UPDATE A SINGLE CELL*/
12 DO_CUB_CELL {LET FL_DATA,@INDEX(RAW_DATA,C_COL,C_ROW)}    /*GET CURRENT DATA FROM TEMPLATE*/
13             {IF FL_DATA>@MAX(FL_U)#OR#FL_DATA<LOWEST}{PUT CONC_DATA,C_COL,C_ROW,@NA}{RETURN}
14             {LET C_OFST,@VLOOKUP(FL_DATA,FL_U,2)-1}        /*FIND OFFSET TO USE FOR EQN*/
15             {FOR TRM,0,3,1,ASSEMBLE}                   /*ASSEMBLE THE VALUE*/
16             {IF @ISERR(@SUM(TERMS))}{PUT CONC_DATA,C_COL,C_ROW,@NA}{RETURN}
17             {PUT CONC_DATA,C_COL,C_ROW,@SUM(TERMS)}       /*PUT THE VALUE INTO TEMPLATE*/
18
19             /*ASSEMBLE THE CONCENTRATION VALUE*/
20 ASSEMBLE    {LET C_BUF,1}                           /*INITIALIZE BUFFER FOR VALUE*/
21             {FOR FACT,0,3,1,MK_TERM}                 /*MAKE ONE OF THE EQN TERMS*/
22             {PUT TERMS,0,TRM,C_BUF*@INDEX(DNA,0,C_OFST+TRM)}      /*PUT TERM IN BUFFER*/
23
24             /*MAKE ONE OF THE TERMS IN THE CUBIC EQUATION*/
25 MK_TERM     {IF TRM=FACT}{RETURN}     /*SKIP "-Zi" IF FACTOR SAME SUBSCRIPT AS TERM*/
26             {LET C_BUF,C_BUF*(FL_DATA-@INDEX(FL_U,0,C_OFST+FACT))}
27             {LET C_BUF,C_BUF/(@INDEX(FL_U,0,C_OFST+TRM)-@INDEX(FL_U,0,C_OFST+FACT))}
28
29                       /*SCRATCH PAD*/
30 NUM_C_COLS        7 /*NUMBER OF COLUMNS IN THE FLUORESCENT DATA RANGE*/
31 C_ROW             9 /*BUFFER FOR ROW COUNTER*/
32 C_COL             7 /*BUFFER FOR COLUMN COUNTER*/
33 C_OFST            9 /*INTERPOLATION OFFSET FOR THE LOOKUP TABLE*/
34 FL_DATA        2354 /*FLUORESCENT UNIT VALUE FOR CURRENT CELL*/
35 C_BUF      0.000170 /*BUFFER FOR CURRENT CUBIC TERM*/
36 TRM               4 /*BUFFER FOR CURRENT TERM BEING ASSEMBLED*/
37 FACT              4 /*CURRENT (Z-Zi)*/
38                       /*TABLE OF CUBIC INTERPOLATION TERMS*/
39          -12.9813 /*FIRST TERM*/
40          53.15330 /*SECOND TERM*/
41          1057.210 /*THIRD TERM*/
42                 0 /*FOURTH TERM*/
```

Figure   9-4

This command calls the ASSEMBLE macro four times, once for each part of the equation. ASSEMBLE begins by initializing a buffer (called C__BUF) to 1. All further mathematics performed on this buffer are either multiplications or divisions. Therefore, 1 is required as a neutral starting value. Next, ASSEMBLE repeatedly calls MK__TERM. If you recall from the above equations, a multiplication and division are made for each subscript that is not equal to the current one for v_i. For example, when $i=0$ the following needs to be performed for the numerator and denominator of the first term, respectively:

$$\text{Numerator: } (z - z_1)(z - z_2)(z - z_3)$$

$$\text{Denominator: } (z_0 - z_1)(z_0 - z_2)(z_0 - z_3)$$

For this reason, MK__TERM begins by issuing the following command to test whether the prerequisite condition exists:

{IF TRM=FACT}

If the condition exists, the macro returns to the calling {FOR} command, skipping the multiplication and division. For all other terms, the appropriate subtractions are made for the numerator and denominator and then the multiplication and division are performed, respectively.

After each of the terms have been appended to the equation, the equation is multiplied by v_i and returned to the TERMS table. Once ASSEMBLE has assembled the four parts of the equation, the DO__CUB__CELL macro tests the value to ensure that a valid value has been achieved. If not, an @NA is placed in the CONC__DATA cell and the macro returns to the calling {FOR} command. If a valid value has been achieved, a summation is performed on TERMS and the value is placed into the CONC__DATA cell.

It would be good practice for you to trace through the program and review the values that correspond to each of the subscripts in the @INDEX functions. That is, you should see which values in the various templates are being targeted and returned. Keeping the correct data table positions during the development of a macro program can become a very trying experience, as you shall see in the next section. However, with a firm understanding about how a program executes and some simple tools (provided at the end of this chapter), you can greatly simplify your programming efforts.

Cubic interpolation usually decreases error by an order of magnitude over Linear Interpolation, but not always (especially not with "S" shaped curves that have very flat plateaus at either end). As an enhancement to Cubic Interpolation, you could use higher degree polynomials. However, higher degree polynomials tend to fluctuate wildly. Because of the unpredictability in polynomial interpolation, a more sophisticated method (such as the Spline method) is recommended.

THE SPLINE METHOD

The *Spline method* is the mathematical, computerized version of a spline (flexicurve). A *spline* is a mechanical device that a draftsperson uses to draw aesthetically pleasing curves. The draftsperson does so by fixing a set of points (called *knots*) on the drawing and bending a flexible strip of wood or plastic around them. The draftsperson then traces the strip to produce the curve.

The spline method is an extension of the point-to-point and Polynomial Interpolation family. The curve is described by a cubic polynomial at any given point and is divided into several local regions tied together. The curve is forced to be continuous by smoothing the connections at the points where segments join.

There are $N - 1$ different cubic polynomials in the cubic spline method (where N is the number of values in the table). The polynomials describe the curve according to the following set of equations:

$$s_i(z) = a_0 + a_1 z_i + a_2 z_i^2 + a_3 z_i^3; \; i = 1, 2, \cdots, N - 1$$

with $s_i(z)$ defined as the cubic polynomial to be used in the interval between z_i and z_{i+1}. Creating a spline involves calculating the a_j coefficients using the following criteria:

- The spline must touch the knots.
- The spline must curve smoothly around the knots with no sharp bends or kinks.

Mathematically, the second criterion means that the first and second derivatives of the spline polynomials must be equal at the knots. These criteria give a total of $4N - 6$ equations in the $4(N - 1)$ unknown coefficients. The most convenient way of solving these equations is to prepare matrices and to solve the matrices by the method of Gauss elimination. The solution of these matrices is a matrix of the coefficients.

Because many spline conditions are redundant, typically not all of them are included in calculations. These omissions provide increased program speed.

To make macro programming easier, a canonical equation will be used to express the equations for the spline segments. If we define t as $(z - z_i)/(z_{i+1} - z_i)$, the spline equation can be expressed as

$$s_i(t) = tv_{i+1} + (1 - t)v_i + (z_{i+1} - z_i)^2((t^3 - t)p_{i+1} + ((1 - t)^3 - (1 - t))p_i)/6$$

where, p_i and p_{i+1} are the second derivatives at z_i and z_{i+1}, respectively. This form of the equation contains fewer unknown coefficients and expresses the coefficients in terms of the second derivative values (p). In the equation, either t or $(1 - t)$ is zero at the endpoints of the spline interval. Because the first derivatives of the spline segments must be equal at the endpoints, the values for p_i and p_{i+1} can be found. The first derivative of the above equation is

$$s_i(t) = r_i + (z_{i+1} - z_i)((3t^2 - 1)p_{i+1} + (3(1 - t)^2 - 1)p_i)/6$$

where, $r_i = (v_{i+1} - v_i)/(z_{i+1} - z_i)$. Making the first derivatives equal at the endpoint yields the following series of $N - 2$ equations:

$$(z_i - z_{i-1})p_{i-1} + 2(z_{i+1} - z_{i-1})p_i + (z_{i+1} - z_i)p_{i+1} = 6(r_i - r_{i-1}).$$

These equations form a simple tridiagonal matrix that is easily solved using Lotus® inversion and multiplication routines. The following is an example matrix for $N = 7$:

$$
\begin{vmatrix}
d_2 & u_2 & 0 & 0 & 0 \\
u_2 & d_3 & u_3 & 0 & 0 \\
0 & u_3 & d_4 & u_4 & 0 \\
0 & 0 & u_4 & d_5 & u_5 \\
0 & 0 & 0 & u_5 & d_6
\end{vmatrix}
\begin{vmatrix}
p_2 \\
p_3 \\
p_4 \\
p_5 \\
p_6
\end{vmatrix}
=
\begin{vmatrix}
w_2 \\
w_3 \\
w_4 \\
w_5 \\
w_6
\end{vmatrix}
$$

where,

$$u_i = z_{i+1} - z_i, \quad d_i = 2(z_{i-1} - z_{i-1}), \text{ and } w_i = 6(r_i - r_{i-1}).$$

Figures 9-5 through 9-7 show a macro program that uses the above equations to perform a spline fit on the data in Figure 9-1. To understand this program, start by creating it. Type the program into the spreadsheet. Issue the /Range Name Labels Right command and specify the labels in Column EV to activate the macro program.

```
--------EV---------EW-------EX-------EY-------EZ-------FA-------FB-------FC-------FD-------FE----
 1                 /*AUTO-EXECUTING MACRO FOR LOTUS 1-2-3*/
 2 \0              {LET VERSION,0}   /*IDENTIFY SPREADSHEET AS 1-2-3*/
 3
 4                 /*AUTO-EXECUTING MACRO FOR LOTUS 1-2-3*/
 5 AUTOEXEC        {LET VERSION,1}   /*IDENTIFY SPREADSHEET AS SYMPHONY*/
 6                 {IF @ISERR(@APP('STAT',''))}{ATTACH}/*SEE IF STAT.APP ATTACHED*/
 7
 8 ATTACH          {S}AASTAT~Q       /*ATTACH THE STATISTICS ADD-IN*/
 9
10                 /*PROGRAM TO UPDATE THE DATA SECTION USING CUBIC SPLINE*/
11 SPLINE          {LET NUM,@COUNT(DNA)}       /*DETERMINE THE NUMBER OF STANDARDS*/
12                 {BLANK CONC_DATA}         /*REMOVE OLD DATA FROM CONCENTRATION SECTION*/
13                 {LET NUM_S_COLS,@COUNT(RAW_DATA)/9} /*DETERMINE HOW MANY COLUMNS OF DATA*/
14                 {SET RNG}         /*INITIALIZE THE MATRIX RANGES*/
15                 {CR_BUF_MTX}      /*CREATE A BUFFER MATRIX WITH DERIVS*/
16                 {MULT_MTX}        /*MULTIPLY THE INVERTED SPLINE MATRIX BY THE W MATRIX*/
17                 {FOR S_COL,0,NUM_S_COLS-1,1,DO_SPL_COL}     /*UPDATE COLUMNS*/
18                 {CALC}                     /*FORCE SPREADSHEET RECALCULATION*/
19
20                 /*SET UP A BUFFER MATRIX AS A SCRATCH-PAD FOR INVERTING*/
21 CR_BUF_MTX {BLANK BUF_MTX}         /*BLANK OUT THE BUFFER MATRIX*/
22                 {FOR S_ROW,1,NUM-2,1,MK_D_MTX}
23                 {FOR S_ROW,0,NUM-2,1,MK_U_MTX}
24                 {FOR S_ROW,0,NUM-2,1,MK_R_MTX}
25                 {FOR S_ROW,1,NUM-2,1,MK_W_MTX}
26                 {INIT_SPL}        /*ZERO OUT THE SPLINE MATRIX*/
27                 {FOR S_ROW,0,NUM-3,1,TFR_D_VALS}     /*TRANSFER THE INFO TO THE SPLINE MATRIX*/
28                 {FOR S_ROW,0,NUM-4,1,TFR_U_VALS}
29                 {INVERT_MTX}      /*INVERT THE SPLINE MATRIX*/
30
31                 /*SET THE RANGES ON THE SPLINE, INVERTED, AND W MATRICES*/
32                 /*THIS IS REQUIRED BY LOTUS MATRIX HANDLING ROUTINES*/
33 SET_RNG         {BLANK SPL_MTX}{BLANK INV_MTX}{BLANK W_MTX}
34                 {GOTO}SPL_MTX~
35                 /RNCSPL_MTX~{BACKSPACE}.{DOWN NUM-3}{RIGHT NUM-3}~
36                 {GOTO}INV_MTX~
37                 /RNCINV_MTX~{BACKSPACE}.{DOWN NUM-3}{RIGHT NUM-3}~
38                 {GOTO}W_MTX~
39                 /RNCW_MTX~{BACKSPACE}.{DOWN NUM-3}~
40                 {HOME}
41
```

Figure   9-5

If you are using Symphony®, issue the {SERVICES} Settings Auto-Execute Set command and specify AUTOEXEC. To the right of the program, issue the /Range Name Create command to create the following ranges:

Range name	Cells
BUF__ MTX	EW3..FB26
SPL__ MTX	FF3..FX26
INV__ MTX	FF53..FF76
P__ MTX	FA4..FA26
W__ MTX	EZ4..EZ26

The starting position of the W__ MTX matrix is very important and **must** be offset by one cell from the top row of the BUF__ MTX. (Otherwise, a blank cell will appear within the matrix and the Lotus inversion program will abort.) The offset in the starting

```
---------EV---------EW-------EX-------EY-------EZ-------FA-------FB-------FC-------FD-------FE---------FF----
42
43                  /*THESE MACROS BUILD THE COMPONENTS FOR THE SPLINE MATRIX*/
44 MK_D_MTX         {PUT BUF_MTX,0,S_ROW,2*(@INDEX(FL_U,0,S_ROW+1)-@INDEX(FL_U,0,S_ROW-1))}
45
46 MK_U_MTX         {PUT BUF_MTX,1,S_ROW,@INDEX(FL_U,0,S_ROW+1)-@INDEX(FL_U,0,S_ROW)}
47
48 MK_R_MTX         {PUT BUF_MTX,2,S_ROW,(@INDEX(DNA,0,S_ROW+1)-@INDEX(DNA,0,S_ROW))/@INDEX(BUF_MTX,1,S_ROW)}
49
50 MK_W_MTX         {PUT BUF_MTX,3,S_ROW,6*(@INDEX(BUF_MTX,2,S_ROW)-@INDEX(BUF_MTX,2,S_ROW-1))}
51
52                  /*INITIALIZE THE SPLINE MATRIX*/
53                  /*LOTUS REQUIRES ALL CELLS IN MATRIX TO HAVE VALUES PRIOR TO INVERTING*/
54 INIT_SPL         {BLANK SPL_MTX}
55                  {FOR S_ROW,0,NUM-3,1,SPL_ROW}
56
57                  /*INITIALIZE A COLUMN OF THE SPLINE MATRIX*/
58 SPL_ROW          {FOR S_COL,0,NUM-3,1,SPL_COL}
59
60                  /*PLACE A ZERO INTO ONE OF THE CELLS OF THE SPLINE MATRIX*/
61                  /*LOTUS MATRIX-HANDLING ROUTINES REQUIRE FULL MATRICES*/
62 SPL_COL          {PUT SPL_MTX,S_COL,S_ROW,0}
63
64                  /*THESE MACROS TRANSFER THE VALUES FROM THE BUFFER MATRIX TO THE SPLINE MATRIX*/
65
66 TFR_D_VALS       {PUT SPL_MTX,S_ROW,S_ROW,@INDEX(BUF_MTX,0,S_ROW+1)}
67
68 TFR_U_VALS       {PUT SPL_MTX,S_ROW,S_ROW+1,@INDEX(BUF_MTX,1,S_ROW+1)}
69                  {PUT SPL_MTX,S_ROW+1,S_ROW,@INDEX(BUF_MTX,1,S_ROW+1)}
70
71                  /*THIS MACRO INVERTS THE MATRIX USING LOTUS MATRIX INVERSION PROGRAM*/
72 INVERT_MTX       {IF VERSION=0}/DMISPL_MTX~INV_MTX~      /*LOTUS 1-2-3 COMMAND*/
73                  {IF VERSION=1}/RMISPL_MTX~INV_MTX~      /*SYMPHONY COMMAND*/
74
75                  /*THIS MACRO MULTIPLIES THE INVERTED MATRIX BY THE W MATRIX*/
76 MULT_MTX         {IF VERSION=0}/DMMINV_MTX~W_MTX~P_MTX~  /*LOTUS 1-2-3 COMMAND*/
77                  {IF VERSION=1}/RMMINV_MTX~W_MTX~P_MTX~  /*SYMPHONY COMMAND*/
78                  {PUT BUF_MTX,4,0,0}                     /*SET P(0)=0*/
79                  {PUT BUF_MTX,4,NUM-1,0}                 /*SET P(N)=0*/
80
81                  /*SCRATCH PAD*/
82 NUM_S_COLS            7 /*TOTAL NUMBER OF DATA COLUMNS TO BE CONVERTED USING SPLINE*/
83 NUM                  12 /*TOTAL NUMBER OF STANDARDS*/
84 S_COL                 7 /*COLUMN COUNTER FOR {FOR} LOOPS*/
85 S_ROW                 9 /*ROW COUNTER FOR {FOR} LOOPS*/
86 VERSION               1 /*0=LOTUS 1-2-3; 1=SYMPHONY*/
87 S_BUF          1055.178 /*BUFFER FOR ASSEMBLING VALUE TO BE PLACED INTO SPREADSHEET*/
88 S2_BUF         -0.01523 /*BUFFER FOR SECOND HALF OF SPLINE EQUATION*/
89 SP_DATA            2354 /*CURRENT DATA VALUE FROM THE SAMPLE FLUORESCENCE DATA SECTION OF TEMPLATE*/
90 S_OFST               10 /*THE ROW OFFSET OF SP_DATA IN THE LOOKUP TABLE*/
91 T_VAL          0.565217 /*THE VALUE OF T IN THE SPLINE EQUATION*/
```

Figure   9-6

position of the P__MTX is also important because the values for P_0 and P_N are not calculated from the matrix calculations and need to be inserted directly. (Both values are equal to zero in a "natural" spline.) The sizes of SPL__MTX, INV__MTX, and W__MTX are unimportant as long as they are larger than two cells. This size requirement exists because a macro program automatically re-adjusts matrix sizes to correspond to the number of data points in the standards section of a template. However, when you create your own application, be careful to include enough room

```
--------EV---------EW-------EX-------EY-------EZ-------FA-------FB-------FC-------FD-------FE-----
91              /*UPDATE A COLUMN OF DATA*/
92 DO_SPL_COL {FOR S_ROW,0,8,1,PUT_SPLINE}
93
94              /*UPDATE A CELL USING CUBIC SPLINE FIT---BASED ON FLUORESCENT VALUES*/
95 PUT_SPLINE {LET SP_DATA,@INDEX(RAW_DATA,S_COL,S_ROW)}   /*GET CURRENT DATA FROM TEMPLATE*/
96           {IF SP_DATA>@MAX(FL_U)#OR#SP_DATA<LOWEST}{PUT CONC_DATA,S_COL,S_ROW,@NA}{RETURN}
97           {LET S_OFST,@VLOOKUP(SP_DATA,FL_U,2)}        /*FIND OFFSET TO USE FOR EQN*/
98           {LET T_VAL,SP_DATA-@INDEX(FL_U,0,S_OFST)}
99           {LET T_VAL,T_VAL/@INDEX(BUF_MTX,1,S_OFST)}
100          {LET S_BUF,T_VAL*@INDEX(DNA,0,S_OFST+1)}
101          {LET S_BUF,S_BUF+(1-T_VAL)*@INDEX(DNA,0,S_OFST)}
102          {LET S2_BUF,(T_VAL^3-T_VAL)*@INDEX(BUF_MTX,4,S_OFST+1)}
103          {LET S2_BUF,S2_BUF+((1-T_VAL)^3-(1-T_VAL))*@INDEX(BUF_MTX,4,S_OFST)}
104          {LET S_BUF,S_BUF+(@INDEX(BUF_MTX,1,S_OFST)^2*S2_BUF/6)}
105          {IF @ISERR(S_BUF)}{PUT CONC_DATA,S_COL,S_ROW,@NA}{RETURN}
106          {PUT CONC_DATA,S_COL,S_ROW,S_BUF}            /*PUT THE VALUE INTO TEMPLATE*/
107
```

Figure   9-7

```
--------EV-------EW-------EX-------EY-------EZ-------FA-------FB-------FC-------FD-------FE----
1                        BUFFER MATRIX
2           D_MTX    U_MTX    R_MTX    W_MTX    P_MTX
3      0                     5.025  15.90049                    0
4      1     116.3   53.125  1.882352 -108.108 -1.11760
5      2     170.4   32.075  3.117692  7.412039 0.411652
6      3     280.8  108.325  0.923147 -13.1672 -0.10477
7      4     437.25   110.3  0.906618 -0.09917  0.028152
8      5     651.7   215.55  0.463929 -2.65613 -0.00960
9      6     928.5    248.7  0.402090 -0.37103  0.002300
10     7    1159.2    330.9  0.302206 -0.59930 -0.00175
11     8    1518.15 428.175  0.233549 -0.41194  0.002620
12     9    1423.55   283.6  0.352609  0.714359 -0.00889
13    10     613.2       23  4.347826  23.97130 0.043205
14    11                                                       0
15    12
16    13
17    14
18    15
19    16
20    17
21    18
22    19
23    20
24    21
25    22
26    23
27
```

Figure   9-8

so that the matrices don't collide with each other or portions of a template or macro program. Remember—these matrices can become VERY large, because they will occupy the same number of rows AND columns as the number of z data points in your lookup table.

Figure 9-8 shows the nested BUF__ MTX/P__ MTX/W__ MTX arrangement for this example. The BUF__ MTX is actually a scratch-pad staging area for the d_i and u_i

values that will later be placed into SPL__MTX. Once SPL__MTX has been constructed and inverted (and the results placed in INV__MTX), INV__MTX will be multiplied by the values in W__MTX. The result of this multiplication will be placed in P__MTX. The values in P__MTX are the coefficients used for calculating the values to be placed into the CONC__DATA template.

The SPLINE macro is the main Spline program. After determining the number of cells being occupied in DNA and RAW__DATA, SPLINE calls SET__RNG. The SET__RNG macro resets the size of the SPL__MTX, INV__MTX, and W__MTX ranges. This adjustment is necessary because the Lotus matrix inversion and multiplication programs require N by N (square) matrices for inversion and matrices with the same number of rows for multiplication. Therefore, you must reset the sizes of the matrices or the macro will abort and an error message will be displayed. Note that NUM is used in the specification for the number of columns and rows to be used in the matrices. NUM was determined by @COUNT(DNA) in the main program.

Next, CR__BUF__MTX is called to create the buffer (scratch-pad) matrix and transfer the pertinent data to an initialized spline matrix. CR__BUF__MTX uses a series of {FOR} commands to perform the necessary calculations. These calculations are quite straightforward. The equations used were defined earlier in this section. Specifying fluorescence units as the z variables and DNA concentration as the v variables,

Matrix name	Equation
D__MTX	$d_i = 2(z_{i+1} - z_{i-1})$
U__MTX	$u_i = z_{i+1} - z_i$
R__MTX	$r_i = (v_{i+1} - v_i)/(z_{i+1} - z_i)$
W__MTX	$w_i = 6(r_i - r_{i-1})$

CR__BUF__MTX then calls INIT__SPL. The Lotus matrix inversion and multiplication programs require a full matrix. That is, the matrix program will abort if there are any unfilled cells in the matrix being manipulated. For this reason, INIT__SPL places a 0 into all cells of the spline matrix (SPL__MTX).

Once cells have been zeroed out, the D and U values can be placed into the matrix. The D values are placed down the diagonal of the matrix using the (S__ROW, S__ROW) designation in the {PUT} command. The U values are placed into the matrix at positions (S__ROW, S__ROW+1) and (S__ROW+1, S__ROW). Now that the spline matrix has been completed, it is inverted (by INVERT__MTX).

Next, the main program calls MULT__MTX to multiply the inverted spline matrix by the W matrix. The output of this multiplication is placed into P__MTX. The values in P__MTX can then be used to convert fluorescence units data (in RAW__DATA) into DNA concentrations.

This conversion is performed by the PUT__SPLINE macro in Figure 9-7. PUT__SPLINE is similar to DO__INT__CELL and DO__CUB__CELL because it uses an @VLOOKUP function to determine the offset to be used in a series of @INDEX functions. These @INDEX functions pick out the appropriate values from the matrices. Specifying fluorescence units as z variables and DNA concentration as the v variable, the sequence of calculations is

$$T\_VAL = (z - z_i)/(z_{i+1} - z_i) = (z - z_i)/u_i$$

$$S\_BUF = tv_{i+1} + (1 - t)v_i$$

$$S2\_BUF = ((t^3 - t)p_{i+1} + ((1 - t)^3 - (1 - t))p_i)$$

$$S\_BUF = S\_BUF + (z_{i+1} - z_i)^2*S2\_BUF/6$$

It is appropriate at this point to reiterate that Spline Interpolations can go in either direction. In the next section, I will show you how to convert the spline macro program so that the DNA concentration is the v variable and the fluorescence units are the z variables. But first, let us discuss some of the strengths and weaknesses of the spline fit.

The advantages of the Spline method are that it

- provides an excellent fit, particularly for asymmetric curves of unspecified shape and "camel hump" shaped curves.
- uses a natural response rather than a transformation.
- shows a good uniformity of variance throughout the range of values.

The disadvantages of the Spline method are that it

- can have ambiguity in the compromise between smoothness and Goodness-of-Fit.
- does not provide reliable descriptive statistics.
- has difficulty providing quality control parameters.

SWITCHING INDEPENDENT AND DEPENDENT VARIABLE DEFINITIONS

The above programs were based on fluorescence units being the z variables and DNA concentration being the v variable. However, converting programs to reverse this definition is quite straightforward. You merely exchange the range names in the programs (with the possible exception of the range name in the @COUNT function). For example, in the spline programs, you would exchange all references to FL__U with references to DNA and all references to DNA (except the one in the @COUNT(DNA) at the beginning of the program) with references to FL__U.

However, one other important change needs to be made. It is in the @VLOOKUP function of the PUT__SPLINE macro. The column that returns the offset must be re-specified. Because Column H is at offset 7 from the DNA range, this information also needs to be added. Figure 9-9 illustrates the changes necessary for the PUT__SPLINE macro.

Alternatively, you could merely switch the definitions of the FL__U and DNA ranges. To do so, you would first delete the ranges with the /Range Name Delete

```
--------EV---------EW-------EX-------EY-------EZ-------PA-------PB-------PC-------PD-------PE----
91              /*UPDATE A COLUMN OF DATA*/
92 DO_SPL_COL {FOR S_ROW,0,8,1,PUT_SPLINE}
93
94              /*UPDATE BASED ON SPLINE FIT---BASED ON DNA VALUES*/
95 PUT_SPLINE {LET SP_DATA,@INDEX(RAW_DATA,S_COL,S_ROW)}
96             {IF SP_DATA>@MAX(DNA)#OR#SP_DATA<@MIN(DNA)}{PUT CONC_DATA,S_COL,S_ROW,@NA}{RETURN}
97             {LET S_OFST,@VLOOKUP(SP_DATA,DNA,7)}
98             {LET T_VAL,SP_DATA-@INDEX(DNA,0,S_OFST)}
99             {LET T_VAL,T_VAL/@INDEX(BUF_MTX,1,S_OFST)}
100            {LET S_BUF,T_VAL*@INDEX(FL_U,0,S_OFST+1)}
101            {LET S_BUF,S_BUF+(1-T_VAL)*@INDEX(FL_U,0,S_OFST)}
102            {LET S2_BUF,(T_VAL^3-T_VAL)*@INDEX(BUF_MTX,4,S_OFST+1)}
103            {LET S2_BUF,S2_BUF+((1-T_VAL)^3-(1-T_VAL))*@INDEX(BUF_MTX,4,S_OFST)}
104            {LET S_BUF,S_BUF+(@INDEX(BUF_MTX,1,S_OFST)^2*S2_BUF/6)}
105            {IF @ISERR(S_BUF)}{PUT CONC_DATA,S_COL,S_ROW,@NA}{RETURN}
106            {PUT CONC_DATA,S_COL,S_ROW,S_BUF}
107
```

Figure   9-9

command and then re-assign the opposite names to the ranges with the /Range Name Create command. Next, update only the @COUNT and @VLOOKUP functions accordingly. This process could be performed at runtime as part of the macro program and would produce a high degree of flexibility. However, this extra capability is not normally needed and may make the program harder to interpret and debug, should a problem occur.

DETERMINING MATRIX OFFSETS

Keeping matrix offsets straight during programming can be a confusing and frustrating task. Matrix offsets can be especially baffling if you are using several subscripted variables that begin at different starting points. The PUT_SPLINE macro is an example of this problem. There are eight @INDEX functions that retrieve data in the PUT_SPLINE macro. A mistake in any of these @INDEX functions will lead to erroneous final results. Finding the @INDEX that is the culprit can be very a difficult task.

If you are having trouble with a program that uses @INDEX functions within {PUT} and/or {LET} commands, there is a simple method that can be an invaluable aid to identifying the problem. First, copy the {PUT} and {LET} commands to an open section of the spreadsheet. Then use the {EDIT} key (F2) to edit out all of the characters other than the actual @INDEX function. In the cell to the right of each @INDEX function, place the name of the variable that is being retrieved (e.g., p_{i+1}). Next, place a {CALC} command and a {GETLABEL "PAUSE",TXT} command at the end of the macro subroutine (creating a cell named TXT for the {GETLABEL}). The {CALC} command will cause a recalculation of the @INDEX functions and the {GETLABEL} command will pause program execution until you press Return.

Next, run the program and write down the values for each @INDEX function. Press Return when you are ready for the next set of values. Compare to the expected values. If you cannot pinpoint which datum is being withdrawn from the matrix (e.g., because the matrix contains repeating values), then you should put some fake data into the matrix, see when they are retrieved, and compare the time the data is retrieved with when you thought they should have been retrieved.

END CONDITIONS

Figures 9-10 through 9-12 illustrate the differences in the efficiency of the three interpolation techniques when working with "S" shaped curves. Figure 9-10 shows the entire curve and Figures 9-11 and 9-12 show the lower and upper regions, respectively. Depending on the type of curve being fit, there can be some substantial differences in the fitting accuracy of the three methods. If differences are found, they typically occur at the ends of curves (as illustrated in Figures 9-11 and 9-12).

For this reason, it is important to test all three interpolation methods with data that span the range of the curve that you plan to use and then choose the technique that fits the data best. Do so by filling the RAW__DATA range with values that extend from the lowest to the highest standard value. Use the /Data Fill (Lotus 1-2-3®) or /Range Fill (Symphony) command to fill the range. Choose an increment size that will place as many points along the curve as you can. Then convert the raw data using the three interpolation techniques. Copy the raw data values and the converted values to

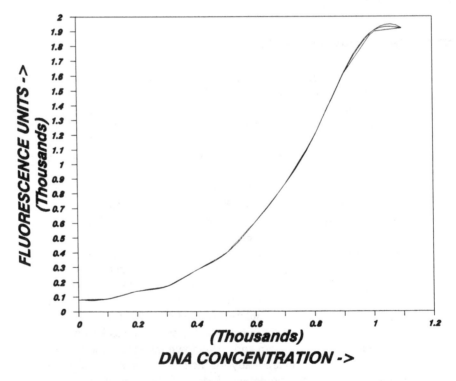

Figure   9-10

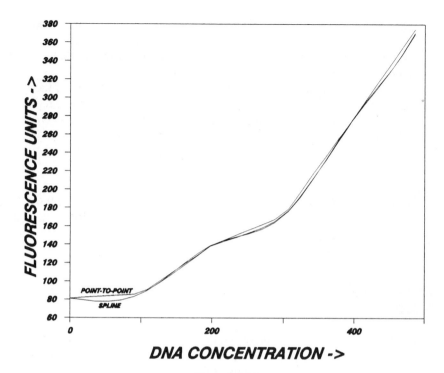

Figure   9-11

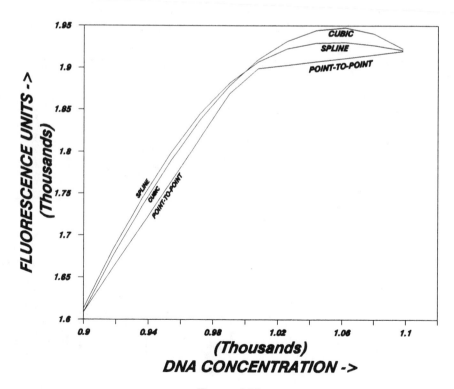

Figure   9-12

columns in an open portion of the spreadsheet. Next, use the graphics environment to plot curves corresponding to Figures 9-10 through 9-12.

If you find unacceptable fluctuation at one end or the other, there are a number of end condition procedures available to provide a better fit in these regions when using the spline method. The spline method program presented in this chapter is called a *natural spline* and defines

$$P_0 = 0$$

$$P_N = 0$$

using the {PUT} commands in the MULT__ MTX macro of Figure 9-6. This procedure is equivalent to assuming that the spline approaches linearity at the ends. For good discussions on other end conditions, smoothing methods, and end condition programming, see Gerald and Wheatley.

EXECUTION TIME VS. COMPUTER CONFIGURATION AND VERSION OF LOTUS

Execution time for the macro programs in Chapters 8 and 9 make a convenient way to compare various computer configurations (Table 9-1) and versions of Lotus (Table 9-2). Table 9-1 was generated using Lotus Symphony, Release 2.0, while Table 9-2 was generated using an IBM PC/AT running at 8 Mhz, with and without EMS memory. As you can see, there can be a substantial difference in the run time for these macro programs. For example, Lotus 1-2-3, Release 3, is by far the slowest program.

				LOTUS SYMPHONY (2.0) MACROS (SECONDS)							
Computer	XT	XT	XT	XT	XT	AT	AT	80	80	80	80
EMS?	YES	YES	NO	NO	YES	NO	YES	NO	YES	NO	YES
Math Coprocessor	NO	NO	NO	80387	80387	80287	80287	80387	80387	NO	NO
Coprocessor Board	NO	ORCHID	INBOARD	INBOARD	INBOARD	NO	NO	NO	NO	NO	NO
MAIN (4P)	204.58	33.63	23.49	23.26	30.74	43.77	60.00	20.64	27.37	20.87	27.79
P TO P	72	12.37	8.02	7.96	10.53	15.51	20.88	7.29	9.62	7.30	9.55
CUBIC	384.34	64.16	41.17	40.91	56.16	78.78	110.89	36.79	49.95	36.71	49.91
SPLINE	162.81	28.26	21.05	20.49	28.26	37.18	49.46	17.5	24.87	19.74	25.12

Table 9-1

	WITHOUT EMS MEMORY				WITH EMS MEMORY			
PROGRAM:	1-2-3	1-2-3	1-2-3	SYMPHONY	1-2-3	1-2-3	1-2-3	SYMPHONY
RELEASE:	2.01	2.2	3.0	2.0	2.01	2.2	3.0	2.0
		(SECONDS)				(SECONDS)		
MAIN (4P)	33.49	31.02	67.71	43.86	33.68	31.24	99.63	61.01
P_TO_P	28.32	26.63	48.45	39.44	28.67	27.17	81.34	51.38
CUBIC	13.07	12.31	26.43	15.37	13.28	12.47	39.84	20.93
SPLINE	67.28	63.73	146.88	78.76	68.06	64.77	229.36	109.66

Table 9-2

Also note that there is a substantial increase in execution time when EMS memory is added to both Lotus 1-2-3, Release 3, and Symphony, Release 2. Thus, choosing the correct computer configuration and Lotus program for your data reduction is very important and can have far-reaching effects on how fast your macro programs run and your spreadsheet recalculates.

SUMMARY OF WHAT HAS BEEN ACCOMPLISHED

This chapter presented three interpolation methods for fitting nonlinear curves. The following is a ranking of the methods in Chapters 8 and 9 in order of increasing complexity and increasing informativeness of the statistical analysis:

- Point-to-point Linear interpolation
- Cubic interpolation
- Cubic spline
- Iteration

The interpolation methods begin by finding an appropriate segment based on the value of the data point being interpolated. A line is then calculated for the segment and the predicted value is determined from the line. The choice of the segment is based on the Lotus @VLOOKUP function. Because @VLOOKUP identifiers must be in ascending order, the z_i coordinates of the curve must be ever-increasing (although the v_i portion need not be).

This chapter added some clever programming techniques to our growing repertoire. Of particular note was the building and solving of matrices. With the ability to access the Lotus matrix inversion and multiplication programs, solving matrices is genuinely simple and very fast.

Finally, a method was presented to help you locate offset problems in the @INDEX functions of the {PUT} and {LET} commands.

WHAT'S NEXT?

This chapter concludes our three chapter discussion of curve fitting. The next chapter will return to statistics. The statistical tests discussed in the next chapter relate to data that do not necessarily follow normal (Gaussian) distributions. The chapter also contains some more examples of macro programming, although much less sophisticated than those presented in this and the last chapter.

REFERENCES

GERALD, CURTIS, and WHEATLEY, PATRICK, *Applied Numerical Analysis*, *4th ed.* Reading, MA: Addison-Wesley Publishing Co., Inc., 1989.

HAMMING, R.W., *Numerical Methods for Scientists and Engineers*. New York, NY: Dover Publications, Inc., 1973.

10

Nonparametric Statistical Tests

Prior to this chapter, the emphasis was on parametric statistical techniques. These techniques

- require that the level of measurement attained on the collected data be in the form of an interval scale or ratio scale.
- involve hypothesis testing of specified parameters.
- require very stringent assumptions and are valid only if these assumptions hold.

Among these assumptions are

- the observations are independent of each other.
- the sample data are randomly drawn from a population that is normally distributed.
- if a comparison is being made, the samples are drawn from normal populations having equal variances.

Some parametric test procedures are said to be *robust* because they are relatively tolerant of slight violations in the assumptions and/or criteria.

However, not all assay data meet these criteria or assumptions well enough to provide valid results for parametric tests. For example,

- Some sets of data are not normally distributed.
- Some experiments are performed on a very limited number of samples (e.g., only 3 or 4 data points), so that the distribution of the data cannot be known.

In these situations, procedures that are free of distribution restrictions must be used. These techniques are referred to as *distribution-free* methods because they compare distributions without specifying their form. That is, this group of tests make no assumptions about the distribution of the observations.

Because the comparison is between distributions without the use of parameters of the normal distribution and because a comparison is not being made between parameters, these procedures are frequently called *nonparametric statistics*, though this term can be misleading. Nonparametric tests may be applied regardless of the kind of distribution the data exhibit and are, therefore, applicable to samples from any population. Such distribution-free tests are somewhat less efficient in some particular instances than tests that make full use of assumptions about the form of the population distribution; but they provide a quick and, in general, conservative means for data evaluation.

We have already encountered a nonparametric test, the chi-square test for Goodness-of-Fit. This chapter describes several other nonparametric tests. However, a large number of nonparametric tests exist and a book of this nature cannot cover all of them. Therefore, the emphasis of this chapter is to show you how to develop your own tools that will allow you to implement nonparametric testing templates of your own choosing.

THE RUNS TEST FOR RANDOMNESS

The Wald-Wolfowitz Runs test is used to determine whether a set of observations is distributed randomly in time or follows a pattern (e.g., a steady increase, cyclical, etc.). The set of data to be tested should contain from 10 to 50 values.

Figure 10-1 shows a revised version of Figure 5-7. The template in Figure 10-1 demonstrates how to perform the Runs test. The template utilizes the median and data (in order of occurrence) from Figure 5-7 (cells A78..A113). This template exhibits several techniques that you will need to know to implement your nonparametric tests, so let us take a moment to examine them.

Column I has formulae that identify whether the value in Column A is greater than, equal to, or less than the median. Although this column is not actually required for the calculation of the Runs test, it can be quite informative to the person reviewing the template. The formulae used in Column I are of the following format:

```
@IF(A78=$MEDIAN,"    0",@IF(A78>$MEDIAN,"    +","    -"))
```

This formula is a good example of nesting @IF functions to sequentially test for more than one condition. Using nested @IF functions, you can determine the appropriate value to be returned to the cell. The formula in this example begins by determining whether cell A78 is equal to the median. If it is, then a zero is returned for the cell. If it is not, then the second @IF test is performed. The second @IF function tests whether cell A78 is greater than the median. If it is, a "+" is returned to the cell; if it is not, a "−" is returned.

```
--------A--------B----------C------D------E--------F--------G--------H--------I--------J--------K--------L--------M-------
 73            ****SKEWNESS, KURTOSIS, MEDIAN, AND MODE ANALYSIS****
 74
 75        TEST FOR SKEWNESS                  TEST FOR KURTOSIS        ***********RUNS TEST ANALYSIS***************
 76 RAW DATA SORTED DATA OCCURENCES    DIFF   DIFF^2   DIFF^3   DIFF^4 HI/MED/LO  TALLY   RUN?   MORE?   LESS?
 77
 78   221.1    198.4    1.0 -         10.6    112.1   1186.3  12558.7     +       221.1    1      1       0
 79   224.2    202.5    1.0 -         13.7    187.3   2563.5  35084.9     +       224.2    0      1       0
 80   217.8    203.4    1.0 -          7.3     53.1    386.8   2818.3     +       217.8    0      1       0
 81   208.8    203.7    1.0 -         -1.7      2.9     -5.0      8.6     -       208.8    1      0       1
 82   212.6    205.8    1.0 -          2.1      4.4     18.9     18.9     +       212.6    1      1       0
 83   213.3    205.9    1.0 -          2.8      7.8     21.6     60.3     +       213.3    0      1       0
 84   203.4    206.0    1.0 -         -7.1     50.6   -360.0   2561.1     -       203.4    1      0       1
 85   206.4    206.4    1.0 -         -4.1     16.9    -69.6    286.4     -       206.4    0      0       1
 86   208.0    206.9    1.0 -         -2.5      6.3    -15.9     39.9     -       208.0    0      0       1
 87   208.8    207.7    1.0 -         -1.7      2.9     -5.0      8.6     -       208.8    0      0       1
 88   206.9    208.0    1.0 -         -3.6     13.1    -47.2    170.6     -       206.9    0      0       1
 89   205.9    208.0    2.0 -         -4.6     21.3    -98.2    453.2     -       205.9    0      0       1
 90   211.4    208.3    1.0 -          0.9      0.8      0.7      0.6     +       211.4    1      1       0
 91   205.8    208.8    1.0 -         -4.7     22.2   -104.7    493.8     -       205.8    1      0       1
 92   202.5    208.8    2.0 -         -8.0     64.2   -514.7   4124.5     -       202.5    0      0       1
 93   214.5    208.8    3.0 MODE       4.0     15.9     63.3    252.5     +       214.5    1      1       0
 94   212.6    208.9    1.0 -          2.1      4.4      9.1     18.9     +       212.6    0      1       0
 95   215.6    209.1    1.0 -          5.1     25.9    131.6    669.2     +       215.6    0      1       0
 96   211.1    209.8    1.0 -          0.6      0.3      0.2      0.1     +       211.1    0      1       0
 97   198.4    210.2    1.0 -        -12.1    146.7  -1777.7  21534.5     -       198.4    1      0       1
 98   213.0    211.1    1.0 -          2.5      6.2     15.4     38.2     +       213.0    1      1       0
 99   208.9    211.4    1.0 -         -1.6      2.6     -4.2      6.8     -       208.9    1      0       1
100   206.0    212.6    1.0 -         -4.5     20.4    -92.0    415.1     -       206.0    0      0       1
101   209.8    212.6    2.0 -         -0.7      0.5     -0.4      0.3     +       209.8    1      1       0
102   208.8    212.7    1.0 -         -1.7      2.9     -5.0      8.6     -       208.8    1      0       1
103   210.2    213.0    1.0 -         -0.3      0.1     -0.0      0.0     +       210.2    1      1       0
104   208.0    213.3    1.0 -         -2.5      6.3    -15.9     39.9     -       208.0    1      0       1
105   208.3    214.1    1.0 -         -2.2      4.9    -10.9     24.0     -       208.3    0      0       1
106   214.1    214.5    1.0 -          3.6     12.9     46.1    165.4     +       214.1    1      1       0
107   209.1    214.6    1.0 -         -1.4      2.0     -2.8      4.0     -       209.1    1      0       1
108   207.7    215.6    1.0 -         -2.8      7.9    -22.3     62.7     -       207.7    0      0       1
109   216.2    216.2    1.0 -          5.7     32.3    183.8   1045.3     +       216.2    1      1       0
110   203.7    217.8    1.0 -         -6.8     46.4   -316.4   2155.7     -       203.7    1      0       1
111   218.3    218.3    1.0 -          7.8     60.6    472.0   3675.2     +       218.3    1      1       0
112   212.7    221.1    1.0 -          2.2      4.8     10.4     22.8     +       212.7    0      1       0
113   214.6    224.2    1.0 -          4.1     16.7     68.2    278.8     +       214.6    0      1       0
114
115 AVERAGE    210.5            MOM1     MOM2     MOM3     MOM4              # RUNS #GREATER #LESSER
116 STD DEV      5.309           0.0     27.4     47.2   2475.2               19       18      18
117 SKEWNESS     0.3156
118 MEDIAN     209.45                  KURTOSIS   3.1147       ****RUNS TEST SUMMARY****
119                                    (K-3)      0.1147       LO_CUTOFF     12
120                                                            HI_CUTOFF     26
121                                                            STATUS   DON'T REJECT RANDOM HYPOTHESIS
122
```

Figure 10-1

However, spreadsheet analysis can mean more to you than just displaying plus and minus signs. If possible, you should strive to take advantage of the full power of the spreadsheet to perform the entire analysis, including both numerical summaries and a display of whether the test passes or fails.

To understand how to discern the methods and formulae to use for your statistical test, consider the following plan. Begin by reviewing how the statistical test is performed manually. Most statistical books list the following protocol to calculate the Runs test:

1. Arrange the data in order of magnitude and calculate the median.
2. List the data in order of occurrence.
3. Score a plus ("+") to values greater than the median, or a minus ("−") to values less than the median.
4. Ignore values equal to the median.
5. Identify each run. A run is a sequence of signs of the same kind bounded by signs of the other kind. (i.e., A run ends and the next one begins when there is a sign change, excluding values equal to the median.)
6. Count the number of runs, the number of +s, and the number of −s. Set n_1 equal to the smaller number of signs, and n_2 equal to the larger number of signs.
7. Compare the number of runs with the critical value in a table. If the number of runs is less than or greater than the range bounded by the table values for n_1 and n_2, reject the hypothesis that the data are random at the level of significance.

Your task is to mimic this procedure with your template, turning each step into a formula(e) that will provide the correct answer without error under all possible conditions. Therefore, a good place to start is by considering the definition of a run and all possible permutations that can exist. In the following sequence of numbers there are two runs:

$$+ , \ + , \ + , \ + , \ - , \ - , \ - , \ - , \ -$$

In the example, there were four positives and five negatives. All of the positives occurred together and all of the negatives occurred together. The set, therefore, has two runs: One of length four and one of length five. The following sequence has the same number of positives and negatives, but they are alternating:

$$- , \ + , \ - , \ + , \ - , \ + , \ - , \ + , \ -$$

In this example, there are nine runs, each of length 1. Likewise, the following sequence has three runs, two with a single negative and one with three positives:

$$- , \ + , \ + , \ + , \ -$$

This discussion is fairly straightforward. If only plus and minus signs were possible, it would be very easy to create a template. That is, you could create the following formula to test the previous value to see if it was of the opposite sign of the current value. If it was, then the test would be true (1); if not, the test would be false (0).

```
@IF((A79<$MEDIAN#AND#A78>$MEDIAN)
    #OR#(A79>$MEDIAN#AND#A78<$MEDIAN),1,0)
```

Note the parentheses, the #AND# operators, and the #OR# operator in the equation. The parentheses cordon off the testing into two units. Each unit contains its own test. If one **OR** the other unit is true, the @IF evaluates to true and a 1 is returned. On the other hand, if **BOTH** units are false, @IF will return a 0.

Taking this discussion to the unit level, **BOTH** tests **within** a parentheses **MUST** be true before a unit is true. That is, either of the following will cause @IF to be true and return a one:

- If the current cell (A79) is less than the median AND the previous cell (A78) is greater than the median.

OR

- If the current cell (A79) is greater than the median AND the previous cell (A78) is less than the median.

You could then use the @SUM function to add up all of the 1s and use the value to determine whether Runs test criteria are satisfied. However, a problem arises in Steps 4 and 5 of the manual procedure. That is, you must ignore values equal to the median. This criterion considerably muddies the waters. Consider the following possible scenarios and the correct scoring for an @IF test.

Sequence	@IF value
−, 0, −	0
−, 0, +	1
+, 0, +	0
+, 0, −	1

There is a temptation at this realization to create new formulae, each of which nest two @IF tests (which are analogs to the one shown above) into another @IF test. The new @IF test would begin by testing the current cell to see if it was equal to the median. If it was, then the process would back up one row so that the @IF test used the value of the previous cell and the one before that. Conversely, if the value of the current cell was not equal to the median, then the current cell and previous cell would be tested. For example,

```
@IF(A79=$MEDIAN,@IF((A78<$MEDIAN#AND#A77>$MEDIAN)
        #OR#(A78>$MEDIAN#AND#A77<$MEDIAN),1,0),
            @IF((A79<$MEDIAN#AND#A78>$MEDIAN)
                #OR#(A79>$MEDIAN#AND#A78<$MEDIAN),1,0))
```

But, what happens if there are two sequential cells that are equal to the median. Or three? Or ten? For example, consider the following scenarios and the correct scoring for an @IF test:

Sequence	@IF value
−, 0, 0, . . . , 0, −	0
−, 0, 0, . . . , 0, +	1
+, 0, 0, . . . , 0, +	0
+, 0, 0, . . . , 0, −	1

Clearly, multiple zeros are a problem because you cannot test for them, ad infinitum, with multiple-nested @IF tests. You therefore need to be a little clever and add a second, "staging" column to transform the data into a form that addresses the median problem. This realization and the resulting protocol are very important to the solution of many nonparametric problems. The protocol is not readily apparent. In spite of its seeming simplicity, the protocol took nearly three days of intensive thought to conceive.

The implementation is shown in Columns J and K in Figure 10-1. Column J starts with a formula in cell J78 that tests the first cell of the original data (cell A78) to see if they are equal to the median. If they are, then the value of the cell in the **next** row is used (cell A79). If not, then the value of the first cell is used. The following formula is the one used in this example:

```
@IF(A78=$MEDIAN,A79,A78)
```

Each of the subsequent cells in Column J have an equivalent to the following formula (starting at Row 79, not 78):

```
@IF(A79=$MEDIAN,J78,A79)
```

Thus, each succeeding cell tests to see if the corresponding original value (i.e., the value in the same row) is equal to the median. If it is, then the previous value in Column J is used. The previous value in Column A is **NOT** used. In this way, if several cells are equal to the median, the last "non-median" value will be carried along until a value is encountered that is not equal to the median. Whenever a value not equal to the median is obtained, the value of the original data is used.

Except for cell K78 (which contains a 1), the cells of Column K contain formulae to test the corresponding cells of Column J. The formulae used are the ones shown earlier in this section and originally thought to work before the issue of equality to the median was considered. These formulae are amended to refer to the values in Column J. That is,

```
@IF((J79<$MEDIAN#AND#J78>$MEDIAN)
   #OR#(J79>$MEDIAN#AND#J78<$MEDIAN),1,0)
```

As discussed earlier, this formula detects changes from minus to plus or from plus to minus. Because a series of cells that were originally equal to the median now repeat the previous non-median cell's value, that non-median cell's value is preserved for testing and the "median problem" is eliminated.

Create the example in Figure 10-1 as follows:

1. Type the following formula into cell I78 and copy it to cells I79..I113:

```
@IF(A78=$MEDIAN,"   0",@IF(A78>$MEDIAN,"   +","   -"))
```

2. Type the following formula into cell J78:

```
@IF(A78=$MEDIAN,A79,A78)
```

3. Type the following formula into cell J79 and copy it to cells J80..J113:

```
@IF(A79=$MEDIAN,J78,A79)
```

4. Type a 1 into cell K78.

5. Type the following formula into cell K79 and copy it to cells K80..K113:

```
@IF((J79<$MEDIAN#AND#J78>$MEDIAN)
    #OR#(J79>$MEDIAN#AND#J78<$MEDIAN),1,0)
```

6. Find the total number of runs by typing the following formula into cell K116:

```
@SUM(K78..K113)
```

Save the template and then test it. Type the median (209.45) into several cells of Column A. Make sure you test each scenario. For example, type 209.45 into cells A79, A82, A85, A86, A94, and A95 and see how the template reacts. You should also take a moment to review the importance of significant figures on the outcome of statistical tests by typing a 209.45000000001 into cell A87. Upon doing so, you will add 2 to the Runs test value, even though the last digit (1) is far beyond the numerical precision of the assay.

Retrieve the original, un-modified template.

Your next task is to automate a comparison of the number of runs detected by the above procedure to the cutoff values for the Runs test. Unlike most other statistical tests, the Runs test has two cutoff values and the determined value must be between these two values. If the determined value (r) is less than or equal to a lower critical value from one table or greater than or equal to an upper critical value from a second table, the randomness hypothesis is rejected at the significance level (5% for this example).

Again, start by considering what needs to be done manually. To perform the evaluation manually, you must count the number of runs, the number of $+$'s and the number of $-$'s. You then set n_1 and n_2 equal to the number of plus signs and minus signs, respectively. Next, compare the number of runs with critical values in the tables. If the number of runs is less than, or greater than the range listed, for n_1 and n_2, you would reject the hypothesis that the data are random at the level of significance.

Armed with this information, you can design the Lotus® equivalent of the manual method. In this case, you must add two columns to the template. Columns L and M generate the data needed to calculate the two n's for this example. Columns L and M each consist of a series of Lotus @IF tests to determine whether the cell in the original data column (A) meets the criterion of greater than (Column L) or less than (Column M) than the median. The formulae have the following format:

Column L	Column M
@IF(A78>$MEDIAN,1,0)	@IF(A78<$MEDIAN,1,0)

If the criterion in the @IF test is met, then a true (1) is returned. If not, then a zero is returned. A summation on each of these columns is made using the @SUM function (analogous to Column K).

To create the example, type the above formulae into Cells L78 and M78, respectively. Issue the /Copy command to copy cells L78..M78 to L79..L113. Then, copy cell K116 to L116..M116. This action will give summations for Columns L and M.

With the information in cells K116, L116, and M116, you can determine whether the Runs test benchmarks are met. As outlined above, unlike most other tests, two tables are required for the Runs test. The first table gives a low cutoff value and the second table gives a high cutoff. If the value determined is less than or equal to the critical value in the low cutoff table or greater than or equal to the critical value in the high cutoff table, the hypothesis that the data are random is rejected.

As always, you can either look up the values in tables found in statistics texts or automate the lookup using the tools developed in Chapters 2 and 4. If the template will be re-used repeatedly, then you should automate the process with lookup tables.

Figures 10-2 and 10-3 show typical lookup tables for this type of application. They are created by typing in the row and column identifiers (indexes) and the data into the tables. Using the /Range Name Create command, create ranges named RUNS_LO and RUNS_HI for cells V3..AO22 and AU3..BN22, respectively.

Using the /Range Name command, give cells K119 and K120 the names LO_CUTOFF and HI_CUTOFF, respectively. The formulae in these cells work analogously to the lookup functions for the GAUSS table in Chapter 4. That is, an @HLOOKUP is nested in a @VLOOKUP function. By placing an offset of zero in the @HLOOKUP, the number in the identifier (top) row of the table is returned.

```
--------V---W--X--Y--Z--AA-AB-AC-AD-AE-AF-AG-AH-AI-AJ-AK-AL-AM-AN-AO----
 1
 2                    CRITICAL  LOW  VALUES  FOR  THE  RUN  TEST
 3            2   3   4   5   6   7   8   9  10  11  12  13  14  15  16  17  18  19  20
 4       2                                    2   2   2   2   2   2   2   2   2
 5       3                2   2   2   2   2   2   2   3   3   3   3   3   3   3
 6       4            2   2   2   3   3   3   3   3   3   3   4   4   4   4   4
 7       5        2   2   3   3   3   3   4   4   4   4   4   4   5   5   5
 8       6    2   2   3   3   3   4   4   4   4   5   5   5   5   6   6
 9       7    2   2   3   3   3   4   4   5   5   5   5   6   6   6   6   6
10       8    2   3   3   3   4   4   5   5   5   6   6   6   6   7   7   7
11       9    2   3   3   4   4   5   5   6   6   6   7   7   7   7   8   8
12      10    2   3   3   4   5   5   5   6   6   7   7   7   8   8   8   8
13      11    2   3   4   4   5   5   6   6   7   7   7   8   8   8   9   9   9
14      12  2 2   3   4   4   5   6   6   7   7   7   8   8   8   9   9   9  10  10
15      13  2 2   3   4   4   5   6   6   7   7   8   9   9   9  10  10  10  10
16      14  2 2   3   4   5   5   6   7   7   8   8   9   9   9  10  10  10  11
17      15  2 3   3   4   5   6   6   7   7   8   8   9   9  10  10  11  11  11  12
18      16  2 3   4   4   5   6   6   7   8   8   9   9  10  10  11  11  11  12  12
19      17  2 3   4   4   5   6   7   7   8   8   9   9  10  10  11  11  11  12  12  13
20      18  2 3   4   5   5   6   7   8   8   9   9  10  10  11  11  12  12  13  13
21      19  2 3   4   5   6   6   7   8   8   9  10  10  11  11  12  12  13  13  13
22      20  2 3   4   5   6   6   7   8   9   9  10  10  11  12  12  13  13  13  14
```

Figure   10-2

```
-------AU---AV-AW-AX-AY-AZ-BA-BB-BC-BD-BE-BF-BG-BH-BI-BJ-BK-BL-BM-BN----
    1                        CRITICAL HIGH VALUES FOR THE RUN TEST
    2
    3           2   3   4   5   6   7   8   9  10  11  12  13  14  15  16  17  18  19  20
    4     2
    5     3
    6     4               9   9
    7     5           9  10  10  11  11
    8     6           9  10  11  12  12  13  13  13  13
    9     7              11  12  13  13  14  14  14  14  15  15  15
   10     8              11  12  13  14  14  15  15  16  16  16  16  17  17  17  17
   11     9                  13  14  14  15  16  16  16  17  17  18  18  18  18
   12    10                  13  14  15  16  16  17  17  18  18  18  19  19  19  20
   13    11                  13  14  15  16  17  17  18  19  19  19  20  20  20  21
   14    12                  13  14  16  16  17  18  19  19  20  20  21  21  21  22
   15    13                      15  16  17  18  19  19  20  20  21  21  22  22  23
   16    14                      15  16  17  18  19  20  20  21  22  22  23  23  24
   17    15                      15  16  18  18  19  20  21  22  22  23  23  24  25
   18    16                          17  18  19  20  21  21  22  23  23  24  25  25
   19    17                          17  18  19  20  21  22  23  23  24  25  25  26
   20    18                          17  18  19  20  21  22  23  24  25  25  26  27
   21    19                          17  18  20  21  22  23  23  24  25  26  26  27
   22    20                          17  18  20  21  22  23  24  25  25  26  27  28
```

Figure   10-3

One is then subtracted from this value to provide the column offset to the @VLOOKUP. The @VLOOKUP then provides the row offset and the cutoff value is returned. The formulae in cells LO__CUTOFF and HI__CUTOFF are, respectively,

@VLOOKUP (L116, RUNS_LO, @HLOOKUP (M116, RUNS_LO, 0) -1)

@VLOOKUP (L116, RUNS_HI, @HLOOKUP (M116, RUNS_HI, 0) -1)

Type these formulae into the appropriate cells. In this example, cells L116 and M116 are the programming equivalents to n_1 and n_2, respectively, and specify the criteria to find the locations of the cutoff values in the lookup tables.

The status cell (K119) contains a formula that compares the two critical values from the RUNS__LO and RUNS__HI tables to the experimental Runs test value and returns either a "REJECT RANDOM HYPOTHESIS" or "DON'T REJECT RANDOM HYPOTHESIS" message. The formula has the following format:

@IF (K116<=LO_CUTOFF#OR#K116>=HI_CUTOFF,

    "REJECT RANDOM HYPOTHESIS", "DON'T REJECT RANDOM HYPOTHESIS")

That is, if K116 is **EITHER** lower than or equal to the LO__CUTOFF **OR** greater than or equal to the HI__CUTOFF, the message to reject the random hypothesis is displayed. If the value of K116 is between the LO__CUTOFF and HI__CUTOFF, the message displayed instructs the user not to reject the random hypothesis. In this example, the randomness hypothesis cannot be rejected.

THE SIGN TEST

The sign test was first described in the early eighteenth century and is a convenient method to determine

- whether the mean value of a number of observations differs significantly from some standard (or stated) value.
- whether paired samples from two different methods, analysts, etc., are significantly different.

The sign test, therefore, is a nonparametric alternative to the t-tests on unpaired and paired samples discussed in Chapters 5 and 6.

The sign test applies when studying a continuously symmetrical population. Under this condition, the probability of getting an observed value less than the reference value and the probability of getting an observed value greater than the reference value are equal. The sign test gets its name from the method used for scoring results. That is, a minus sign is used if the observed value is smaller than the reference value and a plus sign is used if the observed value is greater.

Manually, the *one-sample sign test* is performed as follows:

1. For every x_i less than the standard (stated) value, record a minus sign ($-$).
2. For every x_i greater than the standard value, record a plus sign ($+$).
3. Discard any x_i exactly equal to the standard value and reduce the value of N (number of observations) accordingly. (If more than 20% of the observations are exactly equal, the results of this test may be misleading.)
4. Count the number of minus signs and plus signs (separately).
5. Using the number of the less frequent sign and a critical value from a text or lookup table, determine whether the difference is significant. A significant difference is concluded if the observed number is less than or equal to the critical value.

As you can see, the sign test is mechanically similar to the Runs test described in the previous section. With Lotus, the formulae and implementation of the template are also very similar. Figure 10-4 illustrates this similarity. The example shows a comparison of the assay values for several vials of DNA to the DNA concentration stated for the lot of vials.

To create this example, you can either start fresh or just modify the existing Runs test template. The following will show you how to create the template from scratch:

1. Type the text and data into the cells. Move the cell-pointer to the cell containing the label "STATED" and issue the /Range Name Label Down command to name the cell below it STATED.
2. Into cell J78, type the formula for displaying the signs and copy it to cells J79..J113:

```
@IF(A78=$STATED, "   0",@IF(A78>$STATED, "   +", "   -"))
```

3. Into cell K78, type the formula for determining whether the observed value is greater than the standard value and copy it to cells K79..K113:

```
@IF(A78>$STATED,1,0)
```

```
--------A--------B----------C------D------E--------F--------G--------H--------I--------J--------K--------L--------M--------
73          ****SKEWNESS, KURTOSIS, MEDIAN, AND MODE ANALYSIS****
74
75          TEST FOR SKEWNESS              TEST FOR KURTOSIS       ************SIGN TEST ANALYSIS****************
76 RAW DATA SORTED DATA OCCURENCES    DIFF   DIFF^2   DIFF^3   DIFF^4      HI/MED/LO   MORE?    LESS?
77
78    221.1      198.4    1.0 -       10.6    112.1   1186.3  12558.7         +          1        0
79    224.2      202.5    1.0 -       13.7    187.3   2563.5  35084.9         +          1        0
80    217.8      203.4    1.0 -        7.3     53.1    386.8   2818.3         +          1        0
81    208.8      203.7    1.0 -       -1.7      2.9     -5.0      8.6         -          0        1
82    212.6      205.8    1.0 -        2.1      4.4      9.1     18.9         -          0        1
83    213.3      205.9    1.0 -        2.8      7.8     21.6     60.3         -          0        1
84    203.4      206.0    1.0 -       -7.1     50.6   -360.0   2561.1         -          0        1
85    206.4      206.4    1.0 -       -4.1     16.9    -69.6    286.4         -          0        1
86    208.0      206.9    1.0 -       -2.5      6.3    -15.9     39.9         -          0        1
87    208.8      207.7    1.0 -       -1.7      2.9     -5.0      8.6         -          0        1
88    206.9      208.0    1.0 -       -3.6     13.1    -47.2    170.6         -          0        1
89    205.9      208.0    2.0 -       -4.6     21.3    -98.2    453.2         -          0        1
90    211.4      208.3    1.0 -        0.9      0.8      0.7      0.6         -          0        1
91    205.8      208.8    1.0 -       -4.7     22.2   -104.7    493.8         -          0        1
92    202.5      208.8    2.0 -       -8.0     64.2   -514.7   4124.5         -          0        1
93    214.5      208.8    3.0 MODE     4.0     15.9     63.3    252.5         -          0        1
94    212.6      208.9    1.0 -        2.1      4.4      9.1     18.9         -          0        1
95    215.6      209.1    1.0 -        5.1     25.9    131.6    669.2         +          1        0
96    211.1      209.8    1.0 -        0.6      0.3      0.2      0.1         -          0        1
97    198.4      210.2    1.0 -      -12.1    146.7  -1777.7  21534.5         -          0        1
98    213.0      211.1    1.0 -        2.5      6.2     15.4     38.2         -          0        1
99    208.9      211.4    1.0 -       -1.6      2.6     -4.2      6.8         -          0        1
100   206.0      212.6    1.0 -       -4.5     20.4    -92.0    415.1         -          0        1
101   209.8      212.6    2.0 -       -0.7      0.5     -0.4      0.3         -          0        1
102   208.8      212.7    1.0 -       -1.7      2.9     -5.0      8.6         -          0        1
103   210.2      213.0    1.0 -       -0.3      0.1     -0.0      0.0         -          0        1
104   208.0      213.3    1.0 -       -2.5      6.3    -15.9     39.9         -          0        1
105   208.3      214.1    1.0 -       -2.2      4.9    -10.9     24.0         -          0        1
106   214.1      214.5    1.0 -        3.6     12.9     46.1    165.4         -          0        1
107   209.1      214.6    1.0 -       -1.4      2.0     -2.8      4.0         -          0        1
108   207.7      215.6    1.0 -       -2.8      7.9    -22.3     62.7         -          0        1
109   216.2      216.2    1.0 -        5.7     32.3    183.8   1045.3         +          1        0
110   203.7      217.8    1.0 -       -6.8     46.4   -316.4   2155.7         -          0        1
111   218.3      218.3    1.0 -        7.8     60.6    472.0   3675.2         +          1        0
112   212.7      221.1    1.0 -        2.2      4.8     10.4     22.8         -          0        1
113   214.6      224.2    1.0 -        4.1     16.7     68.2    278.8         -          0        1
114
115 AVERAGE     210.5            MOM1     MOM2     MOM3     MOM4      STATED     N      #GREATER  #LESSER
116 STD DEV     5.309             0.0     27.4     47.2   2475.2       215      36         6        30
117 SKEWNESS    0.3156
118 MEDIAN      209.45                  KURTOSIS  3.1147        ****SIGN TEST SUMMARY****
119                                     (K-3)     0.1147        # SIGNS       6
120                                                             CUTOFF       11
121                                                             STATUS   SIGNIFICANTLY DIFFERENT
122
```

Figure 10-4

4. Into cell L78, type the formula for determining whether the observed value is less than the standard value and copy it to cells L79..L113:

 @IF (A78<$STATED, 1, 0)

5. Into cells L116 and M116, type the formulae for performing summations on

columns K and L, respectively,

```
@SUM(K78..K113)
@SUM(L78..L113)
```

6. Into cell K116, type the formula to find the number of non-equal values:

```
@SUM(L116..M116)
```

7. Into the "# SIGNS" cell (K119), type the following formula to choose the lesser of cells L116 and M116:

```
@IF(L116<M116,L116,M116)
```

8. Into cell K121, type the following formulae:

```
@IF(K119<=K120,"SIGNIFICANTLY DIFFERENT",
        "NOT SIGNIFICANTLY DIFFERENT")
```

9. The cutoff value for cell K120 can be placed either manually or automatically (using lookup @functions) into the cell. With either scenario, you should use the concepts explained previously to determine the value for cell K120.

As was the case for the Runs test, the two sets of @IF functions placed into the cells during Steps 3 and 4 return a true (1) if the condition is met and a zero if the condition is not met. A summation on each of these columns is made using the @SUM functions entered as part of Step 5. These @SUMs give the total number of values greater than and less than the STATED value, respectively. The two columns of @IFs and the @SUMs thereby quantify the signs test. The @IF test prepared as part of Step 7 returns the lesser of the two @SUM quantities and is used in the comparison with the cutoff value to determine significance.

In the example, the value of 215 ng was entered into the STATED cell. Based on $N = 36$, the cutoff is 11. Because the number of signs (6) is less than 11, the hypothesis is accepted that there is a significant difference in the stated and experimentally found values. Type various values into the STATED cell and watch how the template changes. For example, type the median value 209.45 into the STATED cell. Columns J, K, and L change accordingly and the STATUS cell updates to "NOT SIGNIFICANTLY DIFFERENT". On the other hand, the discrepant value of 300 in the STATED cell updates the STATUS cell to "SIGNIFICANTLY DIFFERENT".

The procedure for the paired-sample sign test is similar to the one for the one-sample sign test. However with the *paired-sample sign test*, a comparison is made between the two component values of data pairs and not to a single standard or stated value. Thus, if the first value of the data pair is greater than the second value of the pair, a plus sign is assigned to the pair; if the value of the second is greater than the first, a minus sign is assigned to the pair.

Figure 10-5 illustrates a template that performs the paired-sample sign test. The formulae are nearly identical to the one-sample sign test formulae. Note that Figure 10-5 contains some new data. The data were generated in an experiment designed to compare the amount of protein found in cerebrospinal fluid by two different protein measurement methods. The first method, the Lowry method, is considered a reference method. The second method, the Coomassie Blue dye-binding method, is an experimental procedure being qualified for use by the laboratory. Creating this template is very similar to the one given above for the one-sample sign test.

```
--------A--------B--------C------D----------E--------F--------G--------H------I------J--------K--------L--------M------
73 ***PROTEIN METHOD***    *************SIGN TEST ANALYSIS****************
74                           HI/MED/LO   MORE?    LESS?
75  LOWRY    CBB                                                      ****PAIRED SAMPLE****
76    132    114                +         1        0                  ****SIGN TEST SUMMARY****
77     48     41                +         1        0
78     63     49                +         1        0                    N            46
79     26     15                +         1        0                  #GREATER       43
80     68     55                +         1        0                  #LESSER         3
81     83     79                +         1        0
82     81     74                +         1        0                  # SIGNS         3
83     93     89                +         1        0                  CUTOFF         15
84    102     94                +         1        0                  STATUS   SIGNIFICANTLY DIFFERENT
85     74     70                +         1        0
86     24     16                +         1        0
87     80     61                +         1        0
88    124    124                0         0        0
89     99     85                +         1        0
90    158    154                +         1        0
91    152    140                +         1        0
92    148    128                +         1        0
93     49     44                +         1        0
94    114    104                +         1        0
95     81     75                +         1        0
96     84     88                -         0        1
97     37     35                +         1        0
98     46     43                +         1        0
99     51     54                -         0        1
100    55     49                +         1        0
101    22     17                +         1        0
102    88     80                +         1        0
103    98     69                +         1        0
104    30     29                +         1        0
105   108     89                +         1        0
106    49     42                +         1        0
107    48     37                +         1        0
108   182    154                +         1        0
109    88     90                -         0        1
110    55     53                +         1        0
111    53     39                +         1        0
112    56     55                +         1        0
113    33     29                +         1        0
114    34     29                +         1        0
115    43     34                +         1        0
116   100     88                +         1        0
117    58     48                +         1        0
118    54     45                +         1        0
119   116     96                +         1        0
120    49     39                +         1        0
121    38     34                +         1        0
122    37     31                +         1        0
```

Figure   10-5

1. Type the text and data into the cells.

2. Into cell E76, type the formula for displaying the signs and copy it to cells E77..E122:

   ```
   @IF(A76=B76,"     0",@IF(A76>B76,"     +","     -"))
   ```

3. Into cell F76, type the formula for determining whether the observed value is greater than the standard value and copy it to cells F77..F122:

   ```
   @IF(A76>B76,1,0)
   ```

4. Into cell G76, type the formula for determining whether the observed value is less than the standard value and copy it to cells G77..G122:

   ```
   @IF(A76<B76,1,0)
   ```

5. Into cells K79 and K80, type the formulae for performing summations on Columns F and G, respectively:

   ```
   @SUM(F76..F122)
   @SUM(G76..G122)
   ```

6. Into cell K78, type the formula to find the number of non-equal values:

   ```
   @SUM(K79..K80)
   ```

7. Into the "# SIGNS" cell (K82), type the following formula to choose the lesser of cells K79 and K80:

   ```
   @IF(K79<K80,K79,K80)
   ```

8. Into cell K84, type the following formulae:

   ```
   @IF(K82<=K83,"SIGNIFICANTLY DIFFERENT",
                "NOT SIGNIFICANTLY DIFFERENT")
   ```

9. The cutoff value for cell K83 can be placed either manually or automatically (using lookup @functions) into the cell. With either scenario, you should use the concepts explained previously to determine the value for cell K83.

For this example, based on $N = 46$, the cutoff is 15. Because the number of signs (3) is less than 15, the hypothesis is accepted that there is a significant difference in the two protein methods. This discrepancy was later traced to a difference in the way that globulin proteins react in the two methods. It would be quite informative (and some good practice) for you to place difference formulae into Column M and use the values of these formulae to perform a t-test. This study will confirm that the sign test and t-test yield the same conclusions. If you need to review how to implement a t-test on paired samples, see the discussion in Chapter 6.

THE MEDIAN TEST

One variation of the one-sample sign test (called the *Median Test*) can be used to determine whether two groups of data are drawn from identically distributed populations. The Median Test performs this judgment by determining the numbers of observations in each group that are above and below the median of the combined samples. This test is easily implemented by entering the data for both groups into the template of Figure 10-4 (each group occupying sequential cells). Into the STATED cell, you would place the following formula:

```
+MEDIAN
```

Four cells are then created to hold the number of values greater than and less than the STATED cell. These cells should be called GTR__0, LSR__0, GTR__1, and LSR__1, corresponding to the number of values greater and lesser than the combined median for groups 0 and 1, respectively. Use @SUM functions to find these four quantities for the appropriate regions of Columns K and L. For example, suppose the data for Group 0 spanned the cells from A78..A92 and you wanted to find the greater than median value. You would then use the following formula in the GTR__0 cell:

```
@SUM(K78..K92)
```

The Median test is a chi-square calculation using the following formula:

```
+N*(@ABS(GTR_0*LSR_1-GTR_1*LSR_0)-N/2)^2
    /((GTR_0+LSR_0)*(GTR_1+LSR_1)*(GTR_0+GTR_1)*(LSR_0+LSR_1))
```

The chi-square value is compared to chi-square table cutoff values to determine whether the samples are drawn from identically distributed populations. For a review of this process, see Chapter 4.

I encourage you to take the time to convert the Sign test template to a Median Test template because it will give you some invaluable practice in tailoring the templates presented in this book to your particular requirements.

THE WILCOXON SIGNED-RANK TEST

The Sign test is a very useful and easily applied test. However, it has one drawback that is important in certain cases. The only material point in the Sign test is whether a measurement is greater or less than the standard value (one-sample) or its corresponding data pair (paired-sample). The Sign test disregards the magnitude of the change.

The Wilcoxon Signed-Rank test is another alternative to the t-test. The *Wilcoxon Signed-Rank test* takes into account not only the direction of the change, but also the size of the change relative to the sizes of changes for the other data. It achieves this capability by ranking each change.

More specifically, the Wilcoxon Signed-Rank test ranks differences without regard to their signs, assigning rank 1 to the smallest difference, rank 2 to the next smallest difference, and so on. As in the case of the Sign test, zero differences are discarded. If two or more differences are numerically equal, each one is assigned the average value of the ranks that they jointly occupy. The test for significance can be based on the sum of the ranks of the positive differences, the sum of the ranks of the negative differences, or the smaller of the positive and negative rank sums (because the magnitude of the final calculated quantities for the positive and negative tallies are identical). In this book, I will use the smaller of the two because it will also help to illustrate a programming technique.

The two criteria of discarding zero differences and assigning mean values to ranks with equal differences (i.e., tied ranks) make it very difficult to create a template for this test. Also, the sorting process requires user involvement, which can be subject to error. For these reasons, a macro program is a good choice for this application. Figures 10-6 through 10-8 show an example template and macro program that automate the process for paired samples (by far the most common use for this test). The figures build upon the concepts given previously for the Runs test, the Sign test, and macro programming.

To understand the implementation of the Wilcoxon Signed-Rank test, start by creating the example template. Type in the text headers for the template. Copy the data from the paired-sample Sign test example (Figure 10-5) into the template and enter 1.96 into the CUTOFF cell. (Note that the data in Figure 10-6 have already been sorted by the program and that the Wilcoxon calculations have been added to the template. You should **NOT** type in the numbers from the Wilcoxon calculations.) Then, issue the /Range Name Label Right command and highlight the NUM, POSITIVES, NEG-ATIVES, SIGNS, SCORE, CUTOFF, and STATUS cells. This action will activate the cells to their right. These cells will be used by the macro program.

Next, you will need to initialize some other named ranges in the template for use by the macro program. Issue the /Range Name Create command to create the following ranges:

Range name	Cells
WIL __ DATA	A76..B122
SORT__ RNG	A75..F122
WILCOX	A76..F122
KEY	D76

To create the macro program, type the text into the spreadsheet as you see it in Figures 10-7 and 10-8. Activate the program by issuing the /Range Name Label Right command and highlighting all of the labels in Column P. Repeat for all of the labels in Column Z. If you are using Symphony®, issue the {Services} Settings Auto-execute Set

```
--------A--------B--------C--------D----------E--------F--------G--------H--------I--------J--------K----
73 ***PROTEIN METHOD***          ******WILCOXON SIGN-RANK TEST ANALYSIS******
74
75  LOWRY    CBB   DIFFERENCE ABS DIFF    RANK SIGN-RANK          ****PAIRED SAMPLE****
76   124     124       0        0          0      0              ****SIGN TEST SUMMARY****
77    56      55       1        1         1.5    1.5
78    30      29       1        1         1.5    1.5
79    88      90      -2        2          4     -4             NUM            47
80    55      53       2        2          4      4             POSITIVES   1059.5
81    37      35       2        2          4      4             NEGATIVES     21.5
82    51      54      -3        3         6.5    -6.5
83    46      43       3        3         6.5    6.5            SIGNS         21.5
84    93      89       4        4          11     11            SCORE      -5.67026
85   158     154       4        4          11     11            CUTOFF        1.96
86    84      88      -4        4          11    -11            STATUS   SIGNIFICANTLY DIFFERENT
87    33      29       4        4          11     11
88    83      79       4        4          11     11
89    38      34       4        4          11     11
90    74      70       4        4          11     11
91    34      29       5        5          16     16
92    22      17       5        5          16     16
93    49      44       5        5          16     16
94    81      75       6        6          19     19
95    55      49       6        6          19     19
96    37      31       6        6          19     19
97    81      74       7        7          22     22
98    49      42       7        7          22     22
99    48      41       7        7          22     22
100   88      80       8        8          25     25
101   24      16       8        8          25     25
102  102      94       8        8          25     25
103   54      45       9        9         27.5   27.5
104   43      34       9        9         27.5   27.5
105   58      48      10       10          30     30
106  114     104      10       10          30     30
107   49      39      10       10          30     30
108   26      15      11       11         32.5   32.5
109   48      37      11       11         32.5   32.5
110  100      88      12       12         34.5   34.5
111  152     140      12       12         34.5   34.5
112   68      55      13       13          36     36
113   53      39      14       14          38     38
114   63      49      14       14          38     38
115   99      85      14       14          38     38
116  132     114      18       18          40     40
117  108      89      19       19         41.5   41.5
118   80      61      19       19         41.5   41.5
119  148     128      20       20         43.5   43.5
120  116      96      20       20         43.5   43.5
121  182     154      28       28          45     45
122   98      69      29       29          46     46
```

Figure   10-6

command and specify AUTOEXEC. Store the spreadsheet and retrieve it. This action will update the VERSION cell and implement the portability system (explained in Chapter 7).

The template and program are now ready for use. Let us examine the program more closely because it demonstrates some techniques that you will find useful for automating other applications.

The program begins with one of the two auto-executing macros. If the spreadsheet is retrieved under Lotus 1-2-3®, \ 0 will execute and place a zero into VERSION.

```
---------P--------Q--------R------S----------T--------U--------V--------W--------X-------
 1           /*LOTUS 1-2-3 AUTOEXECUTING MACRO*/
 2  \0       {LET VERSION,0}   /*IDENTIFY SPREADSHEET AS 1-2-3*/
 3
 4           /*SYMPHONY AUTOEXECUTING MACRO*/
 5  AUTOEXEC {LET VERSION,1}   /*IDENTIFY SPREADSHEET AS SYMPHONY*/
 6
 7           /*PROGRAM TO PERFORM WILCOXON SIGN-RANK TEST*/
 8  \R       {LET NUM,@COUNT(WIL_DATA)/2}/*DETERMINE NUMBER OF DATA PAIRS*/
 9           {GOTO}WILCOX~               /*SET CELL-POINTER*/
10           {BLANK RANK_CTR}            /*INITIALIZE RANK COUNTER*/
11           {FOR INC,0,NUM-1,1,DO_DIFF}/*FIND THE DIFFERENCES IN THE DATA PAIRS*/
12           {SORT}                      /*SORT THE DATA PAIRS, BASED ON DIFFERENCES*/
13           {FOR INC,0,NUM-1,1,SET_RANK}/*ASSIGN AN INITIAL RANK TO EACH DATA PAIR*/
14           {LET COUNTER,1}             /*INITIALIZE REPEAT COUNTER*/
15           {BLANK BUFFER}              /*INITIALIZE BUFFER FOR REPEAT ARCHIVE SYSTEM*/
16           {FOR INC,1,NUM,1,CRCT_RNK}  /*CORRECT FOR REPEAT RANKS*/
17           {FOR INC,0,NUM-1,1,ADD_SIGN}/*ADD THE SIGNS TO THE RANKS*/
18           {DO_SUM}                    /*SUMMARIZE THE RESULTS*/
19           {CALC}                      /*FORCE RECALCULATION OF THE SPREADSHEET*/
20
21           /*FIND THE DIFFERENCE AND ABSOLUTE DIFFERENCE FOR EACH DATA PAIR*/
22  DO_DIFF  {PUT WILCOX,2,INC,@INDEX(WILCOX,0,INC)-@INDEX(WILCOX,1,INC)}
23           {PUT WILCOX,3,INC,@ROUND(@ABS(@INDEX(WILCOX,2,INC)),2)}
24
25           /*SORT THE DATA PAIRS, BASED ON THE ABSOLUTE DIFFERENCES*/
26  SORT     /RNCSORT_RNG~{ESC}.{DOWN NUM}{RIGHT 5}~/*RESET THE SORT RANGE*/
27           {IF VERSION=1}/QSCSBDSORT_RNG~QS1KEY~A~QRAQ   /*SYMPHONY COMMANDS*/
28           {IF VERSION=0}/DSRDSORT_RNG~PKEY~A~G          /*LOTUS 1-2-3 COMMANDS*/
29
30           /*SET AN INITIAL RANK VALUE FOR EACH PAIR WITH NON-ZERO DIFFERENCE*/
31  SET_RANK {IF @INDEX(WILCOX,3,INC)=0}{PUT WILCOX,4,INC,0}{RETURN}
32           {LET RANK_CTR,RANK_CTR+1}
33           {PUT WILCOX,4,INC,RANK_CTR}
34
35           /*CORRECT THE RANKS FOR REPEATS*/
36  CRCT_RNK {IF @INDEX(WILCOX,3,INC)=@INDEX(WILCOX,3,INC-1)}{REPEAT}{RETURN}
37           {IF COUNTER>1}{DO_AVGS}     *IF NOT A REPEAT, CHECK TO SEE IF     *
38           {LET COUNTER,1}             *PREVIOUS ONES WERE AND PERFORM UPDATE*
39           {BLANK BUFFER}              *OF THE RANKS--THEN RESET SCRATCH-PAD *
40
41           /*RANK IS A REPEAT: ARCHIVE INFORMATION*/
42  REPEAT   {IF COUNTER=1}{LET BUFFER,@INDEX(WILCOX,4,INC-1)}  *IF FIRST IN SET, *
43           {LET BUFFER,BUFFER+@INDEX(WILCOX,4,INC)}           *ADD PREVIOUS RANK*
44           {LET COUNTER,COUNTER+1}                            *TO BUFFER AS WELL*
45
46
```

Figure   10-7

If retrieved under Symphony, AUTOEXEC will execute and place a one into VER-
SION. This information is used by the SORT macro to issue the appropriate set of
menu sorting commands for the spreadsheet being used.

The Wilcoxon Signed-Rank program begins by pressing [ALT]-R. The \R macro
is the main Signed-Rank program. It begins by determining how many data pairs are
in the WIL__DATA range (using @COUNT and dividing by 2). This determination
allows flexibility in the number of data pairs being studied. RANK__CTR is a counter
for the actual number of ranks. Recall that zero differences are not assigned a rank.
The {BLANK RANK__CTR} command initializes this counter to zero.

Next, the DO__DIFF macro is called repeatedly to find the differences for each
data pair. DO__DIFF actually places values into each of **two** columns of the WILCOX

```
--------Z--------AA-------AB-----AC---------AD-------AE-------AF-------AG-------AH-------AI-------AJ------
 1
 2              /*FIND THE AVERAGE RANK AND PLACE IT INTO ALL ROWS OF REPEATING RANK*/
 3 DO_AVGS    {LET BUFFER,BUFFER/COUNTER}
 4            {FOR INC2,INC-COUNTER,INC-1,1,PUT_AVGS}
 5
 6              /*PLACE THE AVERAGE RANK INTO ONE OF THE CELLS OF THE REPEAT*/
 7 PUT_AVGS  {PUT WILCOX,4,INC2,BUFFER}
 8
 9              /*CORRECT THE RANK BY ADDING THE SIGN OF THE DIFFERENCES*/
10 ADD_SIGN  {PUT WILCOX,5,INC,@IF(@INDEX(WILCOX,2,INC)<0,-1*@INDEX(WILCOX,4,INC),@INDEX(WILCOX,4,INC))}
11
12              /*SUMMARIZE THE POSITIVE AND NEGATIVE RANKS*/
13 DO_SUM      {BLANK POSITIVES}                        /*INITIALIZE POSITIVES ACCUMULATOR*/
14            {BLANK NEGATIVES}                        /*INITIALIZE NEGATIVES ACCUMULATOR*/
15            {FOR INC,0,NUM-1,1,SUM_LOOP}            /*COMPILE THE POSITIVES AND NEGATIVES*/
16            {LET NEGATIVES,@ABS(NEGATIVES)}         /*CONVERT NEGATIVE SUMATION TO POSITIVE VALUE*/
17            {LET SIGNS,@IF(NEGATIVES<POSITIVES,NEGATIVES,POSITIVES)}
18            {LET SCORE,SIGNS-(RANK_CTR*(RANK_CTR+1)/4)}  /*TRANSFORM SIGN INTO A Z SCORE*/
19            {LET SCORE,SCORE/@SQRT(RANK_CTR*(RANK_CTR+1)*(2*RANK_CTR+1)/24)}
20            {LET STATUS,@IF(@ABS(SCORE)>CUTOFF,'SIGNIFICANTLY DIFFERENT','NOT SIGNIFICANTLY DIFFERENT')}
21
22              /*UPDATE THE APPROPRIATE BUFFER*/
23 SUM_LOOP  {IF @INDEX(WILCOX,5,INC)>0}{LET POSITIVES,POSITIVES+@INDEX(WILCOX,5,INC)}
24            {IF @INDEX(WILCOX,5,INC)<0}{LET NEGATIVES,NEGATIVES+@INDEX(WILCOX,5,INC)}
25
26              /*SCRATCH-PAD*/
27
28 INC           47        /*COUNTER FOR {FOR} LOOPS*/
29 COUNTER        1        /*COUNTER FOR REPEATED RANKS*/
30 BUFFER                  /*BUFFER FOR REPEATING RANK VALUES*/
31 INC2          45        /*COUNTER FOR REPEAT LOOP*/
32 RANK_CTR      46        /*ONGOING COUNTER FOR INITIAL RANK VALUES*/
33 VERSION        1        /*0=LOTUS 1-2-3; 1=SYMPHONY*/
34
```

Figure   10-8

range. Column 2 retains the sign of the differences and Column 3 is the absolute value
of the sign. The values in Column 3 are rounded to 2 decimal places in this example
because sorting and tests for equality will be performed later on this column. If the
@ROUND function were not used, the sorting and testing functions may yield
inaccurate results. (Your application may require a different number of decimal
places.)

After the differences are placed into the cells, SORT is called. SORT begins by
customizing the sort range (SORT__ RNG) to agree with the number of data pairs. The
{RIGHT 5} command is required so that **all** of the data is re-arranged. Next, the two {IF}
commands are evaluated and the appropriate set of menu commands are issued to sort
the data. If you will note, the key that is used for the sorting function is Column 3 (D),
the column with the rounded absolute values.

Next, SET__RANK is called repeatedly to assign preliminary rank values to
the sorted data pairs. If you recall, zero differences are discarded. Therefore,
SET__RANK begins by testing for this condition. If a zero difference is
detected, the corresponding cell in Column 4 is assigned a value of zero and a
{RETURN} command is issued. The {RETURN} command forces a return to the
{FOR} loop for the next iteration, thereby skipping the remaining commands

in SET_RANK. If a non-zero difference is detected, the RANK_CTR is incremented by one and the current value of the rank counter is placed into the appropriate cell of Column 4 of WILCOX. In this case, RANK_CTR has a dual function. It is not only used to assign the values to Column 4, but it also keeps an ongoing tally of the number of non-zero differences for later use in the determination of the Degrees of Freedom for the significance test.

Next, COUNTER and BUFFER are initialized. COUNTER is a cell that keeps track of the number of sequential tied ranks before the string of ties ends. BUFFER is a cell that keeps a running total of the values of the tied ranks so that an average can be taken.

The ranks in Column 3 of WILCOX are then corrected for tied ranks by repeatedly calling CRCT_RNK. CRCT_RNK begins by testing the current and previous ranks for equality. If equality is found, the REPEAT macro is called. REPEAT begins by testing whether COUNTER is one. If COUNTER is one, then it is known that this time is the first that this string of tied ranks has been encountered and BUFFER is assigned the value of the previous rank. This procedure assures that the first value in the series is used in the average. Next, the current rank value is added to BUFFER and the series counter is incremented by one. COUNTER keeps track of the number of tied ranks in the series for later use in the average calculation. Then REPEAT returns to the {IF} command cell and a {RETURN} command returns to the {FOR} command for the next iteration.

If tied ranks are not detected, the COUNTER is tested to see if the previous ranks comprised a string of repeats that had not been updated. If they had not, then DO_AVGS is called upon to calculate the average rank and place it into each of the cells of the string. DO_AVGS accomplishes this task using a {FOR} loop with starting and ending values adjusted for COUNTER and INC. This adjustment provides a correlation between the {PUT} command in PUT_AVGS and the cells in the string of ties. Next, COUNTER and BUFFER are re-initialized to 1 and 0, respectively.

It is also important to note that CRCT_RNK does not contain an adjustment for un-tied ranks. That is, ranks that are not in a series are not corrected.

The main program then proceeds to call ADD_SIGN repeatedly to correct the ranks further. This time, a test is made on the differences in Column 2. If the difference is negative, then the corresponding rank value in Column 4 is multiplied by -1 and transferred to Column 5. If the difference is positive, then the corresponding rank value in Column 4 is transferred to Column 5 as-is. Column 5 now contains a way to separate the positive and negative differences. This separation is accomplished by calling DO_SUM, the next step in the main program.

DO_SUM begins by initializing buffers that will be used for the positive and negative running totals on the ranks. Next, DO_SUM calls SUM_LOOP repeatedly. SUM_LOOP tests each of the ranks in Column 5 and adds the rank value to the appropriate running total. After this process is complete, the negative total is replaced by its absolute value. This replacement allows comparison to the positive total in the next step. The next step is to pick the lesser of the positive and negative totals and place this final value into the SIGNS cell, thus completing Signed-Rank data reduction.

To determine the significance of the number in the SIGNS cells, a comparison is made to cutoff values. For less than 25 data pairs, this cutoff value can be determined from a table, analogous to all of the previous statistical tests presented in this book. However, for more than 25 data pairs, you must compare it to a normal deviate. In the case of a 0.05 level of significance and a two-sided test, the normal deviate is 1.96 (corresponding to the two standard deviations encompassing 95% of the sample variation). The sampling distribution of the rank values is approximately normal with the following mean and standard deviation:

$$\text{MEAN} = \frac{N(N + 1)}{4}$$

$$\text{SD} = \sqrt{\frac{N(N + 1)(2N + 1)}{24}}$$

where, N indicates the number of pairs tested with non-zero differences (RANK__CTR in this case).

To test for significance, SIGNS is transformed in DO__SUM to a normal deviate score using the following equation and the result is placed into the cell called SCORE:

$$\frac{\text{SIGNS} - \text{MEAN}}{\text{SD}}$$

The program concludes by determining whether the absolute value of SCORE is greater than the normal deviate cutoff value of 1.96 at 95% confidence limits. Taking the absolute value simplifies the task of testing for the following rules:

- Accept equivalence if between -1.96 and $+1.96$.
- Reject equivalence if less than -1.96 or greater than $+1.96$.

Based on this test, the STATUS cell is updated with either "SIGNIFICANTLY DIFFERENT" or "NOT SIGNIFICANTLY DIFFERENT". A final {CALC} command in the main program forces a recalculation of the spreadsheet to ensure that all formulae are brought up to date.

Although not common, the Wilcoxon Signed-Rank test can also be used as a one-sample test. The above program can be readily converted to a one-sample test for comparing samplings to a standard (or stated) value by simply exchanging the @INDEX reference to Column 1 in the formula of DO__DIFF with a reference to a STATED cell. That is, use the following formula in DO__DIFF:

{PUT WILCOX, 2, INC, @INDEX (WILCOX, 0, INC) -STATED}

THE RANK SUM TEST

The nonparametric equivalent to the independent sample (unpaired) t-test is called the *Rank Sum test*. The Rank Sum test has several aliases, among which are *The U-test*, *The Wilcoxon Test* and *The Mann-Whitney Test*. With the Rank Sum test, you can test whether two groups of samples come from identical populations. Such groups may contain differing numbers of measurements.

The mechanics of the Rank Sum test are similar to the Wilcoxon Signed-Rank test. With the Rank Sum test, the observations from both groups are combined and ranked from smallest (starting with a 1) to largest. In the event of ties:

- If two or more observations **within the same group** are equal, assign them successive ranks.
- If ties occur **between groups**, assign them average ranks of what would have been the successive ranks.

The ranks from the first group are then added together to provide the rank sum. From that quantity, you either compare the rank sum with a table cutoff value (≤ 20 samples) or compute a normal deviate score and compare the score to a normal deviate cutoff value of $+/-1.96$.

Figures 10-9 through 10-12 show example templates and a macro program that automate the Rank Sum test. The figures build upon the concepts given previously for the Wilcoxon Signed-Rank test. Figure 10-9 shows some example data prior to sorting, while Figure 10-10 shows the same data after sorting. Figures 10-11 and 10-12 show the changes that are required to convert the Wilcoxon Signed-Rank program into a Rank Sum program.

To understand the implementation of the Rank Sum test, start by creating the example template. Type in the text headers for the template. Type the data into the template, making certain to include the 0 and 1 digits in Column B. You do not need to separate the two groups when entering the data (i.e., the data can be intermixed). The program is smart enough to perform the separation task for you. The 0 and 1 digits are important because they will be used later to identify the origin of the group for the data point.

Enter 1.96 into the RS__CUTOFF cell. Then, issue the /Range Name Label Right command and highlight the RS__NUM, GROUP__0, RS__SCORE, RS__CUTOFF, and RS__STATUS cells. Press Return. This action will activate the cells to their right. These cells will be used by the macro program. Next, you will need to initialize some other named ranges in the template for use by the macro program. Issue the /Range Name Create command to create the following ranges:

Range name	Cells
RKS__DATA	A76..A122
SORT__RKS	A75..F122
RNK__SUM	A76..F122
RKS__KEY	A76
RKS__2KEY	B76
DATA__FLD	E76..F122

To create the macro program, type the text into the spreadsheet as you see it in Figures 10-11 and 10-12. You can use the same spreadsheet as the one for the Wilcoxon Signed-Rank program, although both programs will work slower. The macro subroutines for the two programs have different names and so do not interfere with each

```
--------A--------B--------C------D----------E-------F--------G--------H----------I--------J--------K-----
 73 ******GROUPED DATA******* *******EXTRACTED DATA FOR GROUP 0*******
 74
 75    DATA   GROUP   RANK          DATA   RANK
 76   221.1     0                                   ****RANK SUM TEST SUMMARY****
 77   224.2     0
 78   217.8     0                                   RS_NUM
 79   208.8     0                                   GROUP_0
 80   212.6     0
 81   213.3     0                                   RS_SCORE
 82   203.4     0                                   RS_CUTOFF
 83   206.4     0                                   RS_STATUS
 84   208.0     0
 85   208.8     0
 86   206.9     0
 87   205.9     0
 88   211.4     0
 89   205.8     0
 90   202.5     0
 91   214.5     1
 92   212.6     1
 93   215.6     1
 94   211.1     1
 95   198.4     1
 96   213.0     1
 97   208.9     1
 98   206.0     1
 99   209.8     1
100   208.8     1
101   210.2     1
102   208.0     1
103   208.3     1
104   214.1     1
105   209.1     1
106   207.7     1
107   216.2     1
108   203.7     1
109   218.3     1
110   212.7     1
111   214.6     1
112
113
114
115
116
117
118
119
120
121
122
```

Figure 10-9

other. However, the named ranges for the Wilcoxon Signed-Rank template overlap with the ones for the Rank Sum template. For example, WILCOX has the same definition as RNK__SUM. This ambiguity not only creates a potentially dangerous situation (changing the size or position of overlapping named ranges gives confusing and unpredictable results), but also **substantially** slows the performance of both macro programs. For this reason, if you want both programs in the same spreadsheet, you should either create two different templates (and, therefore, separate named ranges) or

```
--------A--------B--------C------D----------E-------F--------G--------H----------I--------J--------K-----
73 ******GROUPED DATA****** ******EXTRACTED DATA FOR GROUP 0******
74
75    DATA    GROUP   RANK         DATA    RANK
76   198.4     1       1          202.5     2          ****RANK SUM TEST SUMMARY****
77   202.5     0       2          203.4     3
78   203.4     0       3          205.8     5          RS_NUM          36
79   203.7     1       4          205.9     6          GROUP_0        250.5
80   205.8     0       5          206.4     8
81   205.9     0       6          206.9     9          RS_SCORE     -0.86635
82   206.0     1       7           208     11.5        RS_CUTOFF      1.96
83   206.4     0       8          208.8     14         RS_STATUS   NOT SIGNIFICANTLY DIFFERENT
84   206.9     0       9          208.8     15.5
85   207.7     1      10          211.4     22
86   208.0     0      11.5        212.6     23.5
87   208.0     1      11.5        213.3     27
88   208.3     1      13          217.8     33
89   208.8     0      14          221.1     35
90   208.8     0      15.5        224.2     36
91   208.8     1      15.5
92   208.9     1      17
93   209.1     1      18
94   209.8     1      19
95   210.2     1      20
96   211.1     1      21
97   211.4     0      22
98   212.6     0      23.5
99   212.6     1      23.5
100  212.7     1      25
101  213.0     1      26
102  213.3     0      27
103  214.1     1      28
104  214.5     1      29
105  214.6     1      30
106  215.6     1      31
107  216.2     1      32
108  217.8     0      33
109  218.3     1      34
110  221.1     0      35
111  224.2     0      36
112
113
114
115
116
117
118
119
120
121
122
```

Figure   10-10

use the same range names for both programs and revise the named ranges in the macro commands of one of the programs accordingly.

Activate the program by issuing the /Range Name Label Right command and highlighting all of the labels in Column P. Repeat for all of the labels in Column Z. (Note that a common Scratch-pad can be used for both programs.)

If you are using Symphony, issue the {Services} Settings Auto-execute Set command and specify AUTOEXEC. Store the spreadsheet and retrieve it. This action

```
--------P--------Q--------R------S----------T--------U--------V--------W--------X--------Y----
  1            /*LOTUS 1-2-3 AUTOEXECUTING MACRO*/
  2 \0         {LET VERSION,0}   /*IDENTIFY SPREADSHEET AS 1-2-3*/
  3
  4            /*SYMPHONY AUTOEXECUTING MACRO*/
  5 AUTOEXEC {LET VERSION,1}   /*IDENTIFY SPREADSHEET AS SYMPHONY*/
  6
  7            /*PROGRAM TO PERFORM RANK SUM U TEST*/
  8 \S         {LET RS_NUM,@COUNT(RKS_DATA)}      /*DETERMINE NUMBER OF DATA POINTS*/
  9            {GOTO}RNK_SUM~              /*SET CELL-POINTER*/
 10            {BLANK DATA_FLD}           /*ERASE OLD EXTRACTED DATA FIELD*/
 11            {BLANK RANK_CTR}           /*INITIALIZE RANK SUM COUNTER*/
 12            {SORT_RS}                  /*SORT THE COMBINED GROUPS*/
 13            {FOR INC,0,RS_NUM-1,1,INIT_RANK}    /*ASSIGN AN INITIAL RANK TO EACH DATA PAIR*/
 14            {LET COUNTER,1}             /*INITIALIZE REPEAT COUNTER*/
 15            {BLANK BUFFER}             /*INITIALIZE BUFFER FOR REPEAT ARCHIVE SYSTEM*/
 16            {FOR INC,1,RS_NUM,1,CRCT_RKS}       /*CORRECT RANKS FOR REPEAT VALUES*/
 17            {BLANK COUNTER}             /*INIT A COUNTER TO DETERMINE N FOR GROUP 0*/
 18            {FOR INC,0,RS_NUM-1,1,CHK_GP_0}     /*EXTRACT GROUP 0'S RANKS INTO COL 4*/
 19            {DO_RS_SUM}                /*SUMMARIZE THE RANK SUM RESULTS*/
 20            {CALC}                     /*FORCE RECALCULATION OF THE SPREADSHEET*/
 21
 22            /*SORT THE DATA, BASED ON MAGNITUDE*/
 23 SORT_RS  /RNCSORT_RKS~{ESC}.{DOWN RS_NUM}{RIGHT 5}~   /*RESET THE SORT RANGE*/
 24            {IF VERSION=1}/QSCSBDSORT_RKS~QS1RKS_KEY~A~S2RKS_2KEY~A~QRAQ
 25            {IF VERSION=0}/DSRDSORT_RKS~PRKS_KEY~A~SRKS_2KEY~A~G
 26
 27            /*SET AN INITIAL RANK VALUE FOR EACH DATA POINT*/
 28 INIT_RANK{PUT RNK_SUM,2,INC,INC+1}
 29
 30            /*TEST TO SEE IF FROM GROUP 0; IF IT IS, EXTRACT INFO*/
 31 CHK_GP_0 {IF @INDEX(RNK_SUM,1,INC)=0}{EXTRACT}
 32
 33            /*EXTRACT THE VALUES AND RANKS FOR GROUP 0 AND PLACE INTO COLS 4 AND 5*/
 34 EXTRACT  {PUT RNK_SUM,4,RANK_CTR,@INDEX(RNK_SUM,0,INC)}
 35            {PUT RNK_SUM,5,RANK_CTR,@INDEX(RNK_SUM,2,INC)}
 36            {LET RANK_CTR,RANK_CTR+1}   /*KEEP RUNNING TOTAL OF DATA EXTRACTED*/
 37
 38            /*CORRECT THE RANKS FOR BETWEEN GROUP REPEAT VALUES*/
 39 CRCT_RKS {LET CUR_0,@INDEX(RNK_SUM,0,INC)}{LET PREV_0,@INDEX(RNK_SUM,0,INC-1)}
 40            {LET CUR_1,@INDEX(RNK_SUM,1,INC)}{LET PREV_1,@INDEX(RNK_SUM,1,INC-1)}
 41            {IF CUR_0=PREV_0#AND#CUR_1<>PREV_1}{RPT_RKS}{RETURN}
 42            {IF COUNTER>1}{DO_RS_AVG}   |*IF NOT A REPEAT, CHECK TO SEE IF    *|
 43            {LET COUNTER,1}             |*PREVIOUS ONES WERE AND PERFORM UPDATE*|
 44            {BLANK BUFFER}             |*OF THE RANKS--THEN RESET SCRATCH-PAD *|
 45
 46
 47
 48
```

Figure   10-11

will update the VERSION cell and implement the portability system (explained in Chapter 7).

The template and program are now ready to use. As you will note, the template contains three sections. The first section has columns for raw data, group "tags," and joint rank data. The second section provides a scratch-pad area to place extracted data. In this case, the extracted data contain ranks for group 0. The third section is a summary section, much like the one described in the previous section for the Wilcoxon Signed-Rank test.

Let us now examine some of the nuances of the program, comparing it (as

```
--------Z--------AA-------AB-----AC---------AD-------AE-------AF-------AG-------AH-------AI-------AJ--------AK-----
1
2          /*DATA IS A REPEAT AND FROM DIFFERENT GROUPS: ARCHIVE INFORMATION*/
3 RPT_RKS  {IF COUNTER=1}{LET BUFFER,@INDEX(RNK_SUM,2,INC-1)} |*IF FIRST IN SET, *|
4          {LET BUFFER,BUFFER+@INDEX(RNK_SUM,2,INC)}          |*ADD PREVIOUS RANK*|
5          {LET COUNTER,COUNTER+1}                            |*TO BUFFER AS WELL*|
6
7          /*FIND THE AVERAGE RANK AND PLACE IT INTO ALL ROWS OF REPEATING RANK*/
8 DO_RS_AVG{LET BUFFER,BUFFER/COUNTER}
9          {FOR INC2,INC-COUNTER,INC-1,1,PUT_RS_AV}
10
11         /*PLACE THE AVERAGE RANK INTO ONE OF THE CELLS OF THE REPEAT*/
12 PUT_RS_AV{PUT RNK_SUM,2,INC2,BUFFER}
13
14         /*SUMMARIZE THE GROUP 0 RANK SUMS*/
15 DO_RS_SUM{BLANK GROUP_0}                   /*INITIALIZE GROUP_0 ACCUMULATOR*/
16         {FOR INC,0,RANK_CTR-1,1,RKS_LOOP}  /*COMPILE DATA FOR GROUP 0*/
17         {LET RS_SCORE,GROUP_0-(RANK_CTR*(RS_NUM+1)/2)}     /*TRANSFORM SIGN INTO A Z SCORE*/
18         {LET RS_SCORE,RS_SCORE/@SQRT(RANK_CTR*(RS_NUM-RANK_CTR)*(RS_NUM+1)/12)}
19         {LET RS_STATUS,@IF(@ABS(RS_SCORE)>RS_CUTOFF,'SIGNIFICANTLY DIFFERENT','NOT SIGNIFICANTLY DIFFERENT')}
20
21         /*ADD A RANK TO THE SUMMATION*/
22 RKS_LOOP {LET GROUP_0,GROUP_0+@INDEX(RNK_SUM,5,INC)}
23
24
25         /*SCRATCH-PAD*/
26 INC         15        /*COUNTER FOR {FOR} LOOPS*/
27 COUNTER               /*COUNTER FOR REPEATED RANKS*/
28 BUFFER                /*BUFFER FOR REPEATING RANK VALUES*/
29 INC2        24        /*COUNTER FOR REPEAT LOOP*/
30 RANK_CTR    15        /*ONGOING COUNTER FOR INITIAL RANK VALUES*/
31 VERSION     1         /*0=LOTUS 1-2-3; 1=SYMPHONY*/
32 CUR_0       0         /*BUFFER FOR CURRENT DATA POINT; VALUE*/
33 CUR_1       0         /*BUFFER FOR CURRENT DATA POINT; GROUP ID*/
34 PREV_0      224.2     /*BUFFER FOR PREVIOUS DATA POINT; VALUE*/
35 PREV_1      0         /*BUFFER FOR PREVIOUS DATA POINT; GROUP ID*/
36
```

Figure   10-12

appropriate) to the one discussed in the previous section for the Wilcoxon Signed-Rank test.

The program begins with the one of the two auto-executing macros. If the spreadsheet is retrieved under Lotus 1-2-3, \0 will execute and place a zero into VERSION. If retrieved under Symphony, AUTOEXEC will execute and place a one into VERSION. This information is used by the SORT_RS macro to issue the appropriate set of menu sorting commands for the spreadsheet being used.

The Rank Sum test program begins by pressing [ALT]-S. The \S macro is the main Rank Sum program. It begins by determining how many data pairs are in the RKS_DATA range (using @COUNT). This determination allows flexibility in the amount of data being studied. The {BLANK DATA_FLD} command erases old extracted Group 0 data in Columns 4 and 5 of RNK_SUM. Old data can remain in these columns and cause user confusion if this {BLANK} command is not used.

In this program, RANK_CTR is a counter for the number of Group 0 ranks. The {BLANK RANK_CTR} command initializes this counter to zero.

Next, SORT_RS is called. SORT_RS begins by customizing the sort range (SORT_RKS) to agree with the actual amount of data in the range. The {RIGHT 5} command is required so that **all** of the data in RNK__ SUM is re-arranged. Next, the two {IF} commands are evaluated and the appropriate set of menu commands are issued to sort the data. If you will note, the primary key that is used for this particular sorting function is Column 0 (A), the column containing the raw data values.

Also contrary to the Wilcoxon Rank-Sum program is the use of a secondary sort key. The secondary sort key is used whenever ties are detected with the primary sort key. If such is the case, Column 1 (B; the group identifier column), is used as the sort key and the data within the tied group is aligned on the basis of the secondary sort key. This secondary key is important because it separates tied values into the two groups. If you recall from the discussion at the beginning of this section, it is important to identify whether ties are within the same group or between groups. If a secondary key were not used, there would be random placement of the group identifiers and a reliable rank assignment would be impossible. For example, in Figure 10-10 there are three samples with a value of 208.8. Two of these samples derive from Group 0 and one from Group 1. If only a primary sort key were to be used, the following possible sequences could occur:

100

001

010

In the first two cases, the transition between groups would yield an averaging of the ranks. In the last case, sequential ranks would be assigned. The secondary sort key eliminates this problem by bringing all of the samples within a group together and storing them in sequential order. Therefore, the following sequence would always be received:

001

Next, INIT__ RANK is called repeatedly to assign preliminary ranks to the sorted data. If you recall, contrary to the Signed-Rank test, the Rank Sum test assigns ranks to all data points. Therefore, the {PUT} command in INIT__ RANK is more straight-forward than the Signed-Rank SET__ RANK macro.

Next COUNTER and BUFFER are initialized. COUNTER is a cell that keeps track of the number of sequential tied ranks before the string ends. BUFFER is a cell that keeps a running total of the values of the between-group tied ranks so that an average can be taken.

The ranks in Column 2 of RNK__ SUM are then corrected for tied ranks by repeatedly calling CRCT__ RKS. CRCT__ RNK begins by placing the data and group values for the current and previous samples into buffers of the scratch-pad. This action makes the remainder of the CRCT__ RKS macro more readable and understandable. The program then uses the following {IF} command:

```
{IF CUR_0=PREV_0#AND#CUR_1<>PREV_1}
```

to test for the conditions set forth at the beginning of this section. That is, an {IF} test is made on the current and previous data value (in Column 0) to see if they are equal. If they are not, the {IF} test is considered FALSE and the program proceeds to the next cell in the program. If the current and previous data values are equal, then a test is made to determine whether the current sample comes from the same group as the previous sample (Column 1). If they are from the same group, the program proceeds to the next cell in the program. If the samples are from different groups, the RPT__RKS macro is called upon.

RPT__RKS begins by testing whether COUNTER is one. If COUNTER is one, then it is known that this time is the first that this string of ties has been encountered and BUFFER is assigned the value of the previous rank. This method of assignment assures that the first value in the series is used in the average. Next, the current rank value is added to BUFFER and the series counter is incremented by one. COUNTER keeps track of the number in the series for later use in the average calculation. Then REPEAT returns to the {IF} command cell and a {RETURN} command returns to the {FOR} command for the next iteration.

If the dual specification of tied data values with different groups were not detected, the COUNTER is tested to see if the previous string of repeating, between-group tied ranks were not updated. If they were not, then DO__RS__AVG is called upon to calculate the average rank and place it into each of the cells of the string. DO__RS__AVG accomplishes this task using a {FOR} loop with starting and ending values adjusted for COUNTER and INC. This adjustment provides a correlation between the {PUT} command in PUT__RS__AV and the cells of the between-group repeats. Next, COUNTER and BUFFER are re-initialized back to 1 and 0, respectively.

It is important to note that CRCT__RKS does not contain an adjustment for unequal samples or equal samples from the same group. That is, ranks that are not from a series of equal data values **AND** in different groups are not corrected.

The main program then proceeds to call CHK __GP__ 0 repeatedly to extract the data for Group 0 and place the data into Columns 4 and 5 of RNK __SUM. As data from Group 0 is identified and placed into these columns, a counter is updated to keep track of the number of samples in these two columns. This information is used in the calculation of the normal deviate score in DO__RS__SUM.

DO__RS__SUM begins by initializing a buffer (GROUP__0) that will be used for the totals of the ranks for Group 0. Next, DO__RS__SUM calls RKS__LOOP repeatedly. RKS__LOOP simply adds each rank value in Group 0 to the running total. This addition completes the Rank Sum data reduction.

To determine the significance of the number in the GROUP__0 cell, a comparison is made to cutoff values. For less than 20 data pairs, this cutoff value can be determined from a table, analogous to all of the previous statistical tests presented in this book. However, for more than 20 data pairs, you need to compare it to a normal deviate. In the case of a 0.05 level of significance and a two-sided test, the normal deviate is 1.96 (corresponding to the "two standard deviations" encompassing 95% of the sample variation). The sampling distribution of the rank values is approximately normal with the following mean and standard deviation:

$$\text{MEAN} = \frac{N_0(N_0 + N_1 + 1)}{2}$$

$$\text{SD} = \sqrt{\frac{N_0 N_1(N_0 + N_1 + 1)}{12}}$$

where N_0 and N_1 indicate the number of data points in Groups 0 and 1, respectively. In this case,

$$N_0 = \text{RANK\_CTR}$$
$$N_1 = \text{RS\_NUM-RANK\_CTR}$$

To test for significance, GROUP_0 is transformed in DO_RS_SUM to a normal deviate score using the following equation. The result is placed into the cell called RS_SCORE.

$$\frac{\text{SIGNS} - \text{MEAN}}{\text{SD}}$$

Note that the relationship between N_0 and N_1 shown in the previous paragraph is used to simplify the calculations in the {LET} commands of DO_RS_SUM.

The program concludes by determining whether the absolute value of RS_SCORE is greater than the normal deviate cutoff value of 1.96. Taking the absolute value simplifies the task of testing for the following rules:

- Accept equivalence if between -1.96 and $+1.96$.
- Reject equivalence if less than -1.96 or greater than $+1.96$.

Based on this test, the RS_STATUS cell is updated with either "SIGNIFICANTLY DIFFERENT" or "NOT SIGNIFICANTLY DIFFERENT". A final {CALC} command in the main program forces a recalculation of the spreadsheet to ensure that all formulae are brought up-to-date.

THE KRUSKAL-WALLIS TEST

The Kruskal-Wallis test is an extension of the Rank Sum procedure to applications that require a comparison of the medians of three or more unmatched groups of samples. As in the case of the Rank Sum procedure, the Kruskal-Wallis test calls for a pooling of all of the samples, sorting (to arrange the samples in ascending order) and assigning of ranks. Tied ranks are averaged according to Rank Sum rules. The groups are then extracted and the rank totals are used in the calculation of a chi-square. The following equation is the one that is used:

$$\text{chi-square} = \frac{12}{N(N+1)} \left(\frac{R_0^2}{n_0} + \frac{R_1^2}{n_1} + \frac{R_2^2}{n_2} + \cdots \right) - 3(N+1)$$

where n_i and R_i are the number of measurements and rank totals for Group i and $N = n_0 + n_1 + n_2 + \cdots + n_i$. The chi-square value is compared with tabulated chi-square values. Experimental values of chi-square that exceed tabulated values allow the hypothesis that the medians of the groups are not significantly different enough to be rejected.

As you can see from this procedure, very few changes need to be made to the Rank Sum program to customize the program for the Kruskal-Wallis test. The fact that only minor changes need to be performed to upgrade the program demonstrates one of the advantages of trying to keep macro programs as generic as possible when they are originally designed. For example, the approaches of the SORT_RS and CRCT_RKS macros in Figure 10-11 allow for detection of repeating tied rank strings for more than two groups of data.

The changes involve upgrading the template and the CHK_GP_0 and EXTRACT subroutines so that all of the groups are extracted. Use the same commands as shown in Figure 10-11, but customize them for the group being extracted. For example, you could replace the {FOR} loop in the main program with a {FOR} loop that repeatedly calls a subroutine containing a {FOR INC, RS_NUM-1,1, CHK_GP} command. You should call the increment counter in the main program INC_GP. The CHK_GP subroutine then tests to see if the @INDEX value is equal to the current value of INC_GP. If it is, then EXTRACT is implemented.

The changes to EXTRACT would also be minor. You would replace the column designations in the {PUT} command and @INDEX function to ones that add the value of INC_GP to a base offset to find the appropriate columns. You would also need to keep a running total for each of the groups.

You would then complete the revision by editing DO_RS_SUM to reflect the chi-square equation given earlier in this section. You should take the time to convert the Rank Sum program to a Kruskal-Wallis test because it would give you some invaluable practice in tailoring the programs presented in this book to your particular requirements.

THE SPEARMAN RANK-CORRELATION

The nonparametric equivalent to the correlation coefficient is called the Spearman Rank-Correlation test. The Rank-Correlation test is advantageous when one or both sets of observations under investigation can be expressed only in terms of a rank rather than quantitative units. This situation occurs whenever qualitative results are obtained for either one or both of the sample groups.

The mechanics of the Rank-Correlation test are similar to the Wilcoxon Signed-Rank test. With the Rank-Correlation test, the first group is sorted and ranked. It is then re-sorted to place the samples back into their original order. The second group is then sorted, ranked, and placed back in original order. Next, the sum of the squares of the differences in the paired ranks are found and substituted into the following formula:

$$r_s = 1 - \frac{6(\Sigma d^2)}{n(n^2 - 1)}$$

where n is the number of pairs. In the event of ties (within a group), each tie is assigned the average value of the ranks it jointly occupies.

When there are no ties, r_s equals the correlation coefficient calculated for the two sets of ranks. When ties exist, there may be a small (but typically insignificant) difference.

The value of r_s provides a way to evaluate the significance of correlation. As was the case for the previous tests, r_s is first converted to a normal deviate score. The following equation is used:

$$r_s \sqrt{(n - 1)}$$

and the result is compared to a normal deviate cutoff value of $+/-1.96$.

Figures 10-13 through 10-16 show example templates and a macro program that automates the Rank-Correlation test. The data in the figures show a comparison between the quantitative Coomassie Blue method (described previously in the Sign

	A	B	C	D	E	F	G	H	I	J	K
73											
74		====GROUP 0====		====GROUP 1====							
75	SAMPLE NUM	CBB DATA		RANK	SSA DATA	RANK	DIFFERENCE	SQR_DIFF			
76	1	4		1					****RANK CORRELATION SUMMARY****		
77	2	5.1		3							
78	3	3.6		2					COR_SUM		
79	4	1.3		2					RNK_COR_COEF		
80	5	2.5		2					COR_SCORE		
81	6	1.3		1					COR_CUTOFF		
82	7	8.3		3					COR_STATUS		
83	8	0.3		2							
84	9	2.5		3					T_VALUE		
85	10	0.9		3							
86	11	0.2		1							
87	12	0.2		1							
88	13	3.9		4							
89	14	5.5		4							
90	15	1.2		2							
91	16										
92	17										
93	18										
94	19										
95	20										
96	21										
97	22										
98	23										
99	24										
100	25										
101	26										
102	27										
103	28										
104	29										
105	30										
106	31										
107	32										
108	33										
109	34										
110	35										
111	36										
112	37										
113	38										
114	39										
115	40										
116	41										
117	42										
118	43										
119	44										
120	45										
121	46										
122	47										

Figure   10-13

```
--------A---------B--------C-----D----------E--------F----------G--------H------------I--------J-------K-----
 73
 74                ====GROUP 0====  ====GROUP 1====
 75 SAMPLE NUM CBB DATA   RANK SSA DATA   RANK DIFFERENCE SQR DIFF
 76      1       4       12    1     2.5     9.5    90.25 ****RANK CORRELATION SUMMARY****
 77      2      5.1      13    3    11.5     1.5     2.25
 78      3      3.6      10    2     7       3        9 COR_SUM        232.00
 79      4      1.3      6.5   2     7      -0.5     0.25 RNK_COR_COEF 0.585714
 80      5      2.5      8.5   2     7       1.5     2.25 COR_SCORE    2.191542
 81      6      1.3      6.5   1     2.5     4       16 COR_CUTOFF     1.96
 82      7      8.3      15    3    11.5     3.5    12.25 COR_STATUS   NOT SIGNIFICANTLY DIFFERENT
 83      8      0.3      3     2     7      -4       16
 84      9      2.5      8.5   3    11.5    -3        9 T_VALUE       2.605523
 85     10      0.9      4     3    11.5    -7.5    56.25
 86     11      0.2      1.5   1     2.5    -1        1
 87     12      0.2      1.5   1     2.5    -1        1
 88     13      3.9      11    4    14.5    -3.5    12.25
 89     14      5.5      14    4    14.5    -0.5     0.25
 90     15      1.2      5     2     7      -2        4
 91     16
 92     17
 93     18
 94     19
 95     20
 96     21
 97     22
 98     23
 99     24
100     25
101     26
102     27
103     28
104     29
105     30
106     31
107     32
108     33
109     34
110     35
111     36
112     37
113     38
114     39
115     40
116     41
117     42
118     43
119     44
120     45
121     46
122     47
```

Figure   10-14

Test section of this chapter) and a qualitative Sulfosalicylic acid method. The qualitative method involves the addition of Sulfosalicylic acid to samples containing protein and the visual scoring of the turbidity produced.

Figures 10-13 through 10-16 build upon the concepts given previously for the Wilcoxon Signed-Rank and Rank Sum tests. Figure 10-13 shows some example data prior to sorting, while Figure 10-14 shows the same data after sorting. Figures 10-15 and 10-16 show the changes that are required to convert the Wilcoxon Signed-Rank and Rank Sum programs into a Rank-Correlation program.

```
--------P--------Q--------R------S----------T--------U--------V--------W--------X-------Y----
 1          /*LOTUS 1-2-3 AUTOEXECUTING MACRO*/
 2 \0       {LET VERSION,0}   /*IDENTIFY SPREADSHEET AS 1-2-3*/
 3
 4          /*SYMPHONY AUTOEXECUTING MACRO*/
 5 AUTOEXEC {LET VERSION,1}   /*IDENTIFY SPREADSHEET AS SYMPHONY*/
 6
 7          /*PROGRAM TO PERFORM RANK CORRELATION TEST*/
 8 \T       {LET COR_NUM,@COUNT(COR_DATA)}      /*DETERMINE NUMBER OF DATA POINTS*/
 9          {GOTO}COR_DATA~          /*SET CELL-POINTER*/
10          {BLANK DIF_FLD}          /*ERASE OLD SUM OF SQUARES DATA FIELDS*/
11          {FOR SORT_INC,0,3,1,SRT_LOOP}      /*SORT DATA 0, RE-ORDER, DATA 1, RE-ORDER*/
12          {DO_COR_SUM}             /*SUMMARIZE THE RANK SUM RESULTS*/
13          {CALC}                   /*FORCE RECALCULATION OF THE SPREADSHEET*/
14
15          /*GENERIC SORT: SPECIFIES SORT KEY COLUMN AND CORRECTS REPEATS IF NEEDED*/
16 SRT_LOOP {LET KEY_POS,@INDEX(KEY_OFFSET,0,SORT_INC)}  /*SPECIFY SORT COLUMN OFFSET*/
17          {COR_SORT}               /*SORT BASED ON THE SPECIFIED COLUMN*/
18          {IF SORT_INC=0#OR#SORT_INC=2}{CRT_COR}    /*IF SORTED ON A DATA COLUMN,*/
19                                                     /*CORRECT RANKS          */
20
21          /*UNIVERSAL SORT, USING COLUMN OFFSET AS SORT KEY*/
22 COR_SORT /RNCSORT COR~{ESC}.{DOWN COR_NUM}{RIGHT 6}~ /*RESET THE SORT RANGE*/
23          {GOTO}KEY_HOME~{DOWN}/RNCCOR_KEY~{BACKSPACE}{RIGHT KEY_POS}~
24          {IF VERSION=1}/QSCSBDSORT_COR~QS1COR_KEY~A~QRAQ       7*SYMPHONY*/
25          {IF VERSION=0}/DSRDSORT_COR~PCOR_KEY~A~G             /*LOTUS 1-2-3*/
26
27          /*CORRECT REPEATING RANKS*/
28 CRT_COR  {LET RNK_COL,SORT_INC+2}    /*SPECIFY THE RANK COLUMN TO BE TESTED AND UPDATED*/
29          {FOR INC,0,COR_NUM-1,1,INIT_COR}    /*ASSIGN AN INITIAL VALUE TO EACH RANK*/
30          {LET COUNTER,1}                     7*INITIALIZE REPEAT COUNTER*/
31          {BLANK BUFFER}               /*INITIALIZE BUFFER FOR REPEAT ARCHIVE SYSTEM*/
32          {FOR INC,1,COR_NUM,1,FIX_RNKS}      /*CORRECT RANKS FOR REPEAT VALUES*/
33
34          /*SET AN INITIAL VALUE FOR EACH RANK*/
35 INIT_COR {PUT RNK_COR,RNK_COL,INC,INC+1}
36
37          /*CORRECT THE RANKS FOR BETWEEN GROUP REPEAT VALUES*/
38 FIX_RNKS {LET CUR,@INDEX(RNK_COR,RNK_COL-1,INC)}{LET PREV,@INDEX(RNK_COR,RNK_COL-1,INC-1)}
39          {IF CUR=PREV}{RPT_COR}{RETURN}
40          {IF COUNTER>1}{DO_COR_AVG} |*IF NOT A REPEAT, CHECK TO SEE IF      *|
41          {LET COUNTER,1}            |*PREVIOUS ONES WERE AND PERFORM UPDATE*|
42          {BLANK BUFFER}            |*OF THE RANKS--THEN RESET SCRATCH-PAD *|
43
```

Figure   10-15

To understand the implementation of the Rank-Correlation test, start by creating the example template. Type in the text headers for the template. Type the data as you see it into the template, making certain to include the digits in Column A. These digits are used by the program to keep track of the starting positions of the data during sorting and to return the data to their original positions.

Enter 1.96 into the COR__CUTOFF cell. Then, issue the /Range Name Label Right command and highlight the COR__ SUM, RNK__ COR__ COEF, COR__ SCORE, COR__ CUTOFF, COR__ STATUS, and T__ VALUE cells. Press Return. This action will activate the cells to their right. These cells will be used by the macro program. Next, you will need to initialize some other named ranges in the template for use by the macro program. Issue the /Range Name Create command to

```
--------Z---------AA-------AB-----AC---------AD-------AE-------AF-------AG-------AH-------AI-------AJ--------AK-------
 1
 2             /*DATA IS A REPEAT: ARCHIVE INFORMATION*/
 3 RPT_COR    {IF COUNTER=1}{LET BUFFER,@INDEX(RNK_COR,RNK_COL,INC-1)}      |*IF FIRST IN SET, *|
 4            {LET BUFFER,BUFFER+@INDEX(RNK_COR,RNK_COL,INC)}              |*ADD PREVIOUS RANK*|
 5            {LET COUNTER,COUNTER+1}                                       |*TO BUFFER AS WELL*|
 6
 7             /*FIND THE AVERAGE RANK AND PLACE IT INTO ALL ROWS OF REPEATING RANK*/
 8 DO_COR_AVG {LET BUFFER,BUFFER/COUNTER}
 9            {FOR INC2,INC-COUNTER,INC-1,1,PUT_COR_AV}
10
11             /*PLACE THE AVERAGE RANK INTO ONE OF THE CELLS OF THE REPEAT*/
12 PUT_COR_AV {PUT RNK_COR,RNK_COL,INC2,BUFFER}
13
14             /*SUMMARIZE THE RANK SUM DIFFERENCES*/
15 DO_COR_SUM {BLANK COR_SUM}               /*INITIALIZE ACCUMULATOR BUFFER*/
16            {FOR INC,0,COR_NUM-1,1,DO_COR_DIF} /*COMPILE SUM OF SQUARES DATA*/
17            {LET RNK_COR_COEF,1-(6*(COR_SUM/(COR_NUM*(COR_NUM^2-1))))}    /*CALCULATE RANK-CORRELATION COEFF*/
18            {LET COR_SCORE,RNK_COR_COEF*@SQRT(COR_NUM-1)}      /*CALCULATE NORMAL DEVIATE SCORE*/
19            {LET COR_STATUS,@IF(@ABS(COR_SCORE)<COR_CUTOFF,"SIGNIFICANTLY DIFFERENT","NOT SIGNIFICANTLY DIFFERENT")}
20            {LET T_VALUE,RNK_COR_COEF*@SQRT(COR_NUM-2)/@SQRT(1-RNK_COR_COEF^2)}   /*CALCULATE T-TEST VALUE*/
21
22             /*CALCULATE DIFFERENCE, DIFFERENCE SQUARED, AND SUM OF SQUARES*/
23 DO_COR_DIF {PUT RNK_COR,5,INC,@INDEX(RNK_COR,2,INC)-@INDEX(RNK_COR,4,INC)}
24            {PUT RNK_COR,6,INC,@INDEX(RNK_COR,5,INC)^2}
25            {LET COR_SUM,COR_SUM+@INDEX(RNK_COR,6,INC)}
26
27             /*SCRATCH-PAD*/
28 INC        15          /*COUNTER FOR {FOR} LOOPS*/
29 COUNTER    1           /*COUNTER FOR REPEATED RANKS*/
30 BUFFER                 /*BUFFER FOR REPEATING RANK VALUES*/
31 INC2       15          /*COUNTER FOR REPEAT LOOP*/
32 KEY_POS                /*COLUMN OFFSET FOR SORT KEY*/
33 COR_NUM    15          /*NUMBER OF DATA POINTS*/
34 VERSION    1           /*0=LOTUS 1-2-3; 1=SYMPHONY*/
35 CUR        0           /*BUFFER FOR CURRENT DATA POINT; RANK VALUE*/
36 PREV       4           /*BUFFER FOR PREVIOUS DATA POINT; RANK VALUE*/
37 KEY_POS    0           /*CURRENT SORT KEY COLUMN OFFSET*/
38 RNK_COL    4           /*CURRENT RANK COLUMN OFFSET*/
39 SORT_INC   4           /*INCREMENT COUNTER FOR SORTING*/
40
41             /*TABLE OF COLUMN OFFSETS FOR SORT KEYS*/
42             KEY_OFFSET          1 /*DATA COLUMN FOR GROUP 0*/
43                                 0 /*COLUMN TO RE-ARRANGE DATA TO ORIGINAL SEQUENCE*/
44                                 3 /*DATA COLUMN FOR GROUP 1*/
45                                 0 /*COLUMN TO RE-ARRANGE DATA TO ORIGINAL SEQUENCE*/
```

Figure   10-16

create the following ranges:

Range name	Cells
COR __ DATA	B76..B122
SORT__ COR	A75..G122
RNK __ COR	A76..G122
COR __ KEY	A78
KEY__ HOME	A77
DIF__ FLD	F76..G122

To create the macro program, type the text into the spreadsheet as you see it in Figures 10-15 and 10-16. You can use the Wilcoxon Signed-Rank or Rank Sum spreadsheets, although the programs will work more slowly. The macro subroutines for the three programs have different names and do not interfere with each other. However, some of the named ranges in the templates overlap. This overlap not only creates a potentially dangerous situation (changing the size or position of overlapping named ranges gives confusing and unpredictable results), but also **substantially** slows the performance of both macro programs. For this reason, if you want more than one program in the same spreadsheet, either create different templates (and, therefore, separate, named ranges) or use the same range names for all of the programs and revise the named ranges in the macro commands accordingly.

Activate the program by issuing the /Range Name Label Right command and highlighting all of the labels in Column P. Repeat for all of the labels in Column Z. (Note that a common Scratch-pad can be used for all of the programs.) Issue the /Range Name Create command and create a named range, called KEY__OFFSET, for cells AD42..AD45. These cells contain column offsets that will be used when specifying sort keys.

If you are using Symphony, issue the {Services} Settings Auto-execute Set command and specify AUTOEXEC. Store the spreadsheet and retrieve it. These actions will update the VERSION cell and implement the portability system (explained in Chapter 7).

The template and program are now ready to use. As you will note, the template contains three sections. The first section contains sequential integers. These integers will be used by the program to determine the starting row positions of the data. When the data are sorted, these digits are sorted too. Then, when it is time to return the data to their original positions, these digits will become the sort keys, and the data will be restored.

The second section of the template has two pairs of columns (one pair for each of the two groups of data). The first column in each pair provides space for the group's raw data and the second provides a scratch-pad for the rank assignment.

The third section is a summary section, much like the one described in the previous section for the Wilcoxon Signed-Rank test. The pair of columns in the third section hold the difference and square of difference data.

Let us now examine the program, focusing on some of the nuances, and comparing the program (as appropriate) to those discussed in the previous sections for the Wilcoxon Signed-Rank and Rank Sum tests.

The Rank-Correlation program begins by pressing [ALT]-T. The \T macro is the main Rank-Correlation program. The \T macro begins by determining how many data pairs are in the COR__DATA range (using @COUNT). This determination allows flexibility in the amount of data being studied. The {BLANK DIF__FLD} command erases old difference data in Columns 5 and 6 of RNK__COR. Old data can remain in these columns and cause user confusion if this {BLANK} command is not used.

The main difference between \T and the previous programs is the way in which the sorting process is undertaken. If you recall, Group 0's data are first sorted, ranked, and then sorted back to their original positions. Then, Group 1's data are sorted, ranked, and re-sorted back to their original positions. Thus, there are four sorts in this process, and the only differences between them are the sort keys being used. Rather than creating four macro programs for these sorts, it is much more efficient to create a generic macro and reset the sort key to the appropriate column just prior to the sort. This generic macro will need to have a way to detect whether it is time to assign ranks and act accordingly.

The {FOR} command in the main program controls this process. In addition to being an increment counter for the {FOR} command, SORT_INC has several other important functions that serve to simplify the task of creating the generic sorting macro. SORT_INC first serves as an offset for an @INDEX in SRT_LOOP. In this way, SORT_INC specifies the number of columns to move the cell-pointer when specifying the sort key in COR_SORT. (The table of column offsets is called KEY_OFFSET.) This specification customizes COR_SORT to the appropriate sort key.

Just as importantly, SORT_INC serves to designate whether it is time to assign ranks to the sorted data. If you recall, ranks are assigned just after the sort is made on the data of a group. If the data are re-sorted back to their original order and ranks assigned again, the ranks would be incorrect. For this reason, a test is made in SRT_LOOP to determine whether SORT_INC is zero or two. Zero and two correspond to the iterations of the {FOR} loop in the main program in which the data values are in their correct orientations for assignment of the ranks.

SORT_INC, and its derivative RNK _COL, also specify

- the column to be used for placement of the ranks by {PUT} commands (INIT_COR and PUT_COR_AV).
- the column where @INDEX values are to be found when testing for ties in the ranks (FIX_RNKS).
- the column specifier for @INDEX functions to be used in average calculations for repeating ranks (RPT_COR).

Other than the incorporation of SORT_INC, placing the sorting function into a {FOR} loop with four passes, and testing for the appropriate time to assign ranks, this portion of the Rank-Correlation program works almost identically to the Wilcoxon Signed-Rank program discussed in detail earlier in this chapter.

The summary macro, DO_COR_SUM, is relatively straightforward. DO_COR_SUM begins by initializing a buffer (COR_SUM) that will be used for the totals of the squared differences. Next, DO_COR_SUM calls DO_COR_DIF repeatedly. DO_COR_DIF calculates the difference and the square of the difference in the ranks for each data pair and places the values into Columns 5 and 6, respectively. DO_COR_DIF also keeps a running total of the square of the differences in the COR_SUM cell.

The value of the Rank-Correlation coefficient (RNK_COR_COEF) is determined in DO_COR_SUM using COR_SUM and the equation presented at the beginning of this section. This determination completes the Rank-Correlation data reduction.

The value of r_s provides two different ways to evaluate the significance of correlation. As was the case for the previous non-parametric tests, r_s can first be converted to a normal deviate score. The following equation is used:

$$r_s \sqrt{(n-1)}$$

and the result is compared to a normal deviate cutoff value. In the case of a 0.05 level of significance and a two-sided test, the cutoff value is 1.96 (corresponding to the two standard deviations that encompass 95% of the sample variation).

The significance of the number in the RNK__COR__COEF cell can also be attained by making a comparison to t-test cutoff values using protocols outlined in Chapter 6 for the correlation coefficient. That is, the Rank-Correlation coefficient can be transformed to the t-test using the following equation:

$$t = r_s \frac{\sqrt{N-2}}{\sqrt{1 - r_s^2}}$$

and the transformed t-value compared to t-test cutoff values. If the experimental values of t are **higher** than the table cutoff values, a significant correlation exists. Both of these methods are shown in DO__COR__SUM.

THE KOLMOGOROV-SMIRNOV TEST FOR GOODNESS-OF-FIT

The Kolmogorov-Smirnov test can be used in situations similar to those presented for the chi-square test in Chapter 4. The chi-square test is useful when testing whether a sample of observations might come from a particular distribution, such as a normal distribution. The chi-square test is very well-suited for this purpose when data are presented as frequencies. However, as stated in Chapter 4, the chi-square test requires at least five observations per bin and is difficult to use with continuous data. The Kolmogorov-Smirnov test, on the other hand, is well-suited to testing Goodness-of-Fit for continuous data and for fewer than five observations per bin. The Kolmogorov-Smirnov test is therefore a more robust test than the chi-square test and its use is encouraged.

The Kolmogorov-Smirnov test involves comparing the cumulative frequency curve of the data to be tested with the cumulative frequency curve of the hypothesized distribution. Figure 10-17 shows a template that implements the Kolmogorov-Smirnov test. Figure 10-17 was created by converting the template shown in Figure 4-20 to the Kolmogorov-Smirnov equations. It may be helpful for you to review the chi-square discussion in Chapter 4. This review will refresh the genesis of the FREQUENCY DISTRIBUTION columns (Columns A and B) and the MEAN value. In short, the frequency distribution data for the chi-square test were generated for the "Number of Viral Plaques Detected" example data in Figure 4-19 using the /Data Distribution (Lotus 1-2-3) or /Range Distribution (Symphony) command. The MEAN was determined using an equation that referred to the data in Figure 4-19. The equation was

@AVG(A2..K10)

```
--------A--------B--------C------D----------E----------F-----------G--------
| 15              ****KOLMOGOROV-SMIRNOV GOODNESS-OF-FIT TEST****
| 16                    (POISSON DISTRIBUTION)
| 17                  MEAN          2.57
| 18                              OBSERVED    RELATIVE    EXPECTED
| 19 FREQUENCY DISTRIBUTION    CUMULATIVE  CUMULATIVE  CUMULATIVE
| 20    BIN      FREQUENCY     FREQUENCY   FREQUENCY   FREQUENCY   OBS-EXP
| 21     0           7              7       0.071       0.074      0.003
| 22     1          17             24       0.242       0.267      0.025
| 23     2          29             53       0.535       0.518      0.017
| 24     3          20             73       0.737       0.736      0.001
| 25     4          16             89       0.899       0.877      0.022
| 26     5           7             96       0.970       0.951      0.019
| 27     6           1             97       0.980       0.983      0.003
| 28     7           2             99       1.000       0.995      0.005
| 29     8           0             99       1.000       0.999      0.001
| 30                 0
| 31
| 32    N=          99                                  MAX_DEV     0.025
```

Figure   10-17

To create the Kolmogorov-Smirnov template, erase the equation in Columns D through G of the chi-square template, type in the text column headings, and add the following equations:

1. Into cell D21, type

   +B21

2. Into cell D22, type the following formula and copy it to cells D23..D29. This action will assign cumulative frequencies to each bin's row.

   +B22+D21

3. Column E is the Relative Cumulative Frequency column. To calculate the Relative Cumulative Frequency, divide the Observed Cumulative Frequency by the total number of observations (cell B32; which is 99 in this example). Therefore, type the following formula into cell E21 and copy it to cells E22..E29:

   +D21/B32

4. The data in the Expected Cumulative Frequency column (Column F) originate from a table of the "Cumulative Poisson Distribution." This table can be found in most statistical textbooks. If you are determining Goodness-of-Fit for other distributions, you should use the appropriate table. For example, if comparing to a normal distribution, you would use a table called "Cumulative Standard Normal Distribution." The values in this column can be either just typed into the template or can be derived from a lookup table. It would be a good review for you to create a two-way lookup table using the concepts of Chapter 2. When creating the @VLOOKUP and @HLOOKUP functions, most cumulative distri-

bution tables list a lambda value down the first column and an *x* value across the first row. These values correspond to the MEAN and BIN values in the template, respectively. Therefore, the test conditions in the @VLOOKUP and @HLOOKUP of cell F21 would be $MEAN and +A21, respectively. Copy your nested formula to cells F22..F29.

5. Into cell G21, type the formula to determine the absolute value of the difference between the Observed and Expected Relative Cumulative Frequencies and copy it to cells G22..G29:

```
@ABS(E21-F21)
```

6. Into cell G32, type the formula to determine the largest discrepancy between the Observed and Expected Relative Cumulative Frequencies.

```
@MAX(G21..G29)
```

The value in cell G32 is compared to a Kolmogorov-Smirnov Goodness-of-Fit test critical value for the sample size and level of significance. In this example, the maximum deviation is 0.025. The critical value for 99 observations and a 0.05 level of significance is

```
+1.36/@SQRT(99)=0.136685
```

Therefore, the hypothesis that the data constitute a random sample from a Poisson population with a mean of 2.57 may not be rejected.

It would be good practice for you to create a macro program that generates a Frequency Distribution table and performs data analysis automatically.

SUMMARY OF WHAT HAS BEEN ACCOMPLISHED

This chapter presented a number of nonparametric tests. In general, for each of the common parametric tests, there is a corresponding nonparametric test. The following will demonstrate this relationship:

Comparison	Standard test	Nonparametric test
Comparison to Reference Value	One-sample *t*-test	One-sample Sign test Signed-Rank test
Compare Means of Independent Samples	Two-sample *t*-test	Mann-Whitney U test
Compare Paired Samples	Paired-sample *t*-test	Paired-sample Sign test Signed-Rank Test
Compare Means of Several Groups	One-way ANOVA	Kruskal-Wallis test
Paired Sample Correlation	Correlation Coeff. with *t*-transformation	Rank-Correlation Coefficient
Compare Distributions	—	Median Test Kolmogorov-Smirnov Chi-square

This chapter also presented macro programs that perform nonparametric testing. These programs demonstrated the commonality of nonparametric procedures. Thus, by studying the programs and using the macro subroutines as a "tool box," you should be able to assemble a program to fit your particular nonparametric application.

Several opportunities were provided in this chapter for you to review concepts presented in previous chapters and to improve your template and macro program design skills.

WHAT'S NEXT?

The next chapter will show you how to implement statistical procedures that allow sampling and testing variations to be separated and their magnitudes estimated. The procedures are called the ANalysis Of VAriance, or ANOVA for short. Because ANOVA relies heavily on table calculations, Lotus provides a powerful environment within which to accomplish these analyses.

_11_

Analysis Of Variance (ANOVA)

Tabular testing procedures that compare three or more population means fall into a broad category of statistics called the *ANalysis Of VAriance* (ANOVA). It is important to recognize at the onset that although the term ANOVA refers to studies of variance, the primary focus is on the homogeneity or heterogeneity of group **means**, not variances. Moreover, the method used for ANOVA splits up sums of squares (SS) and Degrees of Freedom (DF) rather than analyzing variances directly. Variances are calculated only at the end of the procedure and are used in the calculation of a variance ratio (i.e., the F-test).

The use of ANOVA can be applied to a large number of experimental problems. Among them are

- to assess the average purity of a chemical in a bulk process.
- to study the effects of storage under various conditions on the concentration or activity of a chemical (i.e., "stability studies").
- to compare results obtained by several different analysts using the same apparatus or method.
- to compare results obtained by several different analysts using several different apparati or methods.
- to compare the average (mean) results obtained for concentrations of an analyte by several different assay methods.
- to compare the effects of different treatments on a base material(s).
- to compare yields produced by variations in manufacturing processes during maximization studies.
- to compare product manufactured by several machines and verify that the product is consistent over time.

- to compare product manufactured by several different plants.
- to assess the levels of contaminants in raw material samples from several different sources.
- to survey samples from a filling line to ensure uniformity.
- to study the degree of optimization of a reagent or process.
- to appraise the latitude of a set of optimized conditions (i.e., how much components can be changed before affecting the final results).
- to compare the abilities of microorganisms to produce biotechnology products.
- to evaluate day-to-day and within-run precision of an assay or process.
- to survey the results from collaborative laboratory studies.
- to assess the validity of the results being obtained by several laboratories (i.e., after sending the laboratories aliquots of the same sample).

ANOVA is a very powerful data analysis tool that you may find useful in reducing the data from your experiments. This tool will be especially useful if you are in a Development or Quality Control department. More importantly, the preparation of generic Lotus spreadsheet templates make the use of ANOVA convenient and almost routine.

However, to achieve the greatest benefit from ANOVA, you must design your experiments with a little thought. The next section discusses some of the historical perspectives of ANOVA. These perspectives will help you understand the jargon used in books on ANOVA and will give you insight into optimal ANOVA experimental design.

HISTORICAL PERSPECTIVE AND DEFINITIONS

The core of ANOVA techniques originated to fulfill the needs of the agricultural industry. Much of the special terminology relating to modern ANOVA experimental design and data reduction have retained hints of their farming heritage.

For example, plots and fields come straight from the agricultural application. A field was literally an area of ground capable of supporting crop growth. In an experiment, the *field* is all of the observations in the experiment. To perform experiments on the field, the field is divided into plots. Each of these plots must be large enough to produce enough plants whose growth, yield, etc., can be measured. Thus, a *plot* is the basic unit of examination and corresponds to a single observation in an experiment.

In an experiment, *treatments* are applied to plots. Historically, a number of experimental designs were common. Among these experiments,

- different fertilizers were applied to plots sown with the same seeds.
- different varieties of seed were sown in plots; each plot being treated with the same fertilizer.
- different varieties of seed were each sown in several plots; each plot being treated with different fertilizer.

In the last design, there are two independent variables. These independent variables are commonly referred to as *factors*. Each combination of factors is considered a treatment and the effect of the treatments is assessed by measuring the yield of the plots.

To make the design of the experiment a bit more elaborate, different levels (doses, frequencies, etc.) of fertilizer mixtures were applied to each variety of seed. The experiment then studied the effect of several independent variables (*factors*) and was called a *factorial* experiment. A factorial experiment was designed to simultaneously study the effects of seed type, fertilizer type, fertilizer, and time of planting. Each combination of these variables then represented a treatment.

Now, it stands to reason that if a farmer's field is large enough, the natural fertility, drainage, wind and sun exposure, etc., may vary from one plot in the field to the next. These variations may affect the results of the treatments to the various plots, producing a systematic error. To avoid systematic errors, care must be observed when assigning treatments to plots. There are several ways to make assignments. Modern ANOVA jargon retains vestiges of these assignments.

- A *fully randomized design* is defined as a design in which various treatments are appointed to plots in a completely random fashion. However, this design can lead to all of the replicates of a treatment being clustered together.

- A *randomized block design* is a design in which the field is first subdivided into fairly large areas of equal size. These large areas are called *blocks*. Each block is then subdivided again into plots (one for each treatment). If within each block, treatments are assigned randomly, the design is called a *randomized block*. If a block is large enough to contain replicates of treatments, the replicates are defined as *block nested*. In the case of a block nested, a joined group of replicates that is associated with a certain block and treatment is called a *CELL*. (To avoid confusing a *CELL* with a spreadsheet cell, an ANOVA cell will be capitalized and italicized in this chapter.)

- If the field's characteristics are known, a double system of blocking may be applied. For example, suppose that there is a directionality to the wind. If such is the case, a *Latin square* technique can be used to decrease the influence of the directionality effect. For example, suppose there are N treatments. The field could then be divided into an $N \times N$ grid of N north-south rows and N east-west rows and each treatment would be assigned at least one north-south row and one east-west row. The following grid shows an example of one of the many block assignment schemes available with the Latin square technique:

$$A\ B\ C\ D$$
$$B\ C\ D\ A$$
$$C\ D\ A\ B$$
$$D\ A\ B\ C$$

(The Latin square technique is presented for historical purposes only and will not be discussed further in this chapter. For a more comprehensive review of ANOVA techniques using Latin square, see the References at the end of this chapter.)

Let us now apply these concepts and jargon to more conventional laboratory experiments. Suppose you were weighing vials that had been filled with a powdered chemical by four different machines. The following would be the correspondence of this experiment with the above discussion:

Experimental	Agricultural
Each weight observation	Plot
Each filling machine	Treatment
Entire experiment	Field

Suppose that you performed the experiments on several days to include a determination of the day-to-day consistency of the product. Then, each day's weighings would constitute a block. If the samples were randomly drawn from each of the filling machines throughout the day and weighed, the design would be randomized block. If triplicate vials were withdrawn from the filling lines and weighed, the design would be nested within blocks and the triplicate would constitute a *CELL*.

A second example would be the testing of vials from a lyophilizer for moisture content. Suppose that the lyophilizer had several shelves from which samples were removed and an experiment was designed to determine whether there were any differences in the shelves. All of the samples from the lyophilizer would be considered the field, the moisture determination on each vial would be a plot, and the samples from each shelf would constitute a treatment. If the study were repeated on several days, the entire experiment would be considered the field and each day would be a block.

A third example would be in the area of animal testing. A litter of newborn mice would be considered a block and each baby mouse a plot. Because a litter may contain both male and female mice and sex may affect the results, an incomplete randomized block design would be used. On the other hand, a cage containing mice of the same age and sex could also constitute a block.

A fourth example would be a test tube rack. A test tube rack can represent a field, with each row in the rack being a block containing randomized samples. In this case, you might be testing whether any biases were being introduced due to the time it took to prepare the test mixtures.

A final example would be an experiment in which you were studying the effects of time and temperature on a reaction. If you had incubators set to various temperatures and you studied the yield produced as a result of incubator temperature, time, and position in the incubators, then you would have a *three factorial* design experiment.

As you can see, by arranging the data collection of an experiment, you can isolate the treatments that you want to test for equivalence. The following are some important considerations to observe when designing your ANOVA experiments:

- Clearly define the hypothesis you are trying to test.
- Reduce biases by randomizing treatments, blocks, plots, *CELL*s, etc.
- Perform replicates on each treatment.
- Use blocking to take any natural subdivisions into account.
- Divide large fields into several blocks.

With these concepts, definitions, and experimental design features in mind, we are ready to start building ANOVA tables. The next section will give you some general remarks on ANOVA tables and the calculations involved.

GENERAL ANOVA CONCEPTS

As stated at the beginning of this chapter, ANOVA is used to test the difference among several treatment means. To this end, the null hypothesis is made that the sample treatment means are equal and are therefore, all simultaneous estimates of the same population.

The basic ANOVA concept was first developed by R. A. Fisher, and the F-test (which is key to ANOVA) was named in his honor. The F-test was introduced in Chapter 5 and is a test that makes a comparison of variances to determine whether there is a significant difference in precision between two different methods, two analysts using the same method, etc.

There are two broad categories of ANOVA procedures that will be discussed in this chapter. The *one-way ANOVA* uses a row to represent each sample observation (plot) and a separate column to represent each treatment. The *two-way ANOVA* involves addition of a third factor, the blocking variable, which is used to associate some of the variation in sample response. In its simplest form, a set of blocking variables can be viewed as the equivalent of a set of row treatments, each row treatment containing a set of column treatments that are affiliated with some criterion. The grid of column and row treatments produces the fundamental basis for the two-way ANOVA concept.

ANOVA uses an elaborate procedure to compute the F statistics that are used for testing whether the variation in response is explained by (or attributed to) different treatments, blocks, etc., as compared to the variation that is left unexplained.

The next section will introduce you to the simplest ANOVA procedure, the one-way ANOVA.

ONE-WAY ANOVA

One-way ANOVA is concerned with testing the difference among several treatment (column) means when the "plots" are assigned randomly to each treatment group. One-way ANOVA can be considered taking the *t*-test for comparison of the means of two independent samples and extending it to the comparison of the means for several independent samples.

The following provides an overview of the manual ANOVA technique for one-way ANOVA. A table of the equations involved can be found at the end of this section.

1. Arrange the data in matrix form. Treatments should be collected and placed in individual columns; within-treatment plots (individual observations) should extend down the rows of the columns.

2. Compute the mean value for each column (treatment).
3. Determine the variance of the treatment means. This estimate of the population variance is called the *mean square between treatment groups (MST)*.
4. Compute the variance separately within each treatment and with respect to each treatment mean.
5. Pool the within-treatment variances by weighting them according to $(n - 1)$ for each sample. The resulting variance is called the *mean square within groups (MSW)* or, more commonly, the *mean square error (MSE)*.
6. If the treatment means are from the same population, then the MST and MSE will be unbiased and estimators of the same population variance. If the treatment means are not from the same population, then the value of the MST will be significantly larger than the MSE. Therefore, the F-distribution can be used to test the differences between the two variances. A one-tailed test is involved and the following form of the F-test equation is used:

$$F = \frac{MST}{MSE}$$

If the F-ratio is in the region of rejection for the specified level of significance, then the hypothesis that the several sample treatments came from the same population is rejected.

Figure 11-1 shows a template that calculates the sum of squares for a one-way ANOVA table. The data in this example illustrate one common use of ANOVA: The evaluation of the equivalency of results being obtained by collaborative laboratories. In the study, a control sample was sent to each of six different laboratories. Each laboratory was required to perform a single set of replicate cholesterol analyses on its sample and report the results back to the referee.

The template in Figure 11-1 contains four regions. The first region contains the raw data returned from each laboratory (plots in sequential rows of the appropriate treatment column). The second region of the template is a scratch-pad for calculating the squares of the raw data. The third region is a scratch-pad for calculating the sum of squares between treatments and within treatments. The final region is the ANOVA table. The ANOVA table is a summary based on the sum of squares scratch-pad. To understand this template, begin by creating it.

1. Type the column and row text headings into the appropriate cells.
2. Type the raw data into cells B3..G11.
3. Create the formulae for calculating the square of the individual raw data cells in the "Square of Sample Concentration Data" cells by typing the following formula into cell B14 and copying the formula to cells B14..G22:

`+B3^2`

4. Type the formula for calculating the sum of the raw data for Laboratory A (Treatment A) into cell B26 and copy it to cells C26..G26 (thereby finding the raw data sums for each laboratory):

`@SUM(B3..B11)`

```
--------A-----------B---------C-----------D--------E--------F--------G--------H-----------I-----
 1              LAB        LAB        LAB       LAB      LAB      LAB
 2     ASSAY    A          B          C         D        E        F
 3       1     221.1      208.8      211.1     208.3    221.1    224.2
 4       2     224.2      206.9      198.4     214.1    208.8    206.9
 5       3     217.8      205.9      213.0     209.1    211.1    198.4
 6       4     208.8      211.4      208.9     207.7             214.1
 7       5     212.6      205.8      206.0     216.2             212.6
 8       6     213.3      202.5      209.8     203.7             205.8
 9       7     203.4                 208.8     218.3             206.0
10       8     206.4                 210.2                       216.2
11       9     208.0                 208.0
12
13                  ****SQUARE OF SAMPLE CONCENTRATION DATA****
14       1     48885      43597      44563     43389    48885    50266
15       2     50266      42808      39363     45839    43597    42808
16       3     47437      42395      45369     43723    44563    39363
17       4     43597      44690      43639     43139        0    45839
18       5     45199      42354      42436     46742        0    45199
19       6     45497      41006      44016     41494        0    42354
20       7     41372          0      43597     47655        0    42436
21       8     42601          0      44184         0        0    46742
22       9     43264          0      43264         0        0        0
23
24                  ****SUM OF SQUARES SCRATCH-PAD****
25  LABORATORY    A          B          C         D        E        F        TOTALS
26     SUM X    1915.6     1241.3     1874.2    1477.4    641.0   1684.2       8834
27   SUM X SQR  408117     256850     390432    311981   137046   355005    1859431
28       N         9          6          9         7        3        8         42
29   (SUM X)^2  3669523    1540826    3512626   2182711   410881  2836530   14153096
30  (SUM X)^2/n 407725     256804     390292    311816   136960   354566    1858163
31       MEAN   212.8      206.9      208.2     211.1    213.7    210.5       1263
32
33
34                  ****ANALYSIS OF VARIANCE TABLE****
35                      (ONE WAY)
36  SOURCE OF    SUM OF    DEGREES OF ESTIMATE           TABLE
37  VARIATION    SQUARES   FREEDOM    OF STD   F RATIO   VALUE
38  ==========================================================
39 BETWEEN
40 TREATMENTS      205         5       40.95    1.163     2.45
41
42 SAMPLING
43 ERROR          1267        36       35.21
44
45 TOTAL          1472        41
46
47
```

Figure   11-1

5. Type the formula for calculating the sum of the squares of the raw data for Laboratory A into cell B27 and copy it to cells C27..G27 (thereby finding the sum of squares for each laboratory):

    @SUM(B14..B22)

6. Type the formula for determining the number of data points in a treatment column (i.e., the number of plots in Treatment k, n_k) into cell B28 and copy it to cells C28..G28:

    @COUNT(B3..B11)

7. Type the formula for determining the square of the *SUM* of the raw data for Laboratory A (NOT the same as the sum of squares) into cell B29 and copy it to cells C29..G29:

```
+B26^2
```

8. Type the formula for dividing the square of the *SUM* of the raw data by n_k into cell B30 and copy it to cells C30..G30:

```
@IF(B28=0,0,B29/B28)
```

These formulae make the template more generic. If the +B29/B28 formula would have been used alone, ERR would appear in all columns that had no raw data (because n_k would be zero and division by zero leads to ERR). These ERRs would be carried throughout the calculations and would find their way down to the ANOVA summary table.

9. Type the formula for calculating the mean raw data value for Laboratory A into cell B31 and copy it to cells C31..G31 (using @IFs to eliminate ERRs in unused columns):

```
@IF(B28=0,0,@AVG(B3..B11))
```

10. Type the formula to perform a summation on Columns B through G into cell H26 and copy it to cells H27..H31:

```
@SUM(B26..G26)
```

You are now ready to create the ANOVA summary table. The formulae used in the table are created as follows:

1. Into cell B40, type the formula for calculating the sum of squares between column means (i.e., between treatments, laboratories, etc.):

```
+H30-H26^2/H28
```

2. Into cell C40, type the formula for determining the number of treatment Degrees of Freedom by detecting the number of cells containing data in the first row of the raw data section:

```
@COUNT(B3..G3)-1
```

3. Into cell D40, type the following formula to determine the MST:

```
+B40/C40
```

4. Into cell B43, type the formula for determining the Sampling Error sum of squares:

```
+H27-H30
```

5. Into cell C43, type the formula for determining Degrees of Freedom for the Sampling Error:

```
+H28-@COUNT(B3..G3)
```

6. Into cell D43, type the formula for determining the MSE, i.e., the variability associated with sampling error and not including any influences associated with treatments):

```
+B43/C43
```

7. Type the formula for determining the total sum of squares into cell B45 and copy it to the total Degrees of Freedom cell (C45):

```
@SUM(B40..B43)
```

8. Complete the table by placing the formula to determine the F-ratio into cell E40:

```
+D40/D43
```

For five numerator and 36 denominator Degrees of Freedom and a 5% level of significance, the calculated value of F must be less than 2.45. Because the calculated F-ratio of 1.16 is less than 2.45, the hypothesis of no difference between the laboratories cannot be rejected.

The following table summarizes the calculations that would be required for manual calculation of the one-way ANOVA table:

Source of variation	Sum of squares	Degrees of freedom	Mean square	F-ratio
Between-Treatments	$SST = \sum\limits_{k=1}^{K} \dfrac{T_k^2}{n_k} - \dfrac{T^2}{N}$	$K - 1$	$MST = \dfrac{SST}{K-1}$	$F = \dfrac{MST}{MSE}$
Sampling Error	$SSE = \sum\limits_{k=1}^{K} \sum\limits_{i=1}^{n_k} X_{ik}^2 - \sum\limits_{k=1}^{K} \dfrac{T_k^2}{n_k}$	$N - K$	$MSE = \dfrac{SSE}{N-K}$	
Total	$SST + SSE$	$N - 1$		

where, k refers to the current treatment (i.e., the column), K refers to the total number of treatments (columns), n_k is the number of samples in the treatment (column), N is the total number of samples for all treatment groups combined, X_{ik} represents an individual observation, T_k represents the sum (total) of the values in a particular treatment group (column), and T represents the sum of all sample values in all groups.

It would be good practice to look at the formulae in the one-way ANOVA template to confirm that the template formulae agree with the manual formulae.

TWO-WAY ANOVA: RANDOMIZED BLOCK DESIGN

Two-way ANOVA is based on two sets of classifications or treatments. For example, suppose the laboratories running the cholesterol tests were asked to perform a single assay on their sample on specific days to see if there were any deteriorations in the reagents or samples. As defined earlier in this chapter, this scenario yields data that are blocked. That is, all treatments (between laboratory assays) performed on a certain day would form a block of data and would be placed in the same table row. Figure 11-2 illustrates a template designed to perform ANOVA on the data. In two-way ANOVA data tables, the treatments identified in the column headings are often called the *A treatments*; those in the row headings are often called the *B treatments*.

When only single assays are performed, the ANOVA model is referred to as a *randomized block design. Interaction* in a two-factor experiment means that the two treatments are not independent and that the particular effect of the treatment levels in one factor differs according to levels of the other factor. With this example experiment, the laboratories are performing single determinations and the day-to-day testing results are anticipated to be independent of laboratory. Therefore, the ANOVA table is termed *Two-way Analysis Without Interaction (Randomized Block Design)*. The Two-way Randomized Block Design can be considered taking the *t*-test for comparison of the difference between the means of paired observations and extending it to the comparison of several means.

Take a closer look at Figure 11-2. Figure 11-2 was prepared by copying the template of Figure 11-1 to Column J and then making some additions and changes. Note that there have been some formulae added to Column Q. These formulae are used in the calculation of the sum of squares for block data. Column Q contains formulae that merely form summations on the rows. To prepare Column Q, type the following formula into cell Q3 and copy it to cells Q4..Q11:

```
@SUM(K3..P3)
```

The second set of formulae in Column Q calculate the square of the raw data row sums (i.e., the block sums). To prepare these formulae, type the following into cell Q14 and copy the formula to cells Q15..Q22:

```
+Q3^2
```

The ANOVA summary table also has some additions. The following will show you how to make the additions:

1. The formulae for the Sum of Squares Scratch-pad remain unchanged.

2. Issue the /Move command to move the TOTAL summary (Row 45) down to Row 48.

3. Type the following formula to determine the sum of squares for the BLOCKS into cell K46:

```
@SUM(Q14..Q22)/@COUNT(K3..P3)-Q26^2/Q28
```

		LAB	LAB	LAB	LAB	LAB	LAB	
1								
2	DAY	A	B	C	D	E	F	SUM
3	1	224.2	215.6	213.0	218.3	221.1	224.2	1316.4
4	2	221.1	214.5	211.1	216.2	211.4	216.2	1290.5
5	3	217.8	212.6	210.2	214.6	211.1	214.1	1280.4
6	4	213.3	211.4	209.8	214.1	208.9	212.6	1270.1
7	5	212.6	208.8	208.9	212.7	208.8	210.2	1262.0
8	6	208.8	206.9	208.8	209.1	208.8	206.9	1249.3
9	7	208.0	205.9	208.0	208.3	208.3	206.0	1244.5
10	8	206.4	205.8	206.0	207.7	208.0	205.8	1239.7
11	9	203.4	202.5	198.4	203.7	207.7	198.4	1214.1
12								
13	DAY	****SQUARE	OF	SAMPLE	CONCENTRATION	DATA****		SQR SUM
14	1	50266	46483	45369	47655	48885	50266	1732909
15	2	48885	46010	44563	46742	44690	46742	1665390
16	3	47437	45199	44184	46053	44563	45839	1639424
17	4	45497	44690	44016	45839	43639	45199	1613154
18	5	45199	43597	43639	45241	43597	44184	1592644
19	6	43597	42808	43597	43723	43597	42808	1560750
20	7	43264	42395	43264	43389	43389	42436	1548780
21	8	42601	42354	42436	43139	43264	42354	1536856
22	9	41372	41006	39363	41494	43139	39363	1474039
23								
24			****SUM	OF	SQUARES	SCRATCH-PAD****		
25	LABORATORY	A	B	C	D	E	F	TOTALS
26	SUM_X	1915.6	1884.0	1874.2	1904.7	1894.1	1894.4	11367
27	SUM_X_SQR	408117	394542	390432	403275	398765	399190	2394320
28	N	9	9	9	9	9	9	54
29	(SUM_X)^2	3669523	3549456	3512626	3627882	3587615	3588751	21535853
30	(SUM_X)^2/n	407725	394384	390292	403098	398624	398750	2392873
31	MEAN	212.8	209.3	208.2	211.6	210.5	210.5	1263
32								

```
33       ****ANALYSIS OF VARIANCE TABLE****
34  (TWO WAY ANALYSIS WITHOUT INTERACTION; RANDOMIZED BLOCK DESIGN)
35
36  SOURCE OF  SUM OF   DEGREES OF  ESTIMATE           TABLE
37  VARIATION  SQUARES  FREEDOM     OF STD    F RATIO  VALUE
38  ==============================================================
39  BETWEEN
40  TREATMENTS  119.1       5        23.8      4.54     2.45
41
42  SAMPLING
43  ERROR       210.0      40         5.25
44
45  BETWEEN
46  BLOCKS      1238        8       154.7     29.46     2.18
47
48  TOTAL       1567       53        29.6
49
50
```

Figure   11-2

4. Type the formula to determine the Degrees of Freedom for the BLOCKS into cell L46:

```
@COUNT(K3..K11)-1
```

5. Edit the formula in cell K43, which calculates the Sampling Error sum of squares, to subtract the Block Error. The following is the appropriate equation:

```
+Q27-Q30-K46
```

6. Edit the formula in cell K48, which calculates the total sum of squares, to the following:

```
@SUM(K40..K46)
```

252

7. Copy the formula in cell K48 into cell L48.

8. Update the formula for the Sampling Error Degrees of Freedom by typing the following into cell L43:

   +L40*L46

9. Copy the Mean Square formula in cell M40 into cells M43, M46, and M48.

10. Type the F-test formula for the Blocks into cell N46:

   +M46/M43

Thus, there are two F-tests in this example. The first F-test is the same as the one for the one-way ANOVA table and the second F-test determines the significance of block effects. Let us look at each, in turn.

The first F-test indicates that there is a significant difference between treatments (laboratories). That is, the calculated F-test is 4.54 as compared to the cutoff of 2.45 for five numerator and 40 denominator Degrees of Freedom.

However, the F-test for the blocks is even more dramatic. In this case, the calculated F-test value of 29.46 far exceeds the 2.45 cutoff value (DF(num) = 5 and DF(denom) = 40) indicating something very significant is taking place. A closer look at the data indicates that either the sample or the assay reagent is deteriorating significantly.

The following tables summarize the calculations that would be required for manual calculation of the two-way ANOVA Randomized Block Design table:

Source of variation	Sum of squares	Degrees of freedom
Between-Treatments	$SST = \sum_{k=1}^{K} \dfrac{T_k^2}{n_k} - \dfrac{T^2}{N}$	$K - 1$
Sampling Error	$SSE = \sum_{j=1}^{J} \sum_{k=1}^{K} X_{jk}^2 - \sum_{k=1}^{K} \dfrac{T_k^2}{n_k} - SSB$	$(J - 1)(K - 1)$
Between-Blocks	$SSB = \dfrac{1}{K} \sum_{j=1}^{J} T_j^2 - \dfrac{T^2}{N}$	$J - 1$
Total	$SST + SSE + SSB$	$N - 1$

Source of Variation	Mean square	F-ratio
Between-Treatments	$MST = \dfrac{SST}{(K - 1)}$	$F = \dfrac{MST}{MSE}$
Sampling Error	$MSE = \dfrac{SSE}{(J - 1)(K - 1)}$	
Between-Blocks	$MSB = \dfrac{SSB}{(J - 1)}$	$F = \dfrac{MSB}{MSE}$

where, k refers to the current treatment (i.e., the column), K refers to the total number of treatments (columns), j refers to the current block (i.e., the row), J refers to the total number of blocks (rows), n_k is the number of samples in the kth treatment (column), N is the total number of samples for all treatment groups combined, X_{jk} represents an individual observation, T_k represents the sum (total) of the values in a particular treatment group (column), T_j represents the sum (total) of the values in a particular block (row), and T represents the sum of all sample values in all groups.

Again, it would be good practice to look at the formulae in the two-way Randomized Block Design ANOVA template to confirm that the template formulae agree with the manual formulae.

TWO-WAY ANOVA WITH INTERACTION (*N* OBSERVATIONS PER CELL)

When replication is included in a two-way ANOVA, the interaction between factors can be tested. In this way, three different null hypotheses are tested by ANOVA. They are,

- that there are no treatment (column) effects (i.e., that the column means are not significantly different).
- that there are no block effects (i.e., that the row means are not significantly different).
- that there is no interaction between the two factors (i.e., the row and column factors are independent and variation is random).

The third hypothesis is important because a significant interaction effect indicates that the effect of the treatments of one factor varies according to the levels of the other factor. If such is the case, the column and/or row effects may not be meaningful.

For example, suppose the laboratories running the cholesterol tests were asked to perform triplicate determinations on their samples in subsequent assay runs. Figures 11-3 and 11-4 illustrate a template designed to perform ANOVA on the data. The template contains five regions. The "Sample Concentration Raw Data," "Square of the Sample Concentration Data," "Sum of Squares Scratch-pad," and "Analysis of Variance Summary" table are much the same as the previous ANOVA tables. The "Square of Blocks Data" section is a new section that serves as a scratch-pad for data formed by adding the cells in a block of triplicates together and then squaring the result.

Additionally, three columns of formulae have been added to the Raw Data section. These formulae compile the triplicate data across the length of a block. For example, the formula in cell AA5 is an @SUM that refers to cells U4..Z6 and the formula in cell AC5 determines the square of cell AA5.

The F-tests in the ANOVA table are summaries based on the sum of squares scratch-pads located throughout the template. To understand this template, begin by creating it.

```
--------T----------U--------V----------W--------X--------Y--------Z--------AA-------AB-------AC-----
 1  REPLICATES    3              ****SAMPLE CONCENTRATION RAW DATA****
 2                LAB     LAB     LAB     LAB     LAB     LAB       BLOCK    BLOCK   SQUARE OF
 3  DAY/ASSAY     A       B       C       D       E       F         SUM     MEAN  BLOCK SUMS
 4     1/1      221.1   208.8   211.1   208.3   221.1   224.2
 5     1/2      224.2   206.9   198.4   214.1   208.8   206.9      3809     211.6   14510005
 6     1/3      217.8   205.9   213.0   209.1   211.1   198.4
 7     2/1      208.8   211.4   208.9   207.7   208.3   214.1
 8     2/2      212.6   205.8   206.0   216.2   208.8   212.6      3768     209.3   14195563
 9     2/3      213.3   202.5   209.8   203.7   211.4   205.8
10     3/1      203.4   214.5   208.8   218.3   208.9   206.0
11     3/2      206.4   212.6   210.2   212.7   207.7   216.2      3790     210.6   14364858
12     3/3      208.0   215.6   208.0   214.6   208.0   210.2
13     4/1      221.1   224.2   217.8   208.8   212.6   213.3
14     4/2      211.4   205.8   202.5   214.5   212.6   215.6      3816     212.0   14564909
15     4/3      211.1   208.3   208.8   211.4   208.9   207.7
16
17                    ****SQUARE OF SAMPLE CONCENTRATION DATA****
18                LAB     LAB     LAB     LAB     LAB     LAB
19  DAY/ASSAY     A       B       C       D       E       F
20     1/1      48885   43597   44563   43389   48885   50266
21     1/2      50266   42808   39363   45839   43597   42808
22     1/3      47437   42395   45369   43723   44563   39363
23     2/1      43597   44690   43639   43139   43389   45839
24     2/2      45199   42354   42436   46742   43597   45199
25     2/3      45497   41006   44016   41494   44690   42354
26     3/1      41372   46010   43597   47655   43639   42436
27     3/2      42601   45199   44184   45241   43139   46742
28     3/3      43264   46483   43264   46053   43264   44184
29     4/1      48885   50266   47437   43597   45199   45497
30     4/2      44690   42354   41006   46010   45199   46483
31     4/3      44563   43389   43597   44690   43639   43139
32
33                    ****SQUARE OF CELLS DATA****
34                LAB     LAB     LAB     LAB     LAB     LAB
35  DAY/ASSAY     A       B       C       D       E       F     ROW SUMMATION
36     1/1
37     1/2     439702  386387  387506  398792  410881  396270  2419538
38     1/3
39     2/1
40     2/2     402844  384028  390250  393882  395012  400056  2366073
41     2/3
42     3/1
43     3/2     381677  413063  393129  416799  390125  399930  2394723
44     3/3
45     4/1
46     4/2     414221  407427  395767  402844  402083  405260  2427601
47     4/3
48
```

Figure   11-3

1. Type the column and row text headings into the appropriate cells.
2. Issue the /Range Name Label Right command and specify cell T1 to name cell U1 REPLICATES. Enter a 3 into cell U1 to indicate triplicates.
3. Type the raw data into cells U4..Z15.
4. Create formulae in the "Sample Concentration Raw Data" section to summarize the triplicate data of the *CELL*s running the entire length of each block. (If you recall from the beginning of this chapter, the collection of replicate data within

a treatment is called a *"CELL"* in ANOVA jargon. An ANOVA *CELL* is capitalized and italicized in this chapter to avoid confusion with Lotus spreadsheet cells.) Into cell AA5 type the formula for finding the total of the block's triplicate *CELLs* and copy to cells AA8, AA11, and AA14:

@SUM(U4..Z6)

IMPORTANT NOTE: If your ANOVA is being performed on a different number of replicates per cell, you will need to amend the row positions and cell references of the formulae in Steps 4, 5, 6, 8, and 9 accordingly. Amending the formulae in a template may diminish the concept of a template somewhat, but with a little care the template will still be quite useful. Alternatively, you could create a custom template for each scenario. You could also create a macro program to automate the process of providing variable replicate flexibility. The macro program would provide flexibility if you find that you are routinely changing the number of replicates in your ANOVA experiments.

5. Into cell AB5, type the formula for finding the mean of the block's triplicate *CELLs* and copy it to cells AB8, AB11, and AB14:

@IF(@COUNT(U4..Z6)>0,@AVG(U4..Z6),0)

6. Into cell AC5, type the formula for finding the square of the sum of the block's triplicate *CELLs* and copy it to cells AC8, AC11, and AC14:

+AA5^2

7. Create the formulae for calculating the square of the individual raw data cells in the "Square of Sample Concentration Data" cells by typing the following formula into cell U20 and copying the formula to cells U20..Z31:

+U4^2

8. Create a table scratch-pad that calculates the square of the sum of the triplicate *CELLs* for each treatment and block. Into cell U37 type the following formula and copy it to cells U37..Z37, U40..Z40, U43..Z43, and U46..Z46:

@SUM(U4..U6)^2

9. Create a summary column for the *CELL* table scratch-pad created in step 8 by typing the following formula into cell AA37 and copying it to cells AA40, AA43, and AA46:

@SUM(U36..Z38)

10. The "Sum of Squares Scratch-Pad" is created in exactly the same way as for the two previous ANOVA categories. Type the formula for calculating the sum of

the raw data for Laboratory A (Treatment A) into cell U52 and copy it to cells V52..Z52 (thereby finding the raw data sums for each laboratory):

@SUM(U4..U15)

11. Type the formula for calculating the sum of the squares of the raw data for Laboratory A into cell U53 and copy it to cells V53..Z53 (thereby finding the sum of squares for each laboratory):

@SUM(U20..U31)

12. Type the formula for determining the number of data points in a treatment column (i.e., the number of plots in Treatment k, n_k) into cell U54 and copy it to cells V54..Z54:

@COUNT(U4..U15)

13. Type the formula for determining the square of the *SUM* of the raw data for Laboratory A (NOT the same as the sum of squares) into cell U55 and copy it to cells V55..Z55:

+U52^2

14. Type the formula for dividing the square of the *SUM* of the raw data by n_k into cell U56 and copy it to cells V56..Z56:

@IF(U54=0,0,U55/U54)

As noted in previous discussions, these formulae make the template more generic. If the +B55/B54 formula would have been used alone, ERR would appear in all columns that had no raw data (because n_k would be zero and division by zero leads to ERR). These ERRs would be carried throughout the calculations and find their way down to the ANOVA summary table.

15. Type the formula for calculating the mean raw data value for Laboratory A into cell U57 and copy it to cells V57..Z57 (using @IFs to eliminate ERRs in unused columns):

@IF(U54=0,0,@AVG(U4..U15))

16. Type the formula to perform a summation on Columns B through G into cell AA52 and copy it to cells AA53..AA57:

@SUM(U52..Z52)

You are now ready to create the ANOVA summary table. The formulae used in the table are created as follows:

1. The formula for the Between-Columns (Treatments) Sum of Squares calculations is similar to, but **NOT** exactly the same as, the corresponding formula for the previous ANOVA tables. Into cell U67, type the new formula for calculating the sum of squares between-column means (i.e., between-treatments, laboratories, etc.):

```
+AA55/@COUNT(U4..U15)-AA52^2/AA54
```

2. Into cell V67, type the formula for determining the number of treatment Degrees of Freedom by detecting the number of cells containing data in the first row of the raw data section:

```
@COUNT(U9..Z9)-1
```

3. Into cell W67, type the following formula to determine the mean square between treatment groups (MST):

```
+U67/V67
```

4. Into cell U70, type the formula for determining the Sampling Error sum of squares:

```
+AA53-AA56-U73-U75
```

5. Into cell V70, type the formula for determining the Sampling Error Degrees of Freedom:

```
@COUNT(U4..U15)/REPLICATES*@COUNT(U4..Z4)*(REPLICATES-1)
```

6. Into cell W70, type the formula for determining the mean square sampling error (MSE; i.e., the variability associated with sampling error and not including any influences associated with treatments, blocks, and *CELLS*):

```
+U70/V70
```

7. Into cell U73, type the formula for determining the sum of squares variation between the row means:

```
@SUM(AC4..AC15)/(REPLICATES*@COUNT(U4..Z4))-AA52^2/AA54
```

8. Into cell V73, type the formula for determining the Degrees of Freedom for the Between-Row variation:

```
@COUNT(U4..U15)/REPLICATES-1
```

9. Into cell W73, type the formula for determining the Mean Square Row Error:

+U73/V73

10. Into cell U75, type the formula for determining the Interaction Sum of Squares:

@SUM(AA37..AA46)/REPLICATES-U67-U73-AA52^2/AA54

11. Into cell V75, type the formula for determining the Degrees of Freedom for the Interaction Sum of Squares:

(@COUNT(U4..U15)/REPLICATES-1)*(@COUNT(U4..Z4)-1)

12. Into cell W75, type the formula for determining the Mean Square Interaction Error (MSI):

+U75/V75

13. Into cell X67, type the formula to determine the F-ratio for the Between-Column means:

+W67/W70

14. Into cell X73, type the formula to determine the F-ratio for the Between-Row means:

+W73/W70

15. Into cell X75, type the formula to determine the F-ratio for the Interaction:

+W75/W70

16. Type the formula for determining the total Sum of Squares into cell U78 and copy it to the total Degrees of Freedom cell (V78):

@SUM(U67..U75)

The F-distribution table values for the appropriate numerator and denominator Degrees of Freedom are given in the ANOVA table of Figure 11-4. It would be good practice to compare the calculated and table values of F and determine their meanings.

The following tables summarize the calculations that would be required for manual calculation of the two-way ANOVA table with Interaction (*n* observations per cell):

Source of Variation	Sum of Squares
Between-Columns	$\displaystyle \text{SST} = \sum_{k=1}^{K} \frac{T_k^2}{nJ} - \frac{T^2}{N}$
Sampling Error	$\displaystyle \text{SSE} = \sum_{j=1}^{J} \sum_{k=1}^{K} \sum_{i=1}^{n} X_{jki}^2 - \sum_{k=1}^{K} \frac{T_k^2}{nJ} - \text{SSB} - \text{SSI}$
Between-Rows	$\displaystyle \text{SSB} = \sum_{j=1}^{J} \frac{T_j^2}{nK} - \frac{T^2}{N}$
Interaction	$\displaystyle \text{SSI} = \frac{1}{n} \sum_{j=1}^{J} \sum_{k=1}^{K} \left(\sum_{i=1}^{n} X_{jki} \right)^2 - \text{SST} - \text{SSB} - \frac{T^2}{N}$
Total	SST+SSE+SSB+SSI

Source of variation	Degrees of freedom	Mean square	F-ratio
Between-Columns	$K - 1$	$\displaystyle \text{MST} = \frac{\text{SST}}{(K - 1)}$	$\displaystyle F = \frac{\text{MST}}{\text{MSE}}$
Sampling Error	$JK(n - 1)$	$\displaystyle \text{MSE} = \frac{\text{SSE}}{JK(n - 1)}$	
Between-Rows	$J - 1$	$\displaystyle \text{MSB} = \frac{\text{SSB}}{(J - 1)}$	$\displaystyle F = \frac{\text{MSB}}{\text{MSE}}$
Interaction	$(J - 1)(K - 1)$	$\displaystyle \text{MSI} = \frac{\text{SSI}}{(J - 1)(K - 1)}$	$\displaystyle F = \frac{\text{MSI}}{\text{MSE}}$
Total	$N - 1$		

where, k refers to the current treatment (i.e., the column), K refers to the total number of treatments (columns), j refers to the current block (i.e., the group of rows that make up the current *CELL* replicate), J refers to the total number of replicate blocks (i.e., the number of rows divided by the number of replicates), n is the number of replicates per *CELL*, N is the total number of data points, T_k represents the sum (total) of the values in a particular treatment group, T_j represents the sum (total) of the values in a particular block group (i.e., group of rows that make up the replicate), X_{jki} represents an individual observation, and T represents the sum of all sample values in all groups.

As in the previous categories, it would be good practice to look at the formulae in the two-way "n Observations Per *CELL*" ANOVA template to confirm that the template formulae agree with the manual formulae.

```
--------T-----------U--------V----------W--------X--------Y--------Z--------AA-------AB-------AC-----
49                          ****SUM OF SQUARES SCRATCH-PAD****
50                 LAB       LAB       LAB       LAB       LAB       LAB
51                  A         B         C         D         E         F      TOTALS
52      SUM_X     2559.2    2522.3    2503.3    2539.4    2528.2    2531     15183
53    SUM_X_SQR   546256    530550    522472    537573    532801    534309   3203961
54         N        12        12        12        12        12        12       72
55   (SUM_X)^2   6549505   6361997   6266511   6448552   6391795   6405961  38424321
56   (SUM_X)^2/n  545792    530166    522209    537379    532650    533830   3202027
57       MEAN     213.3     210.2     208.6     211.6     210.7     210.9     1265
58
59              ****ANALYSIS OF VARIANCE TABLE****
60              (TWO WAY ANALYSIS WITH INTERACTION)
61                 (n OBSERVATIONS PER CELL)
62
63   SOURCE OF   SUM OF  DEGREES OF ESTIMATE          TABLE
64   VARIATION   SQUARES  FREEDOM   OF STD   F RATIO  VALUE
65   ================================================================
66   BETWEEN
67   COLUMNS      143.0       5        29      1.04    2.45
68
69   SAMPLING
70   ERROR       1316.3      48      27.42
71
72   BETWEEN
73   ROWS          79.2       3      26.4      0.96    2.84
74
75   INTERACTION  539.0      15      35.93     1.31    1.92
76
77
78   TOTAL       2077.5      71
79
```

Figure 11-4

EXPANDING THE SIZE OF A TEMPLATE

The templates presented in this chapter are large enough to accommodate nine plots and six treatments. However, it is very easy to expand the size of the templates to hold more data. If you recall, the sum of squares scratch-pads and ANOVA summary tables contain formulae that calculate summations, counts, averages, etc. If you were to merely add rows below the Raw Data and Square Data sections, you would need to edit all of these formulae to include the new rows. Likewise, addition of columns to the right of the last template column would require similar corrections.

However, if you move the last row of a section down (or the last column of a section to the right) by the number of rows (columns) that you want to add, all formulae will automatically adjust to include the new rows or columns. For example, suppose you wanted to add two more columns to Figure 11-2. You would place the cursor on Column P and issue the /Worksheet Insert Column (Lotus 1-2-3®) or /Insert Column (Symphony®) command. When prompted for the column to insert, specify Columns P and Q. The old Column P would then slide to the right and become R. Old Column Q would become S. More importantly, formulae referring to Columns K through P would now refer to Columns K through R.

Similarly, you could insert rows using a /Worksheet Insert Row (Lotus 1-2-3) or /Insert Row (Symphony) menu command. However, these commands often affect other templates in the spreadsheet. For example, Figures 11-1, 11-2, and 11-3 are all in the same spreadsheet. Inserting a row in Figure 11-2 would affect all three templates. Therefore, you should use the /Move command, specify the last row of the section as the starting point, and press the Page Down key several times to specify the entire template. When prompted for the cell to move to, specify the next lower cell (or cells) in the template. You will also need to repeat this process for the Square Data section(s) because it (they) corresponds one-to-one with the Raw Data section(s).

Whenever you add a row or column to a template, you should look to see if any formulae need to be placed into the template or updated. For example, the addition of columns in Figure 11-2 means that the formulae for calculating the square data in Columns P and Q need to be added. Also, because the last row in the Square Data section was moved, it may refer to an incorrect row in the Raw Data section. Therefore, you should use the /Copy command to re-copy cell K14 to the rest of the cells of the new Square Data section.

It would be a good exercise for you to add rows and columns and confirm that the summary formulae update correctly.

SUMMARY OF WHAT HAS BEEN ACCOMPLISHED

The primary focus of this chapter was to provide an insight into using Lotus® as a tool to make the application of ANOVA to the study of experimental data both convenient and routine. In support of this effort, the chapter provided an introduction to ANOVA.

The chapter stressed taking the time to design experiments in such a way that the format of the data provides maximal ANOVA efficiency and meaningfulness. Based on sound experimental design, the data from a multiple treatment experiment can be divided into distinct types of variation. ANOVA can then be used to perform an analysis on the variation so that you can reveal relationships between the means of alternate treatments. To this end, this chapter provided an overview on how to isolate sources of error and perform sum of squares calculations. Once isolated, the mean sum of squares for the variation source could be compared to the MSE using the F-test.

The information provided in this chapter should give you a strong basis for using ANOVA to address your experimental data reduction problems. However, rather than being a definitive reference on the rather large topic of ANOVA, part of the intention of the chapter was to challenge your imagination and creativity so that you will consider Lotus as a tool for ANOVA. The information given in this chapter should give you a good start. Using the information presented in this chapter, you should find it easy to design ANOVA tables for your particular applications. If you are unsure of how to implement ANOVA for your particular application, I have provided a list of references below. These references provide comprehensive reviews of many experimental designs and ANOVA formulae.

WHAT'S NEXT?

The remainder of this book contains a number of useful appendices. The first three appendices provide summaries for the Lotus arithmetic, relational, and logical operators, the Lotus @functions, and the Lotus macro commands. The remainder of the appendices contain statistical tables that have been referred to throughout this book.

REFERENCES

ANDERSON, ROBERT, *Practical Statistics for Analytical Chemists*. New York, NY.: Van Nostrand Reinhold Co., 1987.

BOX, GEORGE, HUNTER, WILLIAM, and HUNTER, J. STUART, *Statistics for Experimenters: An Introduction to Design, Data Analysis, and Model Building*. New York, NY.: John Wiley & Sons, 1978.

DAVIES, OWEN, and GOLDSMITH, PETER, *Statistical Methods in Research and Production*. London, Eng: Longman Group Limited, 1972.

FREUND, JOHN, *Modern Elementary Statistics*. Englewood Cliffs, NJ.: Prentice Hall, Inc., 1988.

MILLER, IRWIN, and FREUND, JOHN, *Probability and Statistics for Engineers*. Englewood Cliffs, NJ.: Prentice Hall, Inc., 1985.

WEINBERG, GEORGE, SCHUMAKER, JOHN, and OLTMAN, DEBRA, *Statistics: An Intuitive Approach*. Monterey, CA: Brooks/Cole Publishing Co., 1981.

A

Arithmetic, Relational, And Logical Operators

Lotus 1-2-3® and Symphony® use a set of "operators" to depict arithmetic operations. For example, the + operator causes the two values flanking it to be added together.

Lotus has a well-defined hierarchy that is used to determine the progression in which operators will be used. For example, consider the equation:

$$\text{value} = 10.0 + 35.0 * \text{slope/factor}$$

This equation has an addition, a multiplication, and a division. Which operation is done first? Is 10.0 added to 35.0, the result 45.0 then multiplied by slope, and then that result divided by factor? Or, is 35.0 multiplied by slope, the result added to 10.0, and then that result divided by factor?

Clearly, the order of execution for various operations can have a profound influence on the final value. To guard against chaos, Lotus® needs unambiguous rules for choosing what to do first. Lotus sets up a pecking order for its operators. This pecking order is called "precedence" and each operator is assigned a precedence level.

The table shown below gives the precedence for each operator. In the table, division and multiplication have higher precedences than addition and subtraction, so division and multiplication are performed first. If the precedence(s) of two or more operators in a statement is the same, they will be executed according to the order in which they occur in the statement (left to right). If you are unsure of the precedence in an equation (or if you want to change it), you should use parentheses. Any operation enclosed in parentheses is performed first. Within parentheses, the usual rules apply.

Operator	Definition	Precedence
^	Exponentiation	7 (highest)
–	Negation	6
*	Multiplication	5
/	Division	5
+	Addition	4
–	Subtraction	4
=	Equality	3
&	String concatenation	1 (lowest)

In addition to arithmetic operators, Lotus® has a category of operators that can be used for making comparisons. These operators are called "relational" operators. Each relational operator compares the value at its left to the value at its right. A relational expression evaluated from an operator and its two operands has the value 1 if the expression is true and the value 0 if the expression is false.

Relational operators are most commonly used for testing the relationship of two expressions in an {IF} macro command or an @IF function. The following table gives the definition and precedence of each relational operator. The precedences are on the same scale as the arithmetic operators. Because the precedences are lower than those for arithmetic operators, the arithmetic operations occur before relational operations are evaluated.

Operator	Definition	Precedence
<	Less than	3
<=	Less than or equal	3
>	Greater than	3
>=	Greater than or equal	3
<>	Not equal	3

The final category of operators contains the "logical" operators. Logical operators normally take relational expressions as operands. In this context, logical operators are useful for combining two or more relational expressions. The following table gives the definition and precedence of each logical operator. Because the precedences for logical operators are lower than the precedences for both arithmetic and relational operators, logical operators are evaluated last.

Operator	Definition	Precedence
#NOT#	Logical NOT	2
#AND#	Logical AND	1
#OR#	Logical OR	1

A logical expression evaluated from an operator and its operands has the value 1 if the expression is true and the value 0 if the expression is false. The following are brief explanations of each logical operator:

#NOT#: The expression is true if the operand is false, and vice versa.

#AND#: The combined expression is true if *both* operands are true, and false otherwise.

#OR#: The combined expression is true if one *or* both operands are true, and false otherwise.

Logical expressions are evaluated from left to right. The evaluation stops as soon as the expression is rendered false. The following examples illustrate the use of logical operands:

6>2#AND#3>1   is true

#NOT#(6>1#AND#10>4)   is false

6>7#OR#6>1   is true

B

@Function Summary

Special-purpose Lotus® formulas are called "@functions". There are a large number of Lotus @functions. Each @function extends your calculating power far beyond simple arithmetic and text-handling operations. There are several broad categories of Lotus @functions: mathematical, statistical, string, special, logical, and date/time.

The following is a list of the @functions used in this book and some others that you will find useful. The definitions are tailored to the primary purpose of this book—to create Lotus spreadsheets that have powerful data reduction templates and macro programs.

MATHEMATICAL @FUNCTIONS

@ABS(number)

Absolute value of *number*.

@ACOS(number)

Arc cosine of *number*.

@ASIN(number)

Arc sine of *number*.

@ATAN(number)

Two-quadrant arc tangent of *number*.

@ATAN2(number1,number2)

Four-quadrant arc tangent of *number2/number1*.

@COS(number)

Cosine of *number*.

@EXP(number)

The number e raised to the *number* power.

@INT(number)

Integer part of *number*.

@LN (number)

Log of *number*, base e.

@LOG(number)

Log of *number*, base 10.

@PI

The number pi (approximately 3.14159).

@SIN(number)

Sine of *number*.

@SQRT(number)

Positive square root of *number*.

@TAN(number)

Tangent of *number*.

STATISTICAL @FUNCTIONS

@AVG(list)

Arithmetic average of values in *list*.

@COUNT(list)

Number of cells containing values in *list*.

@MAX(list)

Maximum value in *list*.

@MIN(list)

Minimum value in *list*.

@STD(list)

Standard deviation of values in *list*.

@SUM(list)

Sum of values in *list*.

@VAR(list)

Variance of values in *list*.

STRING @FUNCTIONS

@CHAR(number)

Returns the ASCII/LICS character represented by *number*.

@CLEAN(string)

Removes the control characters from *string*.

@CODE(string)

Returns the ASCII/LICS code of first character in *string*.

@EXACT(string1,string2)

Tests whether *string1* and *string2* have exactly the same characters.

@LEFT(string,number)

Left-most *number* of characters in *string*.

@LENGTH(string)

Total number of characters in *string*.

@LOWER(string)

Converts all upper-case letters in *string* to lower-case.

@MID(string,startingposition,numberofchars)

Extracts *numberofchars* characters from *string*, beginning at offset *startingposition*.

@N(cell)

Numeric value of *cell*.

@RIGHT(string,number)

Right-most *number* of characters in *string*.

@S(cell)

String value of *cell*.

@STRING(cell,places)

Converts the number in *cell* into a string with *places* decimal places.

@TRIM(string)

Removes the leading and trailing spaces from *string* and compresses multiple spaces within *string* into single spaces.

@UPPER(string)

Converts all lower-case letters in *string* to upper-case.

@VALUE(string)

Converts a *string* that looks like a number into that number.

SPECIAL @FUNCTIONS

@CELLPOINTER(spec)

Attribute of the cell currently highlighted by the cell pointer.

@CHOOSE(selector,val1,val2,val3,...)

Selects value based on its position in the list.

@HLOOKUP(selector,rowrange,offsetnumber)

Table lookup, comparing *selector* value with a row of values. The value of *selector* may be either a string or a number.

@INDEX(range,coloffset,rowoffset)

Lookup based on position in *range*.

@VLOOKUP(selector,colrange,offsetnumber)

Table lookup, comparing *selector* value with a column of values. The value of *selector* may be either a string or a number.

LOGICAL @FUNCTIONS

@IF(criterion,value1,value2)

If criterion is non-zero, *value1* is returned; if criterion is zero, *value2* is returned.

@ISERR(value)

If *value* has the value ERR, then 1 (TRUE) is returned; otherwise 0 (FALSE).

@ISNUMBER(value)

If *value* has a numeric value, then 1 (TRUE) is returned; otherwise 0 (FALSE).

@ISSTRING(value)

If *value* is a string, then 1 (TRUE) is returned; otherwise 0 (FALSE).

DATE AND TIME @FUNCTIONS

@DATE(year,month,day)

Serial number of specified date (Jan 1, 1900 = 1).

@DATEVALUE(datestring)

Serial number of date specified by *datestring*.

@NOW

Serial number of current moment.

@TIME(hour,minute,second)

Serial number of specified time of day.

@TIMEVALUE(timestring)

Serial number of time of day specified by *timestring*.

@DAY(serialnumber)

Day number of *serialnumber*.

@HOUR(serialnumber)

Hour number of *serialnumber*.

@MINUTE(serialnumber)

Minute number of *serialnumber*.

@MONTH(serialnumber)

Month number of *serialnumber*.

@SECOND(serialnumber)

Second number of *serialnumber*.

@YEAR(serialnumber)

Year number of *serialnumber*.

@END(appendix)

At the end of this book's appendix. (Not a real @function.)

C

Macro Command Summary

A set of instructions in a format and language that Lotus® can understand is called a Lotus "macro" program. The language that is used is called the "Lotus Command Language" and the instructions are called "macro commands."

There are six broad categories of Lotus macro commands:

- **System commands**, which allow you to control the screen display and the computer's speaker.
- **Interaction commands**, which allow you to create interactive macros that pause for the user to enter data from the keyboard.
- **Program flow commands**, which let you include branching and looping in your program.
- **Cell commands**, which transfer data between specified cells and/or change cell values.
- **File commands**, which work with data in DOS files or communications ports.
- **Menu commands**, which allow you to manage menus of your own design.

The following is a list, by category, of the macro commands used in this book. The definitions are tailored to the primary purpose of this book—to create Lotus spreadsheets that have powerful data reduction macro programs.

SYSTEM COMMANDS

{BEEP tone}

Sounds one of four different tones (1–4).

{INDICATE string}

Replaces the standard mode indicator message at the upper right corner of the display with the characters specified by *string*. Omitting *string* restores control of the mode indicator to Lotus.

{PANELOFF}

Suppresses updating of the control panel during macro execution.

{PANELON}

Restores standard updating of panel, canceling a {PANELOFF} command.

{WINDOWSOFF}

Suppresses updating of the display during macro execution.

{WINDOWSON}

Restores standard updating of display, canceling a {WINDOWSOFF} command.

INTERACTION COMMANDS

{?}

Halts macro execution temporarily, allowing the user to type in data, move the cell-pointer, and access the Lotus menus. Macro execution resumes when Return is pressed.

{GETLABEL prompt,cell}

Halts macro execution, displays a *prompt* for the user to type in a line of characters, and stores the response as a label in the specified *cell*.

{GETNUMBER prompt,cell}

Halts macro execution, displays a *prompt* for the user to type in a number or numeric expression, and stores the response as a number in the specified *cell*.

PROGRAM FLOW COMMANDS

{macro}

Calls "*macro*" as a subroutine.

{\R}

Calls \R as a subroutine. Same as pressing [ALT]-R.

{BRANCH macro}

Transfers control of the program to the location called "*macro*".

{IF condition}

Conditionally executes the commands that follow the IF command in the same cell if *condition* evaluates to true.

{ONERROR macro,messagecell}

Branches to *macro* if an otherwise fatal error occurs. Optionally stores the error message that Lotus would have displayed in *messagecell*.

{FOR counter,start,stop,stepsize,subroutine}

Repeatedly executes the commands, beginning at the cell called "*subroutine*," as many times as indicated by the values of *start*, *stop*, and *stepsize*.

{QUIT}

Immediately terminates macro processing, returning to manual keyboard control.

{WAIT time}

Causes macro execution to pause until the computer's clock time matches or exceeds *time*.

CELL COMMANDS

{BLANK range}

Erases the cells within the specified *range*.

{LET cell,number}

Places *number* into the specified *cell*.

{LET cell,string}

Places *string* into the specified *cell*.

{PUT range,column,row,number}

Places *number* into the cell in *range* represented by offset coordinates (*column,row*).

{PUT range,column,row,string}

Places *string* into the cell in *range* represented by offset coordinates (*column,row*).

{RECALC range}

Recalculates the formulae in *range*, proceeding column by column.

FILE COMMANDS

{CLOSE}

Closes a disk file or, more commonly, a communications port.

{OPEN filename,mode}

Opens *filename* (disk file or communications port) for read only (R), write only (W), append (A), or both read/write (M).

{READ bytes,buffer}

Reads the number of characters specified by *bytes* from a file or communications port and stores the characters as a string in the cell specified by *buffer*.

{READLN buffer}

Reads a line of characters (until a carriage return is encountered) from a file or communications port and stores the characters as a string in the cell specified by *buffer*.

{WRITE string}

Writes a string of characters (without a carriage return) to a file or communications port.

{WRITELN string}

Writes a line of characters (appending a carriage return) to a file or communications port.

MENU COMMANDS

{MENUBRANCH menu}

Halts macro execution temporarily, branches to a customized menu whose instructions are found starting at the cell called "*menu*," prompts the user to make a choice, then continues execution based on the choice.

{MENUCALL menu}

Similar to {MENUBRANCH}. However, processes *menu* and its associated steps as a subroutine.

MACRO COMMANDS FOR SPECIAL KEYS

In addition to the above macro commands, there are commands which represent special Lotus keys. The cursor control commands allow you to specify the number of times your macro should "press" the named key. For example, {RIGHT 5} is equivalent to manually pressing the right arrow five times. The following is a table of these commands:

{EDIT}	{PGUP}
{GOTO location}	{PGDN}
{WINDOW}	{END}
{CALC}	{ESC}
{UP}	{BACKSPACE}
{DOWN}	{BIGLEFT}
{LEFT}	{BIGRIGHT}
{RIGHT}	{MENU}
{HOME}	~(CARRIAGE RETURN)
{SERVICES} or {S} (SYMPHONY ONLY)	

D

Critical Values for *t*-Distribution

	Single-sided test					Double-sided test			
	P						P			
DF	0·005	0·01	0·05	0·1	DF	0·005	0·01	0·05	0·1
1	63·7	31·8	6·31	3·08	1	127	63·7	12·7	6·31
2	9·92	6·96	2·92	1·89	2	14·1	9·92	4·30	2·92
3	5·84	4·54	2·35	1·64	3	7·45	5·84	3·18	2·35
4	4·60	3·75	2·13	1·53	4	5·60	4·60	2·78	2·13
5	4·03	3·36	2·01	1·48	5	4·77	4·03	2·57	2·01
6	3·71	3·14	1·94	1·44	6	4·32	3·71	2·45	1·94
7	3·50	3·00	1·89	1·42	7	4·03	3·50	2·36	1·89
8	3·36	2·90	1·86	1·40	8	3·83	3·36	2·31	1·86
9	3·25	2·82	1·83	1·38	9	3·69	3·25	2·26	1·83
10	3·17	2·76	1·81	1·37	10	3·58	3·17	2·23	1·81
11	3·11	2·72	1·80	1·36	11	3·50	3·11	2·20	1·80
12	3·05	2·68	1·78	1·36	12	3·43	3·05	2·18	1·78
13	3·01	2·65	1·77	1·35	13	3·37	3·01	2·16	1·77
14	2·98	2·62	1·76	1·34	14	3·33	2·98	2·14	1·76
15	2·95	2·60	1·75	1·34	15	3·29	2·95	2·13	1·75
16	2·92	2·58	1·75	1·34	16	3·25	2·92	2·12	1·75
17	2·90	2·57	1·74	1·33	17	3·22	2·90	2·11	1·74
18	2·88	2·55	1·73	1·33	18	3·20	2·88	2·10	1·73
19	2·86	2·54	1·73	1·33	19	3·17	2·86	2·09	1·73
20	2·85	2·53	1·72	1·32	20	3·15	2·85	2·09	1·72

Critical Values for *t*-Distribution (*continued*)

	Single-sided test					Double-sided test			
	P					**P**			
DF	0·005	0·01	0·05	0·1	**DF**	0·005	0·01	0·05	0·1
21	2·83	2·52	1·72	1·32	21	3·14	2·83	2·08	1·72
22	2·82	2·51	1·72	1·32	22	3·12	2·82	2·07	1·72
23	2·81	2·50	1·71	1·32	23	3·10	2·81	2·07	1·71
24	2·80	2·49	1·71	1·32	24	3·09	2·80	2·06	1·71
25	2·79	2·48	1·71	1·32	25	3·08	2·79	2·06	1·71
26	2·78	2·48	1·71	1·32	26	3·07	2·78	2·06	1·71
27	2·77	2·47	1·70	1·31	27	3·06	2·77	2·05	1·70
28	2·76	2·47	1·70	1·31	28	3·05	2·76	2·05	1·70
29	2·76	2·46	1·70	1·31	29	3·04	2·76	2·05	1·70
30	2·75	2·46	1·70	1·31	30	3·03	2·75	2·04	1·70
40	2·70	2·42	1·68	1·30	40	2·97	2·70	2·02	1·68
60	2·66	2·39	1·67	1·30	60	2·91	2·66	2·00	1·67
120	2·62	2·36	1·66	1·29	120	2·86	2·62	1·98	1·66
∞	2·58	2·33	1·64	1·28	∞	2·81	2·58	1·96	1·64

Source: Reprinted from Anderson, R.L., *Practical Statistics for Analytical Chemists,* Van Nostrand Reinhold, New York, 1987. By permission of the author and publishers.

E

Area Under the Normal (Gaussian) Curve

z	0	1	2	3	4	5	6	7	8	9
0.0	.0000	.0040	.0080	.0120	.0160	.0199	.0239	.0279	.0319	.0359
0.1	.0398	.0438	.0478	.0517	.0557	.0596	.0636	.0675	.0714	.0754
0.2	.0793	.0832	.0871	.0910	.0948	.0987	.1026	.1064	.1103	.1141
0.3	.1179	.1217	.1255	.1293	.1331	.1368	.1406	.1443	.1480	.1517
0.4	.1554	.1591	.1628	.1664	.1700	.1736	.1772	.1808	.1844	.1879
0.5	.1915	.1950	.1985	.2019	.2054	.2088	.2123	.2157	.2190	.2224
0.6	.2258	.2291	.2324	.2357	.2389	.2422	.2454	.2486	.2518	.2549
0.7	.2580	.2612	.2642	.2673	.2704	.2734	.2764	.2794	.2823	.2852
0.8	.2881	.2910	.2939	.2967	.2996	.3023	.3051	.3078	.3106	.3133
0.9	.3159	.3186	.3212	.3238	.3264	.3289	.3315	.3340	.3365	.3389
1.0	.3413	.3438	.3461	.3485	.3508	.3531	.3554	.3577	.3599	.3621
1.1	.3643	.3665	.3686	.3708	.3729	.3749	.3770	.3790	.3810	.3830
1.2	.3849	.3869	.3888	.3907	.3925	.3944	.3962	.3980	.3997	.4015
1.3	.4032	.4049	.4066	.4082	.4099	.4115	.4131	.4147	.4162	.4177
1.4	.4192	.4207	.4222	.4236	.4251	.4265	.4279	.4292	.4306	.4319
1.5	.4332	.4345	.4357	.4370	.4382	.4394	.4406	.4418	.4429	.4441
1.6	.4452	.4463	.4474	.4484	.4495	.4505	.4515	.4525	.4535	.4545
1.7	.4554	.4564	.4573	.4582	.4591	.4599	.4608	.4616	.4625	.4633
1.8	.4641	.4649	.4656	.4664	.4671	.4678	.4686	.4693	.4699	.4706
1.9	.4713	.4719	.4726	.4732	.4738	.4744	.4750	.4756	.4761	.4767

Area Under the Normal (Gaussian) Curve (*continued*)

z	0	1	2	3	4	5	6	7	8	9
2.0	.4772	.4778	.4783	.4788	.4793	.4798	.4803	.4808	.4812	.4817
2.1	.4821	.4826	.4830	.4834	.4838	.4842	.4846	.4850	.4854	.4857
2.2	.4861	.4864	.4868	.4871	.4875	.4878	.4881	.4884	.4887	.4890
2.3	.4893	.4896	.4898	.4901	.4904	.4906	.4909	.4911	.4913	.4916
2.4	.4918	.4920	.4922	.4925	.4927	.4929	.4931	.4932	.4934	.4936
2.5	.4938	.4940	.4941	.4943	.4945	.4946	.4948	.4949	.4951	.4952
2.6	.4953	.4955	.4956	.4957	.4959	.4960	.4961	.4962	.4963	.4964
2.7	.4965	.4966	.4967	.4968	.4969	.4970	.4971	.4972	.4973	.4974
2.8	.4974	.4975	.4976	.4977	.4977	.4978	.4979	.4979	.4980	.4981
2.9	.4981	.4982	.4982	.4983	.4984	.4984	.4985	.4985	.4986	.4986
3.0	.4987	.4987	.4987	.4988	.4988	.4989	.4989	.4989	.4990	.4990
3.1	.4990	.4991	.4991	.4991	.4992	.4992	.4992	.4992	.4993	.4993
3.2	.4993	.4993	.4994	.4994	.4994	.4994	.4994	.4995	.4995	.4995
3.3	.4995	.4995	.4995	.4996	.4996	.4996	.4996	.4996	.4996	.4997
3.4	.4997	.4997	.4997	.4997	.4997	.4997	.4997	.4997	.4997	.4998
3.5	.4998	.4998	.4998	.4998	.4998	.4998	.4998	.4998	.4998	.4998
3.6	.4998	.4998	.4999	.4999	.4999	.4999	.4999	.4999	.4999	.4999
3.7	.4999	.4999	.4999	.4999	.4999	.4999	.4999	.4999	.4999	.4999
3.8	.4999	.4999	.4999	.4999	.4999	.4999	.4999	.4999	.4999	.4999
3.9	.5000	.5000	.5000	.5000	.5000	.5000	.5000	.5000	.5000	.5000

Source: Reprinted from Spiegel, M.R., *Schaum's Outline Series: Theory and Problems of Statistics,* McGraw-Hill Book Company, New York, 1961. By permission of the author and publishers.

F

Critical Values of the *F*-Distribution

Critical Values of the *F* Distribution, $\alpha = 0.05$.

	\multicolumn{18}{c}{Degrees of freedom for the numerator (n)}																	
	1	2	3	4	5	6	7	8	9	10	11	12	13	14	15	16	17	18
1	161	200	216	225	230	234	237	239	241	242	243	244	245	245	246	246	247	247
2	18.5	19.0	19.2	19.2	19.3	19.3	19.4	19.4	19.4	19.4	19.4	19.4	19.4	19.4	19.4	19.4	19.4	19.4
3	10.1	9.55	9.28	9.12	9.01	8.94	8.89	8.85	8.81	8.79	8.76	8.74	8.73	8.71	8.70	8.69	8.68	8.67
4	7.71	6.94	6.59	6.39	6.26	6.16	6.09	6.04	6.00	5.96	5.94	5.91	5.89	5.87	5.86	5.84	5.83	5.82
5	6.61	5.79	5.41	5.19	5.05	4.95	4.88	4.82	4.77	4.74	4.70	4.68	4.66	4.64	4.62	4.60	4.59	4.58
6	5.99	5.14	4.76	4.53	4.39	4.28	4.21	4.15	4.10	4.06	4.03	4.00	3.98	3.96	3.94	3.92	3.91	3.90
7	5.59	4.74	4.35	4.12	3.97	3.87	3.79	3.73	3.68	3.64	3.60	3.57	3.55	3.53	3.51	3.49	3.48	3.47
8	5.32	4.46	4.07	3.84	3.69	3.58	3.50	3.44	3.39	3.35	3.31	3.28	3.26	3.24	3.22	3.20	3.19	3.17
9	5.12	4.26	3.86	3.63	3.48	3.37	3.29	3.23	3.18	3.14	3.10	3.07	3.05	3.03	3.01	2.99	2.97	2.96
10	4.96	4.10	3.71	3.48	3.33	3.22	3.14	3.07	3.02	2.98	2.94	2.91	2.89	2.86	2.85	2.83	2.81	2.80
11	4.84	3.98	3.59	3.36	3.20	3.09	3.01	2.95	2.90	2.85	2.82	2.79	2.76	2.74	2.72	2.70	2.69	2.67
12	4.75	3.89	3.49	3.26	3.11	3.00	2.91	2.85	2.80	2.75	2.72	2.69	2.66	2.64	2.62	2.60	2.58	2.57
13	4.67	3.81	3.41	3.18	3.03	2.92	2.83	2.77	2.71	2.67	2.63	2.60	2.58	2.55	2.53	2.51	2.50	2.48
14	4.60	3.74	3.34	3.11	2.96	2.85	2.76	2.70	2.65	2.60	2.57	2.53	2.51	2.48	2.46	2.44	2.43	2.41
15	4.54	3.68	3.29	3.06	2.90	2.79	2.71	2.64	2.59	2.54	2.51	2.48	2.45	2.42	2.40	2.38	2.37	2.35

Critical Values of the F-Distribution (*continued*)

		Degrees of freedom for the numerator (n)																
	1	2	3	4	5	6	7	8	9	10	11	12	13	14	15	16	17	18
16	4.49	3.63	3.24	3.01	2.85	2.74	2.66	2.59	2.54	2.49	2.46	2.42	2.40	2.37	2.35	2.33	2.32	2.30
17	4.45	3.59	3.20	2.96	2.81	2.70	2.61	2.55	2.49	2.45	2.41	2.38	2.35	2.33	2.31	2.29	2.27	2.26
18	4.41	3.55	3.16	2.93	2.77	2.66	2.58	2.51	2.46	2.41	2.37	2.34	2.31	2.29	2.27	2.25	2.23	2.22
19	4.38	3.52	3.13	2.90	2.74	2.63	2.54	2.48	2.42	2.38	2.34	2.31	2.28	2.26	2.23	2.21	2.20	2.18
20	4.35	3.49	3.10	2.87	2.71	2.60	2.51	2.45	2.39	2.35	2.31	2.28	2.25	2.22	2.20	2.18	2.17	2.15
21	4.32	3.47	3.07	2.84	2.68	2.57	2.49	2.42	2.37	2.32	2.28	2.25	2.22	2.20	2.18	2.16	2.14	2.12
22	4.30	3.44	3.05	2.82	2.66	2.55	2.46	2.40	2.34	2.30	2.26	2.23	2.20	2.17	2.15	2.13	2.11	2.10
23	4.28	3.42	3.03	2.80	2.64	2.53	2.44	2.37	2.32	2.27	2.23	2.20	2.18	2.15	2.13	2.11	2.09	2.07
24	4.26	3.40	3.01	2.78	2.62	2.51	2.42	2.36	2.30	2.25	2.21	2.18	2.15	2.13	2.11	2.09	2.07	2.05
25	4.24	3.39	2.99	2.76	2.60	2.49	2.40	2.34	2.28	2.24	2.20	2.16	2.14	2.11	2.09	2.07	2.05	2.04
26	4.23	3.37	2.98	2.74	2.59	2.47	2.39	2.32	2.27	2.22	2.18	2.15	2.12	2.09	2.07	2.05	2.03	2.02
27	4.21	3.35	2.96	2.73	2.57	2.46	2.37	2.31	2.25	2.20	2.17	2.13	2.10	2.08	2.06	2.04	2.02	2.00
28	4.20	3.34	2.95	2.71	2.56	2.45	2.36	2.29	2.24	2.19	2.15	2.12	2.09	2.06	2.04	2.02	2.00	1.99
29	4.18	3.33	2.93	2.70	2.55	2.43	2.35	2.28	2.22	2.18	2.14	2.10	2.08	2.05	2.03	2.01	1.99	1.97
30	4.17	3.32	2.92	2.69	2.53	2.42	2.33	2.27	2.21	2.16	2.13	2.09	2.06	2.04	2.01	1.99	1.98	1.96
32	4.15	3.29	2.90	2.67	2.51	2.40	2.31	2.24	2.19	2.14	2.10	2.07	2.04	2.01	1.99	1.97	1.95	1.94
34	4.13	3.28	2.88	2.65	2.49	2.38	2.29	2.23	2.17	2.12	2.08	2.05	2.02	1.99	1.97	1.95	1.93	1.92
36	4.11	3.26	2.87	2.63	2.48	2.36	2.28	2.21	2.15	2.11	2.07	2.03	2.00	1.98	1.95	1.93	1.92	1.90
38	4.10	3.24	2.85	2.62	2.46	2.35	2.26	2.19	2.14	2.09	2.05	2.02	1.99	1.96	1.94	1.92	1.90	1.88
40	4.08	3.23	2.84	2.61	2.45	2.34	2.25	2.18	2.12	2.08	2.04	2.00	1.97	1.95	1.92	1.90	1.89	1.87
42	4.07	3.22	2.83	2.59	2.44	2.32	2.24	2.17	2.11	2.06	2.03	1.99	1.96	1.93	1.91	1.89	1.87	1.86
44	4.06	3.21	2.82	2.58	2.43	2.31	2.23	2.16	2.10	2.05	2.01	1.98	1.95	1.92	1.90	1.88	1.86	1.84
46	4.05	3.20	2.81	2.57	2.42	2.30	2.22	2.15	2.09	2.04	2.00	1.97	1.94	1.91	1.89	1.87	1.85	1.83
48	4.04	3.19	2.80	2.57	2.41	2.29	2.21	2.14	2.08	2.03	1.99	1.96	1.93	1.90	1.88	1.86	1.84	1.82
50	4.03	3.18	2.79	2.56	2.40	2.29	2.20	2.13	2.07	2.03	1.99	1.95	1.92	1.89	1.87	1.85	1.83	1.81
55	4.02	3.16	2.77	2.54	2.38	2.27	2.18	2.11	2.06	2.01	1.97	1.93	1.90	1.88	1.85	1.83	1.81	1.79
60	4.00	3.15	2.76	2.53	2.37	2.25	2.17	2.10	2.04	1.99	1.95	1.92	1.89	1.86	1.84	1.82	1.80	1.78
65	3.99	3.14	2.75	2.51	2.36	2.24	2.15	2.08	2.03	1.98	1.94	1.90	1.87	1.85	1.82	1.80	1.78	1.76
70	3.98	3.13	2.74	2.50	2.35	2.23	2.14	2.07	2.02	1.97	1.93	1.89	1.86	1.84	1.81	1.79	1.77	1.75
80	3.96	3.11	2.72	2.49	2.33	2.21	2.13	2.06	2.00	1.95	1.91	1.88	1.84	1.82	1.79	1.77	1.75	1.73
90	3.95	3.10	2.71	2.47	2.32	2.20	2.11	2.04	1.99	1.94	1.90	1.86	1.83	1.80	1.78	1.76	1.74	1.72
100	3.94	3.09	2.70	2.46	2.31	2.19	2.10	2.03	1.97	1.93	1.89	1.85	1.82	1.79	1.77	1.75	1.73	1.71
125	3.92	3.07	2.68	2.44	2.29	2.17	2.08	2.01	1.96	1.91	1.87	1.83	1.80	1.77	1.75	1.72	1.70	1.69
150	3.90	3.06	2.66	2.43	2.27	2.16	2.07	2.00	1.94	1.89	1.85	1.82	1.79	1.76	1.73	1.71	1.69	1.67
200	3.89	3.04	2.65	2.42	2.26	2.14	2.06	1.98	1.93	1.88	1.84	1.80	1.77	1.74	1.72	1.69	1.67	1.66
300	3.87	3.03	2.63	2.40	2.24	2.13	2.04	1.97	1.91	1.86	1.82	1.78	1.75	1.72	1.70	1.68	1.66	1.64
500	3.86	3.01	2.62	2.39	2.23	2.12	2.03	1.96	1.90	1.85	1.81	1.77	1.74	1.71	1.69	1.66	1.64	1.62
1000	3.85	3.00	2.61	2.38	2.22	2.11	2.02	1.95	1.89	1.84	1.80	1.76	1.73	1.70	1.68	1.65	1.63	1.61
∞	3.84	3.00	2.60	2.37	2.21	2.10	2.01	1.94	1.88	1.83	1.79	1.75	1.72	1.69	1.67	1.64	1.62	1.60

Degrees of freedom for the denominator (m)

Example: $P\{F_{46;8;30} < 2.45\} = 95\%$.

$F_{.05;\nu_1,\nu_2} = 1/F_{.95;\nu_2,\nu_1}.$ Example: $F_{.05;8;8} = 1/F_{.95;8,8} = 1/3.15 = 0.317$.

Source: Reprinted from Anderson, R.L., *Practical Statistics for Analytical Chemists,* Van Nostrand Reinhold, New York, 1987. By permission of the author and publishers.

Degrees of freedom for the numerator (n_1)

19	20	22	24	26	28	30	35	40	45	50	60	80	100	200	500	∞	
248	248	249	249	249	250	250	251	251	251	252	252	252	253	254	254	254	1
19.4	19.4	19.5	19.5	19.5	19.5	19.5	19.5	19.5	19.5	19.5	19.5	19.5	19.5	19.5	19.5	19.5	2
8.67	8.66	8.65	8.64	8.63	8.62	8.62	8.60	8.59	8.59	8.58	8.57	8.56	8.55	8.54	8.53	8.53	3
5.81	5.80	5.79	5.77	5.76	5.75	5.75	5.73	5.72	5.71	5.70	5.69	5.67	5.66	5.65	5.64	5.63	4
4.57	4.56	4.54	4.53	4.52	4.50	4.50	4.48	4.46	4.45	4.44	4.43	4.41	4.41	4.39	4.37	4.37	5
3.88	3.87	3.86	3.84	3.83	3.82	3.81	3.79	3.77	3.76	3.75	3.74	3.72	3.71	3.69	3.68	3.67	6
3.46	3.44	3.43	3.41	3.40	3.39	3.38	3.36	3.34	3.33	3.32	3.30	3.29	3.27	3.25	3.24	3.23	7
3.16	3.15	3.13	3.12	3.10	3.09	3.08	3.06	3.04	3.03	3.02	3.01	2.99	2.97	2.95	2.94	2.93	8
2.95	2.94	2.92	2.90	2.89	2.87	2.86	2.84	2.83	2.81	2.80	2.79	2.77	2.76	2.73	2.72	2.71	9
2.78	2.77	2.75	2.74	2.72	2.71	2.70	2.68	2.66	2.65	2.64	2.62	2.60	2.59	2.56	2.55	2.54	10
2.66	2.65	2.63	2.61	2.59	2.58	2.57	2.55	2.53	2.52	2.51	2.49	2.47	2.46	2.43	2.42	2.40	11
2.56	2.54	2.52	2.51	2.49	2.48	2.47	2.44	2.43	2.41	2.40	2.38	2.36	2.35	2.32	2.31	2.30	12
2.47	2.46	2.44	2.42	2.41	2.39	2.38	2.36	2.34	2.33	2.31	2.30	2.27	2.26	2.23	2.22	2.21	13
2.40	2.39	2.37	2.35	2.33	2.32	2.31	2.28	2.27	2.25	2.24	2.22	2.20	2.19	2.16	2.14	2.13	14
2.34	2.33	2.31	2.29	2.27	2.26	2.25	2.22	2.20	2.19	2.18	2.16	2.14	2.12	2.10	2.08	2.07	15
2.29	2.28	2.25	2.24	2.22	2.21	2.19	2.17	2.15	2.14	2.12	2.11	2.08	2.07	2.04	2.02	2.01	16
2.24	2.23	2.21	2.19	2.17	2.16	2.15	2.12	2.10	2.09	2.08	2.06	2.03	2.02	1.99	1.97	1.96	17
2.20	2.19	2.17	2.15	2.13	2.12	2.11	2.08	2.06	2.05	2.04	2.02	1.99	1.98	1.95	1.93	1.92	18
2.17	2.16	2.13	2.11	2.10	2.08	2.07	2.05	2.03	2.01	2.00	1.98	1.96	1.94	1.91	1.89	1.88	19
2.14	2.12	2.10	2.08	2.07	2.05	2.04	2.01	1.99	1.98	1.97	1.95	1.92	1.91	1.88	1.86	1.84	20
2.11	2.10	2.07	2.05	2.04	2.02	2.01	1.98	1.96	1.95	1.94	1.92	1.89	1.88	1.84	1.82	1.81	21
2.08	2.07	2.05	2.03	2.01	2.00	1.98	1.96	1.94	1.92	1.91	1.89	1.86	1.85	1.82	1.80	1.78	22
2.06	2.05	2.02	2.00	1.99	1.97	1.96	1.93	1.91	1.90	1.88	1.86	1.84	1.82	1.79	1.77	1.76	23
2.04	2.03	2.00	1.98	1.97	1.95	1.94	1.91	1.89	1.88	1.86	1.84	1.82	1.80	1.77	1.75	1.73	24
2.02	2.01	1.98	1.96	1.95	1.93	1.92	1.89	1.87	1.86	1.84	1.82	1.80	1.78	1.75	1.73	1.71	25
2.00	1.99	1.97	1.95	1.93	1.91	1.90	1.87	1.85	1.84	1.82	1.80	1.78	1.76	1.73	1.71	1.69	26
1.99	1.97	1.95	1.93	1.91	1.90	1.88	1.86	1.84	1.82	1.81	1.79	1.76	1.74	1.71	1.69	1.67	27
1.97	1.96	1.93	1.91	1.90	1.88	1.87	1.84	1.82	1.80	1.79	1.77	1.74	1.73	1.69	1.67	1.65	28
1.96	1.94	1.92	1.90	1.88	1.87	1.85	1.83	1.81	1.79	1.77	1.75	1.73	1.71	1.67	1.65	1.64	29
1.95	1.93	1.91	1.89	1.87	1.85	1.84	1.81	1.79	1.77	1.76	1.74	1.71	1.70	1.66	1.64	1.62	30
1.92	1.91	1.88	1.86	1.85	1.83	1.82	1.79	1.77	1.75	1.74	1.71	1.69	1.67	1.63	1.61	1.59	32
1.90	1.89	1.86	1.84	1.82	1.80	1.80	1.77	1.75	1.73	1.71	1.69	1.66	1.65	1.61	1.59	1.57	34
1.88	1.87	1.85	1.82	1.81	1.79	1.78	1.75	1.73	1.71	1.69	1.67	1.64	1.62	1.59	1.56	1.55	36
1.87	1.85	1.83	1.81	1.79	1.77	1.76	1.73	1.71	1.69	1.68	1.65	1.62	1.61	1.57	1.54	1.53	38
1.85	1.84	1.81	1.79	1.77	1.76	1.74	1.72	1.69	1.67	1.66	1.64	1.61	1.59	1.55	1.53	1.51	40
1.84	1.83	1.80	1.78	1.76	1.74	1.73	1.70	1.68	1.66	1.65	1.62	1.59	1.57	1.53	1.51	1.49	42
1.83	1.81	1.79	1.77	1.75	1.73	1.72	1.69	1.67	1.65	1.63	1.61	1.58	1.56	1.52	1.49	1.48	44
1.82	1.80	1.78	1.76	1.74	1.72	1.71	1.68	1.65	1.64	1.62	1.60	1.57	1.55	1.51	1.48	1.46	46
1.81	1.79	1.77	1.75	1.73	1.71	1.70	1.67	1.64	1.62	1.61	1.59	1.56	1.54	1.49	1.47	1.45	48
1.80	1.78	1.76	1.74	1.72	1.70	1.69	1.66	1.63	1.61	1.60	1.58	1.54	1.52	1.48	1.46	1.44	50
1.78	1.76	1.74	1.72	1.70	1.68	1.67	1.64	1.61	1.59	1.58	1.55	1.52	1.50	1.46	1.43	1.41	55
1.76	1.75	1.72	1.70	1.68	1.66	1.65	1.62	1.59	1.57	1.56	1.53	1.50	1.48	1.44	1.41	1.39	60
1.75	1.73	1.71	1.69	1.67	1.65	1.63	1.60	1.58	1.56	1.54	1.52	1.49	1.46	1.42	1.39	1.37	65
1.74	1.72	1.70	1.67	1.65	1.64	1.62	1.59	1.57	1.55	1.53	1.50	1.47	1.45	1.40	1.37	1.35	70
1.72	1.70	1.68	1.65	1.63	1.62	1.60	1.57	1.54	1.52	1.51	1.48	1.45	1.43	1.38	1.35	1.32	80
1.70	1.69	1.66	1.64	1.62	1.60	1.59	1.55	1.53	1.51	1.49	1.46	1.43	1.41	1.36	1.32	1.30	90
1.69	1.68	1.65	1.63	1.61	1.59	1.57	1.54	1.52	1.49	1.48	1.45	1.41	1.39	1.34	1.31	1.28	100
1.67	1.65	1.63	1.60	1.58	1.57	1.55	1.52	1.49	1.47	1.45	1.42	1.39	1.36	1.31	1.27	1.25	125
1.66	1.64	1.61	1.59	1.57	1.55	1.53	1.50	1.48	1.45	1.44	1.41	1.37	1.34	1.29	1.25	1.22	150
1.64	1.62	1.60	1.57	1.55	1.53	1.52	1.48	1.46	1.43	1.41	1.39	1.35	1.32	1.26	1.22	1.19	200
1.62	1.61	1.58	1.55	1.53	1.51	1.50	1.46	1.43	1.41	1.39	1.36	1.32	1.30	1.23	1.19	1.15	300
1.61	1.59	1.56	1.54	1.52	1.50	1.48	1.45	1.42	1.40	1.38	1.34	1.30	1.28	1.21	1.16	1.11	500
1.60	1.58	1.55	1.53	1.51	1.49	1.47	1.44	1.41	1.38	1.36	1.33	1.29	1.26	1.19	1.13	1.08	1000
1.59	1.57	1.54	1.52	1.50	1.48	1.46	1.42	1.39	1.37	1.35	1.32	1.27	1.24	1.17	1.11	1.00	∞

Degrees of freedom for the denominator (n_2)

Approximate formula for n_1 and n_2 larger than 30: $\log_{10} F_{.05;\nu_1,\nu_2} \simeq \dfrac{1.4287}{\sqrt{h - 0.95}} - 0.681 \left(\dfrac{1}{n_1} - \dfrac{1}{n_2} \right)$, where $\dfrac{1}{h} = \dfrac{1}{2} \left(\dfrac{1}{n_1} + \dfrac{1}{n_2} \right)$.

Critical Values of the F Distribution, $\alpha = 0.025$.

	1	2	3	4	5	6	7	8	9	10	11	12	13	14	15	16	17	18
1	648	800	864	900	922	937	948	957	963	969	973	977	980	983	985	987	989	990
2	38.5	39.0	39.2	39.2	39.3	39.3	39.4	39.4	39.4	39.4	39.4	39.4	39.4	39.4	39.4	39.4	39.4	39.4
3	17.4	16.0	15.4	15.1	14.9	14.7	14.6	14.5	14.5	14.4	14.4	14.3	14.3	14.3	14.3	14.2	14.2	14.2
4	12.2	10.6	9.98	9.60	9.36	9.20	9.07	8.98	8.90	8.84	8.79	8.75	8.72	8.69	8.66	8.64	8.62	8.60
5	10.0	8.43	7.76	7.39	7.15	6.98	6.85	6.76	6.68	6.62	6.57	6.52	6.49	6.46	6.43	6.41	6.39	6.37
6	8.81	7.26	6.60	6.23	5.99	5.82	5.70	5.60	5.52	5.46	5.41	5.37	5.33	5.30	5.27	5.25	5.23	5.21
7	8.07	6.54	5.89	5.52	5.29	5.12	4.99	4.90	4.82	4.76	4.71	4.67	4.63	4.60	4.57	4.54	4.52	4.50
8	7.57	6.06	5.42	5.05	4.82	4.65	4.53	4.43	4.36	4.30	4.24	4.20	4.16	4.13	4.10	4.08	4.05	4.03
9	7.21	5.71	5.08	4.72	4.48	4.32	4.20	4.10	4.03	3.96	3.91	3.87	3.83	3.80	3.77	3.74	3.72	3.70
10	6.94	5.46	4.83	4.47	4.24	4.07	3.95	3.85	3.78	3.72	3.66	3.62	3.58	3.55	3.52	3.50	3.47	3.45
11	6.72	5.26	4.63	4.28	4.04	3.88	3.76	3.66	3.59	3.53	3.47	3.43	3.39	3.36	3.33	3.30	3.28	3.26
12	6.55	5.10	4.47	4.12	3.89	3.73	3.61	3.51	3.44	3.37	3.32	3.28	3.24	3.21	3.18	3.15	3.13	3.11
13	6.41	4.97	4.35	4.00	3.77	3.60	3.48	3.39	3.31	3.25	3.20	3.15	3.12	3.08	3.05	3.03	3.00	2.98
14	6.30	4.86	4.24	3.89	3.66	3.50	3.38	3.29	3.21	3.15	3.09	3.05	3.01	2.98	2.95	2.92	2.90	2.88
15	6.20	4.76	4.15	3.80	3.58	3.41	3.29	3.20	3.12	3.06	3.01	2.96	2.92	2.89	2.86	2.84	2.81	2.79
16	6.12	4.69	4.08	3.73	3.50	3.34	3.22	3.12	3.05	2.99	2.93	2.89	2.85	2.82	2.79	2.76	2.74	2.72
17	6.04	4.62	4.01	3.66	3.44	3.28	3.16	3.06	2.98	2.92	2.87	2.82	2.79	2.75	2.72	2.70	2.67	2.65
18	5.98	4.56	3.95	3.61	3.38	3.22	3.10	3.01	2.93	2.87	2.81	2.77	2.73	2.70	2.67	2.64	2.62	2.60
19	5.92	4.51	3.90	3.56	3.33	3.17	3.05	2.96	2.88	2.82	2.76	2.72	2.68	2.65	2.62	2.59	2.57	2.55
20	5.87	4.46	3.86	3.51	3.29	3.13	3.01	2.91	2.84	2.77	2.72	2.68	2.64	2.60	2.57	2.55	2.52	2.50
21	5.83	4.42	3.82	3.48	3.25	3.09	2.97	2.87	2.80	2.73	2.68	2.64	2.60	2.56	2.53	2.51	2.48	2.46
22	5.79	4.38	3.78	3.44	3.22	3.05	2.93	2.84	2.76	2.70	2.65	2.60	2.56	2.53	2.50	2.47	2.45	2.43
23	5.75	4.35	3.75	3.41	3.18	3.02	2.90	2.81	2.73	2.67	2.62	2.57	2.53	2.50	2.47	2.44	2.42	2.39
24	5.72	4.32	3.72	3.38	3.15	2.99	2.87	2.78	2.70	2.64	2.59	2.54	2.50	2.47	2.44	2.41	2.39	2.36
25	5.69	4.29	3.69	3.35	3.13	2.97	2.85	2.75	2.68	2.61	2.56	2.51	2.48	2.44	2.41	2.38	2.36	2.34
26	5.66	4.27	3.67	3.33	3.10	2.94	2.82	2.73	2.65	2.59	2.54	2.49	2.45	2.42	2.39	2.36	2.34	2.31
27	5.63	4.24	3.65	3.31	3.08	2.92	2.80	2.71	2.63	2.57	2.51	2.47	2.43	2.39	2.36	2.34	2.31	2.29
28	5.61	4.22	3.63	3.29	3.06	2.90	2.78	2.69	2.61	2.55	2.49	2.45	2.41	2.37	2.34	2.32	2.29	2.27
29	5.59	4.20	3.61	3.27	3.04	2.88	2.76	2.67	2.59	2.53	2.48	2.43	2.39	2.36	2.32	2.30	2.27	2.25
30	5.57	4.18	3.59	3.25	3.03	2.87	2.75	2.65	2.57	2.51	2.46	2.41	2.37	2.34	2.31	2.28	2.26	2.23
32	5.53	4.15	3.56	3.22	3.00	2.84	2.72	2.62	2.54	2.48	2.43	2.38	2.34	2.31	2.28	2.25	2.22	2.20
34	5.50	4.12	3.53	3.19	2.97	2.81	2.69	2.59	2.52	2.45	2.40	2.35	2.31	2.28	2.25	2.22	2.19	2.17
36	5.47	4.09	3.51	3.17	2.94	2.79	2.66	2.57	2.49	2.43	2.37	2.33	2.29	2.25	2.22	2.20	2.17	2.15
38	5.45	4.07	3.48	3.15	2.92	2.76	2.64	2.55	2.47	2.41	2.35	2.31	2.27	2.23	2.20	2.17	2.15	2.13
40	5.42	4.05	3.46	3.13	2.90	2.74	2.62	2.53	2.45	2.39	2.33	2.29	2.25	2.21	2.18	2.15	2.13	2.11
42	5.40	4.03	3.45	3.11	2.89	2.73	2.61	2.51	2.44	2.37	2.32	2.27	2.23	2.20	2.16	2.14	2.11	2.09
44	5.39	4.02	3.43	3.09	2.87	2.71	2.59	2.50	2.42	2.36	2.30	2.26	2.21	2.18	2.15	2.12	2.10	2.07
46	5.37	4.00	3.42	3.08	2.86	2.70	2.58	2.48	2.41	2.34	2.29	2.24	2.20	2.17	2.13	2.11	2.08	2.06
48	5.35	3.99	3.40	3.07	2.84	2.69	2.57	2.47	2.39	2.33	2.27	2.23	2.19	2.15	2.12	2.09	2.07	2.05
50	5.34	3.98	3.39	3.06	2.83	2.67	2.55	2.46	2.38	2.32	2.26	2.22	2.18	2.14	2.11	2.08	2.06	2.03
55	5.31	3.95	3.36	3.03	2.81	2.65	2.53	2.43	2.36	2.29	2.24	2.19	2.15	2.11	2.08	2.05	2.03	2.01
60	5.29	3.93	3.34	3.01	2.79	2.63	2.51	2.41	2.33	2.27	2.22	2.17	2.13	2.09	2.06	2.03	2.01	1.98
65	5.27	3.91	3.32	2.99	2.77	2.61	2.49	2.39	2.32	2.25	2.20	2.15	2.11	2.07	2.04	2.01	1.99	1.97
70	5.25	3.89	3.31	2.98	2.75	2.60	2.48	2.38	2.30	2.24	2.18	2.14	2.10	2.06	2.03	2.00	1.97	1.95
80	5.22	3.86	3.28	2.95	2.73	2.57	2.45	2.36	2.28	2.21	2.16	2.11	2.07	2.03	2.00	1.97	1.95	1.93
90	5.20	3.84	3.27	2.93	2.71	2.55	2.43	2.34	2.26	2.19	2.14	2.09	2.05	2.02	1.98	1.95	1.93	1.91
100	5.18	3.83	3.25	2.92	2.70	2.54	2.42	2.32	2.24	2.18	2.12	2.08	2.04	2.00	1.97	1.94	1.91	1.89
125	5.15	3.80	3.22	2.89	2.67	2.51	2.39	2.30	2.22	2.15	2.10	2.05	2.01	1.97	1.94	1.91	1.89	1.86
150	5.13	3.78	3.20	2.87	2.65	2.49	2.37	2.28	2.20	2.13	2.08	2.03	1.99	1.95	1.92	1.89	1.87	1.84
200	5.10	3.76	3.18	2.85	2.63	2.47	2.35	2.26	2.18	2.11	2.06	2.01	1.97	1.93	1.90	1.87	1.84	1.82
300	5.08	3.74	3.16	2.83	2.61	2.45	2.33	2.23	2.16	2.09	2.04	1.99	1.95	1.91	1.88	1.85	1.82	1.80
500	5.05	3.72	3.14	2.81	2.59	2.43	2.31	2.22	2.14	2.07	2.02	1.97	1.93	1.89	1.86	1.83	1.80	1.78
1000	5.04	3.70	3.13	2.80	2.58	2.42	2.30	2.20	2.13	2.06	2.01	1.96	1.92	1.88	1.85	1.82	1.79	1.77
∞	5.02	3.69	3.12	2.79	2.57	2.41	2.29	2.19	2.11	2.05	1.99	1.94	1.90	1.87	1.83	1.80	1.78	1.75

Degrees of freedom for the numerator (m)

Degrees of freedom for the denominator (n)

Example: $P\{F_{.025;8,20} < 2.91\} = 97.5\%$.

$F_{.975;\nu_1,\nu_2} = 1/F_{.025;\nu_2,\nu_1}$.   Example: $F_{.975;8,20} = 1/F_{.025;20,8} = 1/4.00 = 0.250$.

						Degrees of freedom for the numerator (ν_1)											
19	20	22	24	26	28	30	35	40	45	50	60	80	100	200	500	∞	
992	993	995	997	999	1000	1001	1004	1006	1007	1008	1010	1012	1013	1016	1017	1018	1
39.4	39.4	39.5	39.5	39.5	39.5	39.5	39.5	39.5	39.5	39.5	39.5	39.5	39.5	39.5	39.5	39.5	2
14.2	14.2	14.1	14.1	14.1	14.1	14.1	14.1	14.0	14.0	14.0	14.0	14.0	14.0	13.9	13.9	13.9	3
8.58	8.56	8.53	8.51	8.49	8.48	8.46	8.44	8.41	8.39	8.38	8.36	8.33	8.32	8.29	8.27	8.26	4
6.35	6.33	6.30	6.28	6.26	6.24	6.23	6.20	6.18	6.16	6.14	6.12	6.1?	6.09	6.05	6.03	6.02	5
5.19	5.17	5.14	5.12	5.10	5.08	5.07	5.04	5.01	4.99	4.98	4.96	4.93	4.92	4.88	4.86	4.85	6
4.48	4.47	4.44	4.42	4.39	4.38	4.36	4.33	4.31	4.29	4.28	4.25	4.23	4.21	4.18	4.16	4.14	7
4.02	4.00	3.97	3.95	3.93	3.91	3.89	3.86	3.84	3.82	3.81	3.78	3.76	3.74	3.70	3.68	3.67	8
3.68	3.67	3.64	3.61	3.59	3.58	3.56	3.53	3.51	3.49	3.47	3.45	3.42	3.40	3.37	3.35	3.33	9
3.44	3.42	3.39	3.37	3.34	3.33	3.31	3.28	3.26	3.24	3.22	3.20	3.17	3.15	3.12	3.09	3.08	10
3.24	3.23	3.20	3.17	3.15	3.13	3.12	3.09	3.06	3.04	3.03	3.00	2.97	2.96	2.92	2.90	2.88	11
3.09	3.07	3.04	3.02	3.00	2.98	2.96	2.93	2.91	2.89	2.87	2.85	2.82	2.80	2.76	2.74	2.72	12
2.96	2.95	2.92	2.89	2.87	2.85	2.84	2.80	2.78	2.76	2.74	2.72	2.69	2.67	2.63	2.61	2.60	13
2.86	2.84	2.81	2.79	2.77	2.75	2.73	2.70	2.67	2.65	2.64	2.61	2.58	2.56	2.53	2.50	2.49	14
2.77	2.76	2.73	2.70	2.68	2.66	2.64	2.61	2.58	2.56	2.55	2.52	2.49	2.47	2.44	2.41	2.40	15
2.70	2.68	2.65	2.63	2.60	2.58	2.57	2.53	2.51	2.49	2.47	2.45	2.42	2.40	2.36	2.33	2.32	16
2.63	2.62	2.59	2.56	2.54	2.52	2.50	2.47	2.44	2.42	2.41	2.38	2.35	2.33	2.29	2.26	2.25	17
2.58	2.56	2.53	2.50	2.48	2.46	2.44	2.41	2.38	2.36	2.35	2.32	2.29	2.27	2.23	2.20	2.19	18
2.53	2.51	2.48	2.45	2.43	2.41	2.39	2.36	2.33	2.31	2.30	2.27	2.24	2.22	2.18	2.15	2.13	19
2.48	2.46	2.43	2.41	2.39	2.37	2.35	2.31	2.29	2.27	2.25	2.22	2.19	2.17	2.13	2.10	2.09	20
2.44	2.42	2.39	2.37	2.34	2.33	2.31	2.27	2.25	2.23	2.21	2.18	2.15	2.13	2.09	2.06	2.04	21
2.41	2.39	2.36	2.33	2.31	2.29	2.27	2.24	2.21	2.19	2.17	2.14	2.11	2.09	2.05	2.02	2.00	22
2.37	2.36	2.33	2.30	2.28	2.26	2.24	2.20	2.18	2.15	2.14	2.11	2.08	2.06	2.01	1.99	1.97	23
2.35	2.33	2.30	2.27	2.25	2.23	2.21	2.17	2.15	2.12	2.11	2.08	2.05	2.02	1.98	1.95	1.94	24
2.32	2.30	2.27	2.24	2.22	2.20	2.18	2.15	2.12	2.10	2.08	2.05	2.02	2.00	1.95	1.92	1.91	25
2.29	2.28	2.24	2.22	2.19	2.17	2.16	2.12	2.09	2.07	2.05	2.03	1.99	1.97	1.92	1.90	1.88	26
2.27	2.25	2.22	2.19	2.17	2.15	2.13	2.10	2.07	2.05	2.03	2.00	1.97	1.94	1.90	1.87	1.85	27
2.25	2.23	2.20	2.17	2.15	2.13	2.11	2.08	2.05	2.03	2.01	1.98	1.94	1.92	1.88	1.85	1.83	28
2.23	2.21	2.18	2.15	2.13	2.11	2.09	2.06	2.03	2.01	1.99	1.96	1.92	1.90	1.86	1.83	1.81	29
2.21	2.20	2.16	2.14	2.11	2.09	2.07	2.04	2.01	1.99	1.97	1.94	1.90	1.88	1.84	1.81	1.79	30
2.18	2.16	2.13	2.10	2.08	2.06	2.04	2.00	1.98	1.95	1.93	1.91	1.87	1.85	1.80	1.77	1.75	32
2.15	2.13	2.10	2.07	2.05	2.03	2.01	1.97	1.95	1.92	1.90	1.88	1.84	1.82	1.77	1.74	1.72	34
2.13	2.11	2.08	2.05	2.03	2.00	1.99	1.95	1.92	1.90	1.88	1.85	1.81	1.79	1.74	1.71	1.69	36
2.11	2.09	2.05	2.03	2.00	1.98	1.96	1.93	1.90	1.87	1.85	1.82	1.79	1.76	1.71	1.68	1.66	38
2.09	2.07	2.03	2.01	1.98	1.96	1.94	1.90	1.88	1.85	1.83	1.80	1.76	1.74	1.69	1.66	1.64	40
2.07	2.05	2.02	1.99	1.96	1.94	1.92	1.89	1.86	1.83	1.81	1.78	1.74	1.72	1.67	1.64	1.62	42
2.05	2.03	2.00	1.97	1.95	1.93	1.91	1.87	1.84	1.82	1.80	1.77	1.73	1.70	1.65	1.62	1.60	44
2.04	2.02	1.99	1.96	1.93	1.91	1.89	1.85	1.82	1.80	1.78	1.75	1.71	1.69	1.63	1.60	1.58	46
2.02	2.01	1.97	1.94	1.92	1.90	1.88	1.84	1.81	1.79	1.77	1.73	1.69	1.67	1.62	1.58	1.56	48
2.01	1.99	1.96	1.93	1.91	1.88	1.87	1.83	1.80	1.77	1.75	1.72	1.68	1.66	1.60	1.57	1.55	50
1.99	1.97	1.93	1.90	1.88	1.86	1.84	1.80	1.77	1.74	1.72	1.69	1.65	1.62	1.57	1.54	1.51	55
1.96	1.94	1.91	1.88	1.86	1.83	1.82	1.78	1.74	1.72	1.70	1.67	1.62	1.60	1.54	1.51	1.48	60
1.95	1.93	1.89	1.86	1.84	1.82	1.80	1.76	1.72	1.70	1.68	1.65	1.60	1.58	1.52	1.48	1.46	65
1.93	1.91	1.88	1.85	1.82	1.80	1.78	1.74	1.71	1.68	1.66	1.63	1.58	1.56	1.50	1.46	1.44	70
1.90	1.88	1.85	1.82	1.79	1.77	1.75	1.71	1.68	1.65	1.63	1.60	1.55	1.53	1.47	1.43	1.40	80
1.88	1.86	1.83	1.80	1.77	1.75	1.73	1.69	1.66	1.63	1.61	1.58	1.53	1.50	1.44	1.40	1.37	90
1.87	1.85	1.81	1.78	1.76	1.74	1.71	1.67	1.64	1.61	1.59	1.56	1.51	1.48	1.42	1.38	1.35	100
1.84	1.82	1.79	1.75	1.73	1.71	1.68	1.64	1.61	1.58	1.56	1.52	1.48	1.45	1.38	1.34	1.30	125
1.82	1.80	1.77	1.74	1.71	1.69	1.67	1.62	1.59	1.56	1.54	1.50	1.45	1.42	1.35	1.31	1.27	150
1.80	1.78	1.74	1.71	1.68	1.66	1.64	1.60	1.56	1.53	1.51	1.47	1.42	1.39	1.32	1.27	1.23	200
1.77	1.75	1.72	1.69	1.66	1.64	1.62	1.57	1.54	1.51	1.48	1.45	1.39	1.36	1.28	1.23	1.18	300
1.76	1.74	1.70	1.67	1.64	1.62	1.60	1.55	1.51	1.49	1.46	1.42	1.37	1.34	1.25	1.19	1.14	500
1.74	1.72	1.69	1.65	1.63	1.60	1.58	1.54	1.50	1.47	1.44	1.41	1.35	1.32	1.23	1.16	1.09	1000
1.73	1.71	1.67	1.64	1.61	1.59	1.57	1.52	1.48	1.45	1.43	1.39	1.33	1.30	1.21	1.13	1.00	∞

Degrees of freedom for the denominator (ν_2)

Approximate formula for ν_1 and ν_2 larger than 30: $\log_{10} F_{.025; \nu_1, \nu_2} \simeq \dfrac{1.7023}{\sqrt{h - 1.14}} - 0.846 \left(\dfrac{1}{\nu_1} - \dfrac{1}{\nu_2}\right)$, where $\dfrac{1}{h} = \dfrac{1}{2}\left(\dfrac{1}{\nu_1} + \dfrac{1}{\nu_2}\right)$.

Critical Values of the F Distribution, $\alpha = 0.01$.

	Degrees of freedom for the numerator (n)																	
	1	2	3	4	5	6	7	8	9	10	11	12	13	14	15	16	17	18
	Multiply the numbers of the first row ($n=1$) by 10.																	
1	405	500	540	563	576	586	593	598	602	606	608	611	613	614	616	617	618	619
2	98.5	99.0	99.2	99.2	99.3	99.3	99.4	99.4	99.4	99.4	99.4	99.4	99.4	99.4	99.4	99.4	99.4	99.4
3	34.1	30.8	29.5	28.7	28.2	27.9	27.7	27.5	27.3	27.2	27.1	27.1	27.0	26.9	26.9	26.8	26.8	26.8
4	21.2	18.0	16.7	16.0	15.5	15.2	15.0	14.8	14.7	14.5	14.4	14.4	14.3	14.2	14.2	14.2	14.1	14.1
5	16.3	13.3	12.1	11.4	11.0	10.7	10.5	10.3	10.2	10.1	9.96	9.89	9.82	9.77	9.72	9.68	9.64	9.61
6	13.7	10.9	9.78	9.15	8.75	8.47	8.26	8.10	7.98	7.87	7.79	7.72	7.66	7.60	7.56	7.52	7.48	7.45
7	12.2	9.55	8.45	7.85	7.46	7.19	6.99	6.84	6.72	6.62	6.54	6.47	6.41	6.36	6.31	6.27	6.24	6.21
8	11.3	8.65	7.59	7.01	6.63	6.37	6.18	6.03	5.91	5.81	5.73	5.67	5.61	5.56	5.52	5.48	5.44	5.41
9	10.6	8.02	6.99	6.42	6.06	5.80	5.61	5.47	5.35	5.26	5.18	5.11	5.05	5.00	4.96	4.92	4.89	4.86
10	10.0	7.56	6.55	5.99	5.64	5.39	5.20	5.06	4.94	4.85	4.77	4.71	4.65	4.60	4.56	4.52	4.49	4.46
11	9.65	7.21	6.22	5.67	5.32	5.07	4.89	4.74	4.63	4.54	4.46	4.40	4.34	4.29	4.25	4.21	4.18	4.15
12	9.33	6.93	5.95	5.41	5.06	4.82	4.64	4.50	4.39	4.30	4.22	4.16	4.10	4.05	4.01	3.97	3.94	3.91
13	9.07	6.70	5.74	5.21	4.86	4.62	4.44	4.30	4.19	4.10	4.02	3.96	3.91	3.86	3.82	3.78	3.75	3.72
14	8.86	6.51	5.56	5.04	4.70	4.46	4.28	4.14	4.03	3.94	3.86	3.80	3.75	3.70	3.66	3.62	3.59	3.56
15	8.68	6.36	5.42	4.89	4.56	4.32	4.14	4.00	3.89	3.80	3.73	3.67	3.61	3.56	3.52	3.49	3.45	3.42
16	8.53	6.23	5.29	4.77	4.44	4.20	4.03	3.89	3.78	3.69	3.62	3.55	3.50	3.45	3.41	3.37	3.34	3.31
17	8.40	6.11	5.18	4.67	4.34	4.10	3.93	3.79	3.68	3.59	3.52	3.46	3.40	3.35	3.31	3.27	3.24	3.21
18	8.29	6.01	5.09	4.58	4.25	4.01	3.84	3.71	3.60	3.51	3.43	3.37	3.32	3.27	3.23	3.19	3.16	3.13
19	8.18	5.93	5.01	4.50	4.17	3.94	3.77	3.63	3.52	3.43	3.36	3.30	3.24	3.19	3.15	3.12	3.08	3.05
20	8.10	5.85	4.94	4.43	4.10	3.87	3.70	3.56	3.46	3.37	3.29	3.23	3.18	3.13	3.09	3.05	3.02	2.99
21	8.02	5.78	4.87	4.37	4.04	3.81	3.64	3.51	3.40	3.31	3.24	3.17	3.12	3.07	3.03	2.99	2.96	2.93
22	7.95	5.72	4.82	4.31	3.99	3.76	3.59	3.45	3.35	3.26	3.18	3.12	3.07	3.02	2.98	2.94	2.91	2.88
23	7.88	5.66	4.76	4.26	3.94	3.71	3.54	3.41	3.30	3.21	3.14	3.07	3.02	2.97	2.93	2.89	2.86	2.83
24	7.82	5.61	4.72	4.22	3.90	3.67	3.50	3.36	3.26	3.17	3.09	3.03	2.98	2.93	2.89	2.85	2.82	2.79
25	7.77	5.57	4.68	4.18	3.86	3.63	3.46	3.32	3.22	3.13	3.06	2.99	2.94	2.89	2.85	2.81	2.78	2.75
26	7.72	5.53	4.64	4.14	3.82	3.59	3.42	3.29	3.18	3.09	3.02	2.96	2.90	2.86	2.82	2.78	2.74	2.72
27	7.68	5.49	4.60	4.11	3.78	3.56	3.39	3.26	3.15	3.06	2.99	2.93	2.87	2.82	2.78	2.75	2.71	2.68
28	7.64	5.45	4.57	4.07	3.75	3.53	3.36	3.23	3.12	3.03	2.96	2.90	2.84	2.79	2.75	2.72	2.68	2.65
29	7.60	5.42	4.54	4.04	3.73	3.50	3.33	3.20	3.09	3.00	2.93	2.87	2.81	2.77	2.73	2.69	2.66	2.63
30	7.56	5.39	4.51	4.02	3.70	3.47	3.30	3.17	3.07	2.98	2.91	2.84	2.79	2.74	2.70	2.66	2.63	2.60
32	7.50	5.34	4.46	3.97	3.65	3.43	3.26	3.13	3.02	2.93	2.86	2.80	2.74	2.70	2.66	2.62	2.58	2.55
34	7.44	5.29	4.42	3.93	3.61	3.39	3.22	3.09	2.98	2.89	2.82	2.76	2.70	2.66	2.62	2.58	2.55	2.51
36	7.40	5.25	4.38	3.89	3.57	3.35	3.18	3.05	2.95	2.86	2.79	2.72	2.67	2.62	2.58	2.54	2.51	2.48
38	7.35	5.21	4.34	3.86	3.54	3.32	3.15	3.02	2.92	2.83	2.75	2.69	2.64	2.59	2.55	2.51	2.48	2.45
40	7.31	5.18	4.31	3.83	3.51	3.29	3.12	2.99	2.89	2.80	2.73	2.66	2.61	2.56	2.52	2.48	2.45	2.42
42	7.28	5.15	4.29	3.80	3.49	3.27	3.10	2.97	2.86	2.78	2.70	2.64	2.59	2.54	2.50	2.46	2.43	2.40
44	7.25	5.12	4.26	3.78	3.47	3.24	3.08	2.95	2.84	2.75	2.68	2.62	2.56	2.52	2.47	2.44	2.40	2.37
46	7.22	5.10	4.24	3.76	3.44	3.22	3.06	2.93	2.82	2.73	2.66	2.60	2.54	2.50	2.45	2.42	2.38	2.35
48	7.19	5.08	4.22	3.74	3.43	3.20	3.04	2.91	2.80	2.72	2.64	2.58	2.53	2.48	2.44	2.40	2.37	2.33
50	7.17	5.06	4.20	3.72	3.41	3.19	3.02	2.89	2.79	2.70	2.63	2.56	2.51	2.46	2.42	2.38	2.35	2.32
55	7.12	5.01	4.16	3.68	3.37	3.15	2.98	2.85	2.75	2.66	2.59	2.53	2.47	2.42	2.38	2.34	2.31	2.28
60	7.08	4.98	4.13	3.65	3.34	3.12	2.95	2.82	2.72	2.63	2.56	2.50	2.44	2.39	2.35	2.31	2.28	2.25
65	7.04	4.95	4.10	3.62	3.31	3.09	2.93	2.80	2.69	2.61	2.53	2.47	2.42	2.37	2.33	2.29	2.26	2.23
70	7.01	4.92	4.08	3.60	3.29	3.07	2.91	2.78	2.67	2.59	2.51	2.45	2.40	2.35	2.31	2.27	2.23	2.20
80	6.96	4.88	4.04	3.56	3.26	3.04	2.87	2.74	2.64	2.55	2.48	2.42	2.36	2.31	2.27	2.23	2.20	2.17
90	6.93	4.85	4.01	3.54	3.23	3.01	2.84	2.72	2.61	2.52	2.45	2.39	2.33	2.29	2.24	2.21	2.17	2.14
100	6.90	4.82	3.98	3.51	3.21	2.99	2.82	2.69	2.59	2.50	2.43	2.37	2.31	2.26	2.22	2.19	2.15	2.12
125	6.84	4.78	3.94	3.47	3.17	2.95	2.79	2.66	2.55	2.47	2.39	2.33	2.28	2.23	2.19	2.15	2.11	2.08
150	6.81	4.75	3.92	3.45	3.14	2.92	2.76	2.63	2.53	2.44	2.37	2.31	2.25	2.20	2.16	2.12	2.09	2.06
200	6.76	4.71	3.88	3.41	3.11	2.89	2.73	2.60	2.50	2.41	2.34	2.27	2.22	2.17	2.13	2.09	2.06	2.02
300	6.72	4.68	3.85	3.38	3.08	2.86	2.70	2.57	2.47	2.38	2.31	2.24	2.19	2.14	2.10	2.06	2.03	1.99
500	6.69	4.65	3.82	3.36	3.05	2.84	2.68	2.55	2.44	2.36	2.28	2.22	2.17	2.12	2.07	2.04	2.00	1.97
1000	6.66	4.63	3.80	3.34	3.04	2.82	2.66	2.53	2.43	2.34	2.27	2.20	2.15	2.10	2.06	2.02	1.98	1.95
∞	6.63	4.61	3.78	3.32	3.02	2.80	2.64	2.51	2.41	2.32	2.25	2.18	2.13	2.08	2.04	2.00	1.97	1.93

Degrees of freedom for the denominator (n)

Example: $P\{F_{.01;3;20} < 3.56\} = 99\%$.

$F_{.99;\nu_1,\nu_2} = 1/F_{.01;\nu_2,\nu_1}$.   Example: $F_{.99;3;20} = 1/F_{.01;20;3} = 1/5.36 = 0.187$.

Degrees of freedom for the numerator (n_1)

Multiply the numbers of the first row ($n_1 = 1$) by 10.

19	20	22	24	26	28	30	35	40	45	50	60	80	100	200	500	∞	n_2
620	621	622	623	624	625	626	628	629	630	630	631	633	633	635	636	637	1
99.4	99.4	99.5	99.5	99.5	99.5	99.5	99.5	99.5	99.5	99.5	99.5	99.5	99.5	99.5	99.5	99.5	2
26.7	26.7	26.6	26.6	26.6	26.5	26.5	26.5	26.4	26.4	26.4	26.3	26.3	26.2	26.2	26.1	26.1	3
14.0	14.0	14.0	13.9	13.9	13.9	13.8	13.8	13.7	13.7	13.7	13.7	13.6	13.6	13.5	13.5	13.5	4
9.58	9.55	9.51	9.47	9.43	9.40	9.38	9.33	9.29	9.26	9.24	9.20	9.16	9.13	9.08	9.04	9.02	5
7.42	7.40	7.35	7.31	7.28	7.25	7.23	7.18	7.14	7.11	7.09	7.06	7.01	6.99	6.93	6.90	6.88	6
6.18	6.16	6.11	6.07	6.04	6.02	5.99	5.94	5.91	5.88	5.86	5.82	5.78	5.75	5.70	5.67	5.65	7
5.38	5.36	5.32	5.28	5.25	5.22	5.20	5.15	5.12	5.09	5.07	5.03	4.99	4.96	4.91	4.88	4.86	8
4.83	4.81	4.77	4.73	4.70	4.67	4.65	4.60	4.57	4.54	4.52	4.48	4.44	4.42	4.36	4.33	4.31	9
4.43	4.41	4.36	4.33	4.30	4.27	4.25	4.20	4.17	4.14	4.12	4.08	4.04	4.01	3.96	3.93	3.91	10
4.12	4.10	4.06	4.02	3.99	3.96	3.94	3.89	3.86	3.83	3.81	3.78	3.73	3.71	3.66	3.62	3.60	11
3.88	3.86	3.82	3.78	3.75	3.72	3.70	3.65	3.62	3.59	3.57	3.54	3.49	3.47	3.41	3.38	3.36	12
3.69	3.66	3.62	3.59	3.56	3.53	3.51	3.46	3.43	3.40	3.38	3.34	3.30	3.27	3.22	3.19	3.17	13
3.53	3.51	3.46	3.43	3.40	3.37	3.35	3.30	3.27	3.24	3.22	3.18	3.14	3.11	3.06	3.03	3.00	14
3.40	3.37	3.33	3.29	3.26	3.24	3.21	3.17	3.13	3.10	3.08	3.05	3.00	2.98	2.92	2.89	2.87	15
3.28	3.26	3.22	3.18	3.15	3.12	3.10	3.05	3.02	2.99	2.97	2.93	2.89	2.86	2.81	2.78	2.75	16
3.18	3.16	3.12	3.08	3.05	3.03	3.00	2.96	2.92	2.89	2.87	2.83	2.79	2.76	2.71	2.68	2.65	17
3.10	3.08	3.03	3.00	2.97	2.94	2.92	2.87	2.84	2.81	2.78	2.75	2.70	2.68	2.62	2.59	2.57	18
3.03	3.00	2.96	2.92	2.89	2.87	2.84	2.80	2.76	2.73	2.71	2.67	2.63	2.60	2.55	2.51	2.49	19
2.96	2.94	2.90	2.86	2.83	2.80	2.78	2.73	2.69	2.67	2.64	2.61	2.56	2.54	2.48	2.44	2.42	20
2.90	2.88	2.84	2.80	2.77	2.74	2.72	2.67	2.64	2.61	2.58	2.55	2.50	2.48	2.42	2.38	2.36	21
2.85	2.83	2.78	2.75	2.72	2.69	2.67	2.62	2.58	2.55	2.53	2.50	2.45	2.42	2.36	2.33	2.31	22
2.80	2.78	2.74	2.70	2.67	2.64	2.62	2.57	2.54	2.51	2.48	2.45	2.40	2.37	2.32	2.28	2.26	23
2.76	2.74	2.70	2.66	2.63	2.60	2.58	2.53	2.49	2.46	2.44	2.40	2.36	2.33	2.27	2.24	2.21	24
2.72	2.70	2.66	2.62	2.59	2.56	2.54	2.49	2.45	2.42	2.40	2.36	2.32	2.29	2.23	2.19	2.17	25
2.69	2.66	2.62	2.58	2.55	2.53	2.50	2.45	2.42	2.39	2.36	2.33	2.28	2.25	2.19	2.16	2.13	26
2.66	2.63	2.59	2.55	2.52	2.49	2.47	2.42	2.38	2.35	2.33	2.29	2.25	2.22	2.16	2.12	2.10	27
2.63	2.60	2.56	2.52	2.49	2.46	2.44	2.39	2.35	2.32	2.30	2.26	2.22	2.19	2.13	2.09	2.06	28
2.60	2.57	2.53	2.49	2.46	2.44	2.41	2.36	2.33	2.30	2.27	2.23	2.19	2.16	2.10	2.06	2.03	29
2.57	2.55	2.51	2.47	2.44	2.41	2.39	2.34	2.30	2.27	2.25	2.21	2.16	2.13	2.07	2.03	2.01	30
2.53	2.50	2.46	2.42	2.39	2.36	2.34	2.29	2.25	2.22	2.20	2.16	2.11	2.08	2.02	1.98	1.96	32
2.49	2.46	2.42	2.38	2.35	2.32	2.30	2.25	2.21	2.18	2.16	2.12	2.07	2.04	1.98	1.94	1.91	34
2.45	2.43	2.38	2.35	2.32	2.29	2.26	2.21	2.17	2.14	2.12	2.08	2.03	2.00	1.94	1.90	1.87	36
2.42	2.40	2.35	2.32	2.28	2.26	2.23	2.18	2.14	2.11	2.09	2.05	2.00	1.97	1.90	1.86	1.84	38
2.39	2.37	2.33	2.29	2.26	2.23	2.20	2.15	2.11	2.08	2.06	2.02	1.97	1.94	1.87	1.83	1.80	40
2.37	2.34	2.30	2.26	2.23	2.20	2.18	2.13	2.09	2.06	2.03	1.99	1.94	1.91	1.85	1.80	1.78	42
2.35	2.32	2.28	2.24	2.21	2.18	2.15	2.10	2.06	2.03	2.01	1.97	1.92	1.89	1.82	1.78	1.75	44
2.33	2.30	2.26	2.22	2.19	2.16	2.13	2.08	2.04	2.01	1.99	1.95	1.90	1.86	1.80	1.75	1.73	46
2.31	2.28	2.24	2.20	2.17	2.14	2.12	2.06	2.02	1.99	1.97	1.93	1.88	1.84	1.78	1.73	1.70	48
2.29	2.27	2.22	2.18	2.15	2.12	2.10	2.05	2.01	1.97	1.95	1.91	1.86	1.82	1.76	1.71	1.68	50
2.25	2.23	2.18	2.15	2.11	2.08	2.06	2.01	1.97	1.93	1.91	1.87	1.81	1.78	1.71	1.67	1.64	55
2.22	2.20	2.15	2.12	2.08	2.05	2.03	1.98	1.94	1.90	1.88	1.84	1.78	1.75	1.68	1.63	1.60	60
2.20	2.17	2.13	2.09	2.06	2.03	2.00	1.95	1.91	1.88	1.85	1.81	1.75	1.72	1.65	1.60	1.57	65
2.18	2.15	2.11	2.07	2.03	2.01	1.98	1.93	1.89	1.85	1.83	1.78	1.73	1.70	1.62	1.57	1.54	70
2.14	2.12	2.07	2.03	2.00	1.97	1.94	1.89	1.85	1.81	1.79	1.75	1.69	1.66	1.58	1.53	1.49	80
2.11	2.09	2.04	2.00	1.97	1.94	1.92	1.86	1.82	1.79	1.76	1.72	1.66	1.62	1.54	1.49	1.46	90
2.09	2.07	2.02	1.98	1.94	1.92	1.89	1.84	1.80	1.76	1.73	1.69	1.63	1.60	1.52	1.47	1.43	100
2.05	2.03	1.98	1.94	1.91	1.88	1.85	1.80	1.76	1.72	1.69	1.65	1.59	1.55	1.47	1.41	1.37	125
2.03	2.00	1.96	1.92	1.88	1.85	1.83	1.77	1.73	1.69	1.66	1.62	1.56	1.52	1.43	1.38	1.33	150
2.00	1.97	1.93	1.89	1.85	1.82	1.79	1.74	1.69	1.66	1.63	1.58	1.52	1.48	1.39	1.33	1.28	200
1.97	1.94	1.89	1.85	1.82	1.79	1.76	1.71	1.66	1.62	1.59	1.55	1.48	1.44	1.35	1.28	1.22	300
1.94	1.92	1.87	1.83	1.79	1.76	1.74	1.68	1.63	1.60	1.56	1.52	1.45	1.41	1.31	1.23	1.16	500
1.92	1.90	1.85	1.81	1.77	1.74	1.72	1.66	1.61	1.57	1.54	1.50	1.43	1.38	1.28	1.19	1.11	1000
1.90	1.88	1.83	1.79	1.76	1.72	1.70	1.64	1.59	1.55	1.52	1.47	1.40	1.36	1.25	1.15	1.00	∞

Degrees of freedom for the denominator (n_2)

Approximate formula for n_1 and n_2 larger than 30: $\quad \log_{10} F_{.01;\nu_1,\nu_2} \simeq \dfrac{2.0206}{\sqrt{h-1.40}} - 1.073 \left(\dfrac{1}{\nu} - \dfrac{1}{n}\right)$, where $\dfrac{1}{h} = \dfrac{1}{2}\left(\dfrac{1}{n_1} + \dfrac{1}{n_2}\right)$.

G

Percentile Values for the Chi-Square Distribution (with v Degrees of Freedom)

v	$\chi^2_{.995}$	$\chi^2_{.99}$	$\chi^2_{.975}$	$\chi^2_{.95}$	$\chi^2_{.90}$	$\chi^2_{.75}$	$\chi^2_{.50}$	$\chi^2_{.25}$	$\chi^2_{.10}$	$\chi^2_{.05}$	$\chi^2_{.025}$	$\chi^2_{.01}$	$\chi^2_{.005}$
1	7.88	6.63	5.02	3.84	2.71	1.32	.455	.102	.0158	.0039	.0010	.0002	.0000
2	10.6	9.21	7.38	5.99	4.61	2.77	1.39	.575	.211	.103	.0506	.0201	.0100
3	12.8	11.3	9.35	7.81	6.25	4.11	2.37	1.21	.584	.352	.216	.115	.072
4	14.9	13.3	11.1	9.49	7.78	5.39	3.36	1.92	1.06	.711	.484	.297	.207
5	16.7	15.1	12.8	11.1	9.24	6.63	4.35	2.67	1.61	1.15	.831	.554	.412
6	18.5	16.8	14.4	12.6	10.6	7.84	5.35	3.45	2.20	1.64	1.24	.872	.676
7	20.3	18.5	16.0	14.1	12.0	9.04	6.35	4.25	2.83	2.17	1.69	1.24	.989
8	22.0	20.1	17.5	15.5	13.4	10.2	7.34	5.07	3.49	2.73	2.18	1.65	1.34
9	23.6	21.7	19.0	16.9	14.7	11.4	8.34	5.90	4.17	3.33	2.70	2.09	1.73
10	25.2	23.2	20.5	18.3	16.0	12.5	9.34	6.74	4.87	3.94	3.25	2.56	2.16
11	26.8	24.7	21.9	19.7	17.3	13.7	10.3	7.58	5.58	4.57	3.82	3.05	2.60
12	28.3	26.2	23.3	21.0	18.5	14.8	11.3	8.44	6.30	5.23	4.40	3.57	3.07
13	29.8	27.7	24.7	22.4	19.8	16.0	12.3	9.30	7.04	5.89	5.01	4.11	3.57
14	31.3	29.1	26.1	23.7	21.1	17.1	13.3	10.2	7.79	6.57	5.63	4.66	4.07
15	32.8	30.6	27.5	25.0	22.3	18.2	14.3	11.0	8.55	7.26	6.26	5.23	4.60
16	34.3	32.0	28.8	26.3	23.5	19.4	15.3	11.9	9.31	7.96	6.91	5.81	5.14
17	35.7	33.4	30.2	27.6	24.8	20.5	16.3	12.8	10.1	8.67	7.56	6.41	5.70
18	37.2	34.8	31.5	28.9	26.0	21.6	17.3	13.7	10.9	9.39	8.23	7.01	6.26
19	38.6	36.2	32.9	30.1	27.2	22.7	18.3	14.6	11.7	10.1	8.91	7.63	6.84

Percentile Values for the Chi-Square Distribution (with v Degrees of Freedom) (*continued*)

v	$\chi^2_{.995}$	$\chi^2_{.99}$	$\chi^2_{.975}$	$\chi^2_{.95}$	$\chi^2_{.90}$	$\chi^2_{.75}$	$\chi^2_{.50}$	$\chi^2_{.25}$	$\chi^2_{.10}$	$\chi^2_{.05}$	$\chi^2_{.025}$	$\chi^2_{.01}$	$\chi^2_{.005}$
20	40.0	37.6	34.2	31.4	28.4	23.8	19.3	15.5	12.4	10.9	9.59	8.26	7.43
21	41.4	38.9	35.5	32.7	29.6	24.9	20.3	16.3	13.2	11.6	10.3	8.90	8.03
22	42.8	40.3	36.8	33.9	30.8	26.0	21.3	17.2	14.0	12.3	11.0	9.54	8.64
23	44.2	41.6	38.1	35.2	32.0	27.1	22.3	18.1	14.8	13.1	11.7	10.2	9.26
24	45.6	43.0	39.4	36.4	33.2	28.2	23.3	19.0	15.7	13.8	12.4	10.9	9.89
25	46.9	44.3	40.6	37.7	34.4	29.3	24.3	19.9	16.5	14.6	13.1	11.5	10.5
26	48.3	45.6	41.9	38.9	35.6	30.4	25.3	20.8	17.3	15.4	13.8	12.2	11.2
27	49.6	47.0	43.2	40.1	36.7	31.5	26.3	21.7	18.1	16.2	14.6	12.9	11.8
28	51.0	48.3	44.5	41.3	37.9	32.6	27.3	22.7	18.9	16.9	15.3	13.6	12.5
29	52.3	49.6	45.7	42.6	39.1	33.7	28.3	23.6	19.8	17.7	16.0	14.3	13.1
30	53.7	50.9	47.0	43.8	40.3	34.8	29.3	24.5	20.6	18.5	16.8	15.0	13.8
40	66.8	63.7	59.3	55.8	51.8	45.6	39.3	33.7	29.1	26.5	24.4	22.2	20.7
50	79.5	76.2	71.4	67.5	63.2	56.3	49.3	42.9	37.7	34.8	32.4	29.7	28.0
60	92.0	88.4	83.3	79.1	74.4	67.0	59.3	52.3	46.5	43.2	40.5	37.5	35.5
70	104.2	100.4	95.0	90.5	85.5	77.6	69.3	61.7	55.3	51.7	48.8	45.4	43.3
80	116.3	112.3	106.6	101.9	96.6	88.1	79.3	71.1	64.3	60.4	57.2	53.5	51.2
90	128.3	124.1	118.1	113.1	107.6	98.6	89.3	80.6	73.3	69.1	65.6	61.8	59.2
100	140.2	135.8	129.6	124.3	118.5	109.1	99.3	90.1	82.4	77.9	74.2	70.1	67.3

Source: Reprinted from Thompson, C.M., *Table of percentage points for the χ^2 distribution,* Biometrika, *32* (1941). By permission of the author and publishers.

Cumulative Poisson Distribution

λ \ x	0	1	2	3	4	5	6	7	8	9	10
0.01	0.990										
0.02	0.980										
0.03	0.970										
0.04	0.961	0.999									
0.05	0.951	0.999									
0.06	0.942	0.998									
0.07	0.932	0.998									
0.08	0.923	0.997									
0.09	0.914	0.996									
0.10	0.905	0.995									
0.15	0.861	0.990	0.999								
0.20	0.819	0.982	0.999								
0.25	0.779	0.974	0.998								
0.30	0.741	0.963	0.996								
0.35	0.705	0.951	0.994	0.999							
0.40	0.670	0.938	0.992	0.999							
0.45	0.638	0.925	0.989	0.999							
0.50	0.607	0.910	0.986	0.998							
0.55	0.577	0.894	0.982	0.998							
0.60	0.549	0.878	0.977	0.997							
0.65	0.522	0.861	0.972	0.996	0.999						

Notes: (1) Entries in the table are values of $F(x)$ where

$$F(x) = P(c \leq x) = \sum_{c=0}^{x} e^{-\lambda}\lambda^c/c!$$

(2) Blank spaces to the right of the last entry in any row of the table may be read as 1; blank spaces to the left of the first entry in any row of the table may be read as 0.

Source: Reprinted from Molina, E.C., *Poisson's Eponential Binomial Limit,* Van Nostrand, New York, 1947. By permission of the author and publishers.

λ \ x	0	1	2	3	4	5	6	7	8	9	10
0.70	0.497	0.844	0.966	0.994	0.999						
0.75	0.472	0.827	0.959	0.993	0.999						
0.80	0.449	0.809	0.953	0.991	0.999						
0.85	0.427	0.791	0.945	0.989	0.998						
0.90	0.407	0.772	0.937	0.987	0.998						
0.95	0.387	0.754	0.929	0.984	0.997						
1.00	0.368	0.736	0.920	0.981	0.996	0.999					
1.1	0.333	0.699	0.900	0.974	0.995	0.999					
1.2	0.301	0.663	0.879	0.966	0.992	0.998					
1.3	0.273	0.627	0.857	0.957	0.989	0.998					
1.4	0.247	0.592	0.833	0.946	0.986	0.997	0.999				
1.5	0.223	0.558	0.809	0.934	0.981	0.996	0.999				
1.6	0.202	0.525	0.783	0.921	0.976	0.994	0.999				
1.7	0.183	0.493	0.757	0.907	0.970	0.992	0.998				
1.8	0.165	0.463	0.731	0.891	0.964	0.990	0.997	0.999			
1.9	0.150	0.434	0.704	0.875	0.956	0.987	0.997	0.999			
2.0	0.135	0.406	0.677	0.857	0.947	0.983	0.995	0.999			
2.1	0.122	0.380	0.650	0.839	0.938	0.980	0.994	0.999			
2.2	0.111	0.355	0.623	0.819	0.928	0.975	0.993	0.998			
2.3	0.100	0.331	0.596	0.799	0.916	0.970	0.991	0.997	0.999		
2.4	0.091	0.308	0.570	0.779	0.904	0.964	0.988	0.997	0.999		
2.5	0.082	0.287	0.544	0.758	0.891	0.958	0.986	0.996	0.999		
2.6	0.074	0.267	0.518	0.736	0.877	0.951	0.983	0.995	0.999		
2.7	0.067	0.249	0.494	0.714	0.863	0.943	0.979	0.993	0.998	0.999	
2.8	0.061	0 231	0.469	0.692	0.848	0.935	0.976	0.992	0.998	0.999	
2.9	0.055	0.215	0.446	0.670	0.832	0.926	0.971	0.990	0.997	0.999	
3.0	0.050	0.199	0.423	0.647	0.815	0.916	0.966	0.988	0.996	0.999	

λ \ x	0	1	2	3	4	5	6	7	8	9	10
3.2	0.041	0.171	0.380	0.603	0.781	0.895	0.955	0.983	0.994	0.998	
3.4	0.033	0.147	0.340	0.558	0.744	0.871	0.942	0.977	0.992	0.997	0.999
3.6	0.027	0.126	0.303	0.515	0.706	0.844	0.927	0.969	0.988	0.996	0.999
3.8	0.022	0.107	0.269	0.473	0.668	0.816	0.909	0.960	0.984	0.994	0.998
4.0	0.018	0.092	0.238	0.433	0.629	0.785	0.889	0.949	0.979	0.992	0.997
4.2	0.015	0.078	0.210	0.395	0.590	0.753	0.867	0.936	0.972	0.989	0.996
4.4	0.012	0.066	0.185	0.359	0.551	0.720	0.844	0.921	0.964	0.985	0.994
4.6	0.010	0.056	0.163	0.326	0.513	0.686	0.818	0.905	0.955	0.980	0.992
4.8	0.008	0.048	0.143	0.294	0.476	0.651	0.791	0.887	0.944	0.975	0.990
5.0	0.007	0.040	0.125	0.265	0.440	0.616	0.762	0.867	0.932	0.968	0.986
5.2	0.006	0.034	0.109	0.238	0.406	0.581	0.732	0.845	0.918	0.960	0.982
5.4	0.005	0.029	0.095	0.213	0.373	0.546	0.702	0.822	0.903	0.951	0.977
5.6	0.004	0.024	0.082	0.191	0.342	0.512	0.670	0.797	0.886	0.941	0.972
5.8	0.003	0.021	0.072	0.170	0.313	0.478	0.638	0.771	0.867	0.929	0.965
6.0	0.002	0.017	0.062	0.151	0.285	0.446	0.606	0.744	0.847	0.916	0.957

λ \ x	11	12	13	14	15	16	17	18	19	20	21
3.2											
3.4											
3.6											
3.8	0.999										
4.0	0.999										
4.2	0.999										
4.4	0.998	0.999									
4.6	0.997	0.999									
4.8	0.996	0.999									
5.0	0.995	0.998	0.999								
5.2	0.993	0.997	0.999								
5.4	0.990	0.996	0.999								
5.6	0.988	0.995	0.998	0.999							
5.8	0.984	0.993	0.997	0.999	0.999						
6.0	0.980	0.991	0.996	0.999	0.999						

λ \ x	0	1	2	3	4	5	6	7	8	9	10
6.2	0.002	0.015	0.054	0.134	0.259	0.414	0.574	0.716	0.826	0.902	0.949
6.4	0.002	0.012	0.046	0.119	0.235	0.384	0.542	0.687	0.803	0.886	0.939
6.6	0.001	0.010	0.040	0.105	0.213	0.355	0.511	0.658	0.780	0.869	0.927
6.8	0.001	0.009	0.034	0.093	0.192	0.327	0.480	0.628	0.755	0.850	0.915
7.0	0.001	0.007	0.030	0.082	0.173	0.301	0.450	0.599	0.729	0.830	0.901
7.2	0.001	0.006	0.025	0.072	0.156	0.276	0.420	0.569	0.703	0.810	0.887
7.4	0.001	0.005	0.022	0.063	0.140	0.253	0.392	0.539	0.676	0.788	0.871
7.6	0.001	0.004	0.019	0.055	0.125	0.231	0.365	0.510	0.648	0.765	0.854
7.8		0.004	0.016	0.048	0.112	0.210	0.338	0.481	0.620	0.741	0.835
8.0		0.003	0.014	0.042	0.100	0.191	0.313	0.453	0.593	0.717	0.816
8.2		0.003	0.012	0.037	0.089	0.174	0.290	0.425	0.565	0.692	0.796
8.4		0.002	0.010	0.032	0.079	0.157	0.267	0.399	0.537	0.666	0.774

λ \ x	11	12	13	14	15	16	17	18	19	20	21
6.2	0.975	0.989	0.995	0.998	0.999						
6.4	0.969	0.986	0.994	0.997	0.999						
6.6	0.963	0.982	0.992	0.997	0.999	0.999					
6.8	0.955	0.978	0.990	0.996	0.998	0.999					
7.0	0.947	0.973	0.987	0.994	0.998	0.999					
7.2	0.937	0.967	0.984	0.993	0.997	0.999	0.999				
7.4	0.926	0.961	0.980	0.991	0.996	0.998	0.999				
7.6	0.915	0.954	0.976	0.989	0.995	0.998	0.999				
7.8	0.902	0.945	0.971	0.986	0.993	0.997	0.999				
8.0	0.888	0.936	0.966	0.983	0.992	0.996	0.998	0.999			
8.2	0.873	0.926	0.960	0.979	0.990	0.995	0.998	0.999			
8.4	0.857	0.915	0.952	0.975	0.987	0.994	0.997	0.999			

λ \ x	1	2	3	4	5	6	7	8	9	10	11	12	13	14	15	16	17
8.5	0.002	0.009	0.030	0.074	0.150	0.256	0.386	0.523	0.653	0.763	0.849	0.909	0.949	0.973	0.986	0.993	0.997
9.0	0.001	0.006	0.021	0.055	0.116	0.207	0.324	0.456	0.587	0.706	0.803	0.876	0.926	0.959	0.978	0.989	0.995
9.5	0.001	0.004	0.015	0.040	0.089	0.165	0.269	0.392	0.522	0.645	0.752	0.836	0.898	0.940	0.967	0.982	0.991
10.0		0.003	0.010	0.029	0.067	0.130	0.220	0.333	0.458	0.583	0.697	0.792	0.864	0.917	0.951	0.973	0.986
10.5		0.002	0.007	0.021	0.050	0.102	0.179	0.279	0.397	0.521	0.639	0.742	0.825	0.888	0.932	0.960	0.978
11.0		0.001	0.005	0.015	0.038	0.079	0.143	0.232	0.341	0.460	0.579	0.689	0.781	0.854	0.907	0.944	0.968
11.5		0.001	0.003	0.011	0.028	0.060	0.114	0.191	0.289	0.402	0.520	0.633	0.733	0.815	0.878	0.924	0.954
12.0			0.002	0.008	0.020	0.046	0.090	0.155	0.242	0.347	0.462	0.576	0.682	0.772	0.844	0.899	0.937
12.5			0.002	0.005	0.015	0.035	0.070	0.125	0.201	0.297	0.406	0.519	0.628	0.725	0.806	0.869	0.916
13.0			0.001	0.004	0.011	0.026	0.054	0.100	0.166	0.252	0.353	0.463	0.573	0.675	0.764	0.835	0.890
13.5			0.001	0.003	0.008	0.019	0.041	0.079	0.135	0.211	0.304	0.409	0.518	0.623	0.718	0.798	0.861
14.0				0.002	0.006	0.014	0.032	0.062	0.109	0.176	0.260	0.358	0.464	0.570	0.669	0.756	0.827
14.5				0.001	0.004	0.010	0.024	0.048	0.088	0.145	0.220	0.311	0.413	0.518	0.619	0.711	0.790
15.0				0.001	0.003	0.008	0.018	0.037	0.070	0.118	0.185	0.268	0.363	0.466	0.568	0.664	0.749
16.0					0.001	0.004	0.010	0.022	0.043	0.077	0.127	0.193	0.275	0.368	0.467	0.566	0.659
17.0					0.001	0.002	0.005	0.013	0.026	0.049	0.085	0.135	0.201	0.281	0.371	0.468	0.564
18.0						0.001	0.003	0.007	0.015	0.030	0.055	0.092	0.143	0.208	0.287	0.375	0.469

λ \ x	18	19	20	21	22	23	24	25	26	27	28	29	30	31	32	33	34
8.5	0.999	0.999															
9.0	0.998	0.999															
9.5	0.996	0.998	0.999														
10.0	0.993	0.997	0.998	0.999													
10.5	0.988	0.994	0.997	0.999	0.999												
11.0	0.982	0.991	0.995	0.998	0.999												
11.5	0.974	0.986	0.992	0.996	0.998	0.999											
12.0	0.963	0.979	0.988	0.994	0.997	0.999	0.999										
12.5	0.948	0.969	0.983	0.991	0.995	0.998	0.999										
13.0	0.930	0.957	0.975	0.986	0.992	0.996	0.998	0.999									
13.5	0.908	0.942	0.965	0.980	0.989	0.994	0.997	0.999	0.999								
14.0	0.883	0.923	0.952	0.971	0.983	0.991	0.995	0.997	0.999	0.999							
14.5	0.853	0.901	0.936	0.960	0.976	0.986	0.992	0.996	0.998	0.999	0.999						
15.0	0.819	0.875	0.917	0.947	0.967	0.981	0.989	0.994	0.997	0.998	0.999						
16.0	0.742	0.812	0.868	0.911	0.942	0.963	0.978	0.987	0.993	0.996	0.998	0.999	0.999	0.999			
17.0	0.655	0.736	0.805	0.861	0.905	0.937	0.959	0.975	0.985	0.991	0.995	0.997	0.999	0.999			
18.0	0.562	0.651	0.731	0.799	0.855	0.899	0.932	0.955	0.972	0.983	0.990	0.994	0.997	0.998	0.999		

x / λ	6	7	8	9	10	11	12	13	14	15	16	17
19.0	0.001	0.002	0.004	0.009	0.018	0.035	0.061	0.098	0.150	0.215	0.292	0.378
20.0		0.001	0.002	0.005	0.011	0.021	0.039	0.066	0.105	0.157	0.221	0.297
21.0			0.001	0.003	0.006	0.013	0.025	0.043	0.072	0.111	0.163	0.227
22.0			0.001	0.002	0.004	0.008	0.015	0.028	0.048	0.077	0.117	0.169
23.0				0.001	0.002	0.004	0.009	0.017	0.031	0.052	0.082	0.123
24.0					0.001	0.003	0.005	0.011	0.020	0.034	0.056	0.087
25.0					0.001	0.001	0.003	0.006	0.012	0.022	0.038	0.060

x / λ	18	19	20	21	22	23	24	25	26	27	28	29
19.0	0.469	0.561	0.647	0.725	0.793	0.849	0.893	0.927	0.951	0.969	0.980	0.988
20.0	0.381	0.470	0.559	0.644	0.721	0.787	0.843	0.888	0.922	0.948	0.966	0.978
21.0	0.302	0.384	0.471	0.558	0.640	0.716	0.782	0.838	0.883	0.917	0.944	0.963
22.0	0.232	0.306	0.387	0.472	0.556	0.637	0.712	0.777	0.832	0.877	0.913	0.940
23.0	0.175	0.238	0.310	0.389	0.472	0.555	0.635	0.708	0.772	0.827	0.873	0.908
24.0	0.128	0.180	0.243	0.314	0.392	0.473	0.554	0.632	0.704	0.768	0.823	0.868
25.0	0.092	0.134	0.185	0.247	0.318	0.394	0.473	0.553	0.629	0.700	0.763	0.818

x / λ	30	31	32	33	34	35	36	37	38	39	40	41	42
19.0	0.993	0.996	0.998	0.999	0.999								
20.0	0.987	0.992	0.995	0.997	0.999	0.999							
21.0	0.976	0.985	0.991	0.994	0.997	0.998	0.999	0.999					
22.0	0.959	0.973	0.983	0.989	0.994	0.996	0.998	0.999	0.999				
23.0	0.936	0.956	0.971	0.981	0.988	0.993	0.996	0.997	0.999	0.999			
24.0	0.904	0.932	0.953	0.969	0.979	0.987	0.992	0.995	0.997	0.998	0.999	0.999	
25.0	0.863	0.900	0.929	0.950	0.966	0.978	0.985	0.991	0.994	0.997	0.998	0.999	0.999

Critical Values for the Sign Test

	α for Double-Sided Test				α for Double-Sided Test		
	·10	·05	·01		·10	·05	·01
	α for Single-Sided Test				α for Single-Sided Test		
n	·05	·025	·005	*n*	·05	·025	·005
6	0	0	–	41	14	13	11
7	0	0	–	42	15	14	12
8	1	0	0	43	15	14	12
9	1	1	0	44	16	15	13
10	1	1	0	45	16	15	13
11	2	1	0	46	16	15	13
12	2	2	1	47	17	16	14
13	3	2	1	48	17	16	14
14	3	2	1	49	18	17	15
15	3	3	2	50	18	17	15
16	4	3	2	52	19	18	16
17	4	4	2	54	20	19	17
18	5	4	3	56	21	20	17
19	5	4	3	58	22	21	18
20	5	5	3	60	23	21	19

Critical Values for the Sign Test (*continued*)

n	α for Double-Sided Test			n	α for Double-Sided Test		
	·10	·05	·01		·10	·05	·01
	α for Single-Sided Test				α for Single-Sided Test		
	·05	·025	·005		·05	·025	·005
21	6	5	4	62	24	22	20
22	6	5	4	64	24	23	21
23	7	6	4	66	25	24	22
24	7	6	5	68	26	25	22
25	7	7	5	70	27	26	23
26	8	7	6	72	28	27	24
27	8	7	6	74	29	28	25
28	9	8	6	76	30	28	26
29	9	8	7	78	31	29	27
30	10	9	7	80	32	30	28
31	10	9	7	82	33	31	28
32	10	9	8	84	33	32	29
33	11	10	8	86	34	33	30
34	11	10	9	88	35	34	31
35	12	11	9	90	36	35	32
36	12	11	9	92	37	36	33
37	13	12	10	94	38	37	34
38	13	12	10	96	39	37	34
39	13	12	11	98	40	38	35
40	14	13	11	100	41	39	36

Source: Reprinted from Davis, O.L., and Goldsmith, P.L., *Statistical Methods in Research and Production,* Longman Group Limited, London, 1977. By permission of the authors and publisher.

J

Critical Values for the Run Test

TABLE 1

n_1 \ n_2	2	3	4	5	6	7	8	9	10	11	12	13	14	15	16	17	18	19	20
2											2	2	2	2	2	2	2	2	2
3					2	2	2	2	2	2	2	2	2	3	3	3	3	3	3
4				2	2	2	3	3	3	3	3	3	3	4	4	4	4	4	4
5			2	2	3	3	3	3	3	4	4	4	4	4	4	4	5	5	5
6		2	2	3	3	3	3	4	4	4	4	5	5	5	5	5	5	6	6
7		2	2	3	3	3	4	4	5	5	5	5	5	6	6	6	6	6	6
8		2	3	3	3	4	4	5	5	5	6	6	6	6	6	7	7	7	7
9		2	3	3	4	4	5	5	5	6	6	6	7	7	7	7	8	8	8
10		2	3	3	4	5	5	5	6	6	7	7	7	7	8	8	8	8	9
11		2	3	4	4	5	5	6	6	7	7	7	8	8	8	9	9	9	9
12	2	2	3	4	4	5	6	6	7	7	7	8	8	8	9	9	9	10	10
13	2	2	3	4	5	5	6	6	7	7	8	8	9	9	9	10	10	10	10
14	2	2	3	4	5	5	6	7	7	8	8	9	9	9	10	10	10	11	11
15	2	3	3	4	5	6	6	7	7	8	8	9	9	10	10	11	11	11	12
16	2	3	4	4	5	6	6	7	8	8	9	9	10	10	11	11	11	12	12
17	2	3	4	4	5	6	7	7	8	9	9	10	10	11	11	11	12	12	13
18	2	3	4	5	5	6	7	8	8	9	9	10	10	11	11	12	12	13	13
19	2	3	4	5	6	6	7	8	8	9	10	10	11	11	12	12	13	13	13
20	2	3	4	5	6	6	7	8	9	9	10	10	11	12	12	13	13	13	14

Note: The values of r given in Tables 1 and 2 are various critical values of r associated with selected values of n_1 and n_2. For the one-sample run test, any value of r that is equal to or less than the value shown in Table 1 or equal to or greater than the value shown in Table 2 is significant at the 5 percent level. For the two-sample run test, any value of r that is equal to or less than the value shown in Table 1 is significant at the 5 percent level.

TABLE 2

n_1 \ n_2	2	3	4	5	6	7	8	9	10	11	12	13	14	15	16	17	18	19	20
2																			
3																			
4			9	9															
5		9	10	10	11	11													
6		9	10	11	12	12	13	13	13	13									
7			11	12	13	13	14	14	14	14	15	15	15						
8			11	12	13	14	14	15	15	16	16	16	16	17	17	17	17	17	17
9					13	14	14	15	16	16	16	17	17	18	18	18	18	18	18
10					13	14	15	16	16	17	17	18	18	18	19	19	19	20	20
11					13	14	15	16	17	17	18	19	19	19	20	20	20	21	21
12					13	14	16	16	17	18	19	19	20	20	21	21	21	22	22
13						15	16	17	18	19	19	20	20	21	21	22	22	23	23
14						15	16	17	18	19	20	20	21	22	22	23	23	23	24
15						15	16	18	18	19	20	21	22	22	23	23	24	24	25
16							17	18	19	20	21	21	22	23	23	24	25	25	25
17							17	18	19	20	21	22	23	23	24	25	25	26	26
18							17	18	19	20	21	22	23	24	25	25	26	26	27
19							17	18	20	21	22	23	23	24	25	26	26	27	27
20							17	18	20	21	22	23	24	25	25	26	27	27	28

Source: Reprinted from Swed, F.S., and Eisenhart, C., "Tables for testing randomness of grouping in a sequence of alternatives," *Ann. Math. Stat., 14:* 83–86, 1943. By permission of the Institute of Mathematical Statistics.

K

Critical Values for the Wilcoxon Signed-Rank Test

	Level of Significance for One-Tailed Test		
	.025	.01	.005
	Level of Significance for Two-Tailed Test		
n	.05	.02	.01
6	0	—	—
7	2	0	—
8	4	2	0
9	6	3	2
10	8	5	3
11	11	7	5
12	14	10	7
13	17	13	10
14	21	16	13
15	25	20	16
16	30	24	20
17	35	28	23
18	40	33	28
19	46	38	32
20	52	43	38
21	59	49	43
22	66	56	49
23	73	62	55
24	81	69	61
25	89	77	68

Notes: The values of T given in the table are critical values associated with selected values of n. Any value of T that is less than or equal to the tabulated value is significant at the indicated level of significance.

For $n > 25$, T is approximately normally distributed with mean $n(n + 1)/4$ and variance $n(n + 1)(2n + 1)/24$.

L

Critical Values for the Rank Sum Test

n_2	α for 2-sided Test	α for 1-sided Test	1	2	3	4	5	6	7	8	9	10	11	12	13	14	15	16	17	18	19	20
													n_1 (Smaller Sample)								
3	0·10	0·05			6																	
	0·05	0·025																				
	0·01	0·005																				
4	0·10	0·05			6	11																
	0·05	0·025				10																
	0·01	0·005																				
5	0·10	0·05		3	7	12	19															
	0·05	0·025			6	11	17															
	0·01	0·005					15															
6	0·10	0·05		3	8	13	20	28														
	0·05	0·025			7	12	18	26														
	0·01	0·005				10	16	23														
7	0·10	0·05		3	8	14	21	29	39													
	0·05	0·025			7	13	20	27	36													
	0·01	0·005				10	16	24	32													
8	0·10	0·05		4	9	15	23	31	41	51												
	0·05	0·025		3	8	14	21	29	38	49												
	0·01	0·005				11	17	25	34	43												
9	0·10	0·05		4	10	16	24	33	43	54	66											
	0·05	0·025		3	8	14	22	31	40	51	62											
	0·01	0·005			6	11	18	26	35	45	56											
10	0·10	0·05		4	10	17	26	35	45	56	69	82										
	0·05	0·025		3	9	15	23	32	42	53	65	78										
	0·01	0·005			6	12	19	27	37	47	58	71										
11	0·10	0·05		4	11	18	27	37	47	59	72	86	100									
	0·05	0·025		3	9	16	24	34	44	55	68	81	96									
	0·01	0·005			6	12	20	28	38	49	61	73	87									
12	0·10	0·05		5	11	19	28	38	49	62	75	89	104	120								
	0·05	0·025		4	10	17	26	35	46	58	71	84	99	115								
	0·01	0·005			7	13	21	30	40	51	63	76	90	105								
13	0·10	0·05		5	12	20	30	40	52	64	78	92	108	125	142							
	0·05	0·025		4	10	18	27	37	48	60	73	88	103	119	136							
	0·01	0·005			7	13	22	31	41	53	65	79	93	109	125							

Critical Values for the Rank Sun Test (*continued*)

n_2	α for 2-sided Test	α for 1-sided Test	1	2	3	4	5	6	7	8	9	10	11	12	13	14	15	16	17	18	19	20
										n_1 (Smaller Sample)											
14	0·10	0·05		6	13	21	31	42	54	67	81	96	112	129	147	166						
	0·05	0·025		4	11	19	28	38	50	62	76	91	106	123	141	160						
	0·01	0·005			7	14	22	32	43	54	67	81	96	112	129	147						
15	0·10	0·05		6	13	22	33	44	56	69	84	99	116	133	152	171	192					
	0·05	0·025		4	11	20	29	40	52	65	79	94	110	127	145	164	184					
	0·01	0·005			8	15	23	33	44	56	69	84	99	115	133	151	171					
16	0·10	0·05		6	14	24	34	46	58	72	87	103	120	138	156	176	197	219				
	0·05	0·025		4	12	21	30	42	54	67	82	97	113	131	150	169	190	211				
	0·01	0·005			8	15	24	34	46	58	72	86	102	119	136	155	175	196				
17	0·10	0·05		6	15	25	35	47	61	75	90	106	123	142	161	182	203	225	249			
	0·05	0·025		5	12	21	32	43	56	70	84	100	117	135	154	174	195	217	240			
	0·01	0·005			8	16	25	36	47	60	74	89	105	122	140	159	180	201	223			
18	0·10	0·05		7	15	26	37	49	63	77	93	110	127	146	166	187	208	231	255	280		
	0·05	0·025		5	13	22	33	45	58	72	87	103	121	139	158	179	200	222	246	270		
	0·01	0·005			8	16	26	37	49	62	76	92	108	125	144	163	184	206	228	252		
19	0·10	0·05	1	7	16	27	38	51	65	80	96	113	131	150	171	192	214	237	262	287	313	
	0·05	0·025		5	13	23	34	46	60	74	90	107	124	143	163	183	205	228	252	277	303	
	0·01	0·005		3	9	17	27	38	50	64	78	94	111	129	148	168	189	210	234	258	283	
20	0·10	0·05	1	7	17	28	40	53	67	83	99	117	135	155	175	197	220	243	268	294	320	348
	0·05	0·025		5	14	24	35	48	62	77	93	110	128	147	167	188	210	234	258	283	309	337
	0·01	0·005		3	9	18	28	39	52	66	81	97	114	132	151	172	193	215	239	263	289	315

For larger values of n_1 and n_2, critical values are given to a good approximation by the formula:

$$\frac{n_1}{2}(n_1+n_2+1) - u\left\{\frac{n_1 n_2(n_1+n_2+1)}{12}\right\}^{1/2}$$

where $u = 1·28$ for $\alpha = 0·20$ (double-sided test)　　　　$u = 1·96$ for $\alpha = 0·05$ (double-sided test)
　　　　$u = 1·64$ for $\alpha = 0·10$ (double-sided test)　　　　$u = 2·58$ for $\alpha = 0·01$ (double-sided test)

Source:　Reprinted from Davies, O.L., and Goldsmith, P.L., *Statistical Methods in Research and Production,* Longman Group Limited, London, 1977. By permission of the authors and publisher.

M

Critical Values for the Kolmogorov-Smirnov Goodness-of-Fit Test

Sample Size (n)	Level of Significance for $D = $ Maximum $|F(x) - S_n(x)|$				
	.20	.15	.10	.05	.01
1	.900	.925	.950	.975	.995
2	.684	.726	.776	.842	.929
3	.565	.597	.642	.708	.828
4	.494	.525	.564	.624	.733
5	.446	.474	.510	.565	.669
6	.410	.436	.470	.521	.618
7	.381	.405	.438	.486	.577
8	.358	.381	.411	.457	.543
9	.339	.360	.388	.432	.514
10	.322	.342	.368	.410	.490
11	.307	.326	.352	.391	.468
12	.295	.313	.338	.375	.450
13	.284	.302	.325	.361	.433
14	.274	.292	.314	.349	.418
15	.266	.283	.304	.338	.404
16	.258	.274	.295	.328	.392
17	.250	.266	.286	.318	.381
18	.244	.259	.278	.309	.371
19	.237	.252	.272	.301	.363
20	.231	.246	.264	.294	.356
25	.21	.22	.24	.27	.32
30	.19	.20	.22	.24	.29
35	.18	.19	.21	.23	.27
Over 35	$\dfrac{1.07}{\sqrt{n}}$	$\dfrac{1.14}{\sqrt{n}}$	$\dfrac{1.22}{\sqrt{n}}$	$\dfrac{1.36}{\sqrt{n}}$	$\dfrac{1.63}{\sqrt{n}}$

Note: The values of D given in the table are critical values associated with selected values of n. Any value of D that is greater than or equal to the tabulated value is significant at the indicated level of significance.

Source: Reprinted from Massey, F.J., "The Kolmogorov-Smirnov test for goodness of fit", *J. Am. Stat. Assoc., 46:* 68–78, 1951. By permission of the author and publishers.

Index

Thank you for purchasing this book.

The author would like to offer you the opportunity to obtain the applications and examples in this book on diskette.

To save keyboard input time, all of the applications and examples in this book are available in the two disk set. Each application has the original template(s), named ranges, program(s), etc., described in the book. These applications will work with Lotus 1-2-3®, versions 2.0/2.01/2.2/3 and Lotus Symphony®, versions 1.1, 1.2, and 2.0.

To send for the *Prentice Hall Laboratory Lotus® Series™: Practical Spreadsheet Statistics and Curve Fitting for Scientists and Engineers* disks, just photocopy this page, fill out the information completely and return to the address listed below. Thank you.

Date Purchased: _____

Name: _____

Title: _____

School/Company: _____

Department: _____

Street Address: _____

City: _____ State: _____ Zip: _____

Telephone: (   ) _____

Please enclose a check for $18.00 (US $dollars), payable to: Louis M. Mezei.

Louis M. Mezei
40815 Ondina Ct.
Fremont, CA 94539